S0-CKS-225

Geophysical Monograph Series

Including

IUGG Volumes
Maurice Ewing Volumes
Mineral Physics Volumes

GEOPHYSICAL MONOGRAPH SERIES

Geophysical Monograph Volumes

42 Flow and Transport Through Unsaturated Fractured Rock *Daniel D. Evans and Thomas J. Nicholson (Eds.)*

43 Seamounts, Islands, and Atolls *Barbara H. Keating, Patricia Fryer, Rodey Batiza, and George W. Boehlert (Eds.)*

44 Modeling Magnetospheric Plasma *T. E. Moore, J. H. Waite, Jr. (Eds.)*

45 Perovskite: A Structure of Great Interest to Geophysics and Materials Science *Alexandra Navrotsky and Donald J. Weidner (Eds.)*

46 Structure and Dynamics of Earth's Deep Interior (IUGG Volume 1) *D. E. Smylie and Raymond Hide (Eds.)*

47 Hydrological Regimes and Their Subsurface Thermal Effects (IUGG Volume 2) *Alan E. Beck, Grant Garvin and Lajos Stegena (Eds.)*

IUGG Volumes

1 Structure and Dynamics of Earth's Deep Interior *D. E. Smylie and Raymond Hide (Eds.)*

2 Hydrological Regimes and Their Subsurface Thermal Effects *Alan E. Beck, Grant Garvin and Lajos Stegena (Eds.)*

Maurice Ewing Volumes

1 Island Arcs, Deep Sea Trenches, and Back-Arc Basins *Manik Talwani and Walter C. Pitman III (Eds.)*

2 Deep Drilling Results in the Atlantic Ocean: Ocean Crust *Manik Talwani, Christopher G. Harrison, and Dennis E. Hayes (Eds.)*

3 Deep Drilling Results in the Atlantic Ocean: Continental Margins and Paleoenvironment *Manik Talwani, William Hay, and William B. F. Ryan (Eds.)*

4 Earthquake Prediction—An International Review *David W. Simpson and Paul G. Richards (Eds.)*

5 Climate Processes and Climate Sensitivity *James E. Hansen and Taro Takahashi (Eds.)*

6 Earthquake Source Mechanics *Shamita Das, John Boatwright, and Christopher H. Scholz (Eds.)*

Mineral Physics Volumes

1 Point Defects in Minerals *Robert N. Schock (Ed.)*

2 High Pressure Research in Mineral Physics *Murli H. Manghnani and Yasuhiko Syono (Eds.)*

Geophysical Monograph 48
IUGG Volume 3

T.M

Origin and Evolution of Sedimentary Basins and Their Energy and Mineral Resources

Raymond A. Price
Editor

American Geophysical Union
International Union of Geodesy and Geophysics

Geophysical Monograph/IUGG Series

Library of Congress Cataloging-in-Publication Data

Origin and evolution of sedimentary basins and their energy and mineral resources.

(Geophysical monograph ; 48/IUGG series ; 3)
Papers presented at the 27th International Geological Congress held in Moscow in 1984 and at the 19th General Assembly of the International Union of Geodesy and Geophysics held in Vancouver in 1987.
1. Sedimentary basins—Congresses. 2. Mines and mineral resources—Congresses. 3. Power resources—Congresses. I. Price, Raymond A. II. Series.
QE571.075 1989 551.3 89-6549
ISBN 0-87590-452-1

Printed in the United States of America.

CONTENTS

PREFACE

The International Lithosphere Program was launched in 1981 as a ten-year project of interdisciplinary research in the solid earth sciences. It is a natural outgrowth of the Geodynamics Program of the 1970's, and of its predecessor, the Upper Mantle Project. The Program — "Dynamics and Evolution of the Lithosphere: The Hazards"—is concerned primarily with the current state, origin and development of the lithosphere, with special attention to the continents and their margins. One special goal of the program is the strengthening of interactions between basic research and the applications of geology, geophysics, geochemistry and geodesy to mineral and energy resource exploration and development, to the mitigation of geological hazards, and to protection of the environment; another special goal is the strengthening of the earth sciences and their effective application in developing countries.

The origin and evolution of sedimentary basins is an obvious focus of the International Lithosphere Program because it is fundamentally a problem in the dynamics and evolution of the lithosphere, and moreover, it provides special opportunities for strengthening the interactions between basic research and the applications of geology, geophysics, geochemistry and geodesy to mineral and energy exploration and development. Accordingly, at both the XXVIIth International Geological Congress in Moscow, in 1984, and at the XIXth General Assembly of the International Union of Geodesy and Geophysics in Vancouver, in 1987, the International Lithosphere Program convened special symposia on the subject of the origin and evolution of sedimentary basins and their mineral and energy resources. This special volume presents some of the principal results of those symposia.

The first group of 14 papers is based on presentations made at the 1987 Symposium in Vancouver, which was convened by R. A. Price and A. L. Yanshin. They have been assembled and edited by R. A. Price. They cover a broad spectrum of topics ranging from deep processes controlling the origin of basins, to the hydrodynamics of pore fluids within the basins. A variety of different basin settings are considered as well as several different perspectives on the mechanisms of formation of sedimentary basins.

The second group of five papers is based on presentations at the XXVIIth International Geological Congress in Moscow. It broadens the coverage of the volume to include Precambrian basins as well as additional examples from other parts of the world. These papers were assembled and edited by R. W. Hutchinson and R. W. Macqueen for publication in 1987. Because of the way in which they complement the papers from the 1987 Vancouver symposium, it was decided to combine the two groups of papers for this publication.

The editors gratefully acknowledge the help of the following individuals who critically reviewed one or more of the manuscripts:

J. Toth
R. Stephenson
J. Oliver
R. Macqueen
C. Beaumont
K. Lambeck
B. Wernicke
C. Angevine
L. D. Brown
W. R. Dickinson
P. Ziegler
D. Blundell
T. A. Jordan
G. Bond
P. R. Vail
D. Turcotte
J. S. Bell
L. C. Gerhard
R. W. Hutchinson
D. W. Morrow
P. K. Sims
G. Quinlan
L. H. Royden
C. E. Keen
D. Chapman
K. Furlong
D. Finlayson
N. Kusznir
W. J. Perry, Jr.
J. Dewey
R. I. Walcott
N. H. Sleep
G. Thompson
J. Steidtmann
R. S. Yeats
S. Cloetingh
R. A. Price
T. A. Cross
R. W. Hodder
H. Kent
R. H. Riddler
G. Thorman

Raymond A. Price
Geological Survey of Canada
Ottawa, Ontario

Origin and Evolution of Sedimentary Basins and Their Energy and Mineral Resources

INTRAPLATE STRESSES AND SEDIMENTARY BASIN EVOLUTION

Sierd Cloetingh[1], Henk Kooi[1], and Wim Groenewoud

Vening Meinesz Laboratory, University of Utrecht, The Netherlands

Abstract. Fluctuations in stress levels in the lithosphere can play an important role in basin stratigraphy and may provide a tectonic explanation for Vail's third order cycles in apparent sea levels. The gross onlap/offlap stratigraphic architecture of rifted basins can be described by models with changing horizontal stress fields. We demonstrate the effect of intraplate stress on vertical motions of the lithosphere for a depth-dependent rheology of the lithosphere with brittle fracture in its upper part and ductile flow in its lower part. Comparison of the outcome of the modeling with previous estimates by Cloetingh et al. [1985] of stress-induced subsidence and uplift based on an elastic plate model for the mechanical properties of the lithosphere demonstrates a considerable magnification of the induced vertical motions. These findings have important consequences for the stress levels required to explain the observed onlaps and offlaps at sedimentary basins. Similarly, they bear on our assessment of the relative importance of lithospheric dynamics versus glacio-eustasy as the controlling factor underlying sea-level cycles during periods with a non-icefree world. Modeling of the stratigraphy of the U.S. Atlantic margin demonstrates that the inferred transience in the horizontal stress field is qualitatively consistent with expectations based on what is known about plate kinematics during the same time period. The classic Mid-Oligocene unconformity can be explained by a compressional tectonic phase. The superposition of the stress effect associated with a major plate reorganization and a glacio-eustatic event might explain the exceptional magnitude of the Mid-Oligocene lowering of apparent sea level. Out-of-phase intrabasinal cycles such as relative uplift at the flanks and increased subsidence at the basin center, as observed for the Gulf de Lions margin, are predictable by the models. The large variations in estimates of magnitudes of short-term changes in relative sea level between various basins around the world are in agreement with predictions of the tectonic model.

Introduction

In recent years substantial progress has been made in quantitative modeling of sedimentary basins [e.g., Beaumont and Tankard, 1987]. Modeling studies have highlighted the important role of thermomechanical properties of the lithosphere in models of sedimentary basin evolution [e.g., Sleep, 1971; Watts et al., 1982; Beaumont et al., 1982]. Furthermore, they have quantified the contributions of a variety of lithospheric processes to the vertical motions of lithosphere at sedimentary basins. These processes include thermally induced contraction of the lithosphere amplified by the loading of sediments that accumulate in these basins [Sleep, 1971], isostatic response to crustal thinning and stretching [McKenzie, 1978] and flexural bending in response to vertical loading [Price, 1973; Beaumont, 1978].

Simultaneously, major advances have been made in the study of the stress fields in the plate interiors. Detailed analysis of earthquake focal mechanisms [Wiens and Stein, 1984; Bergman, 1986], in-situ stress measurements and analysis of break-out orientations obtained from wells [Bell and Gough, 1979; Zoback, 1985; Klein and Barr, 1986] have demonstrated the existence of consistently oriented present-day stress patterns in the lithosphere. Studies of paleo-stress fields within the plates by the application of analysis of microstructures [Letouzey, 1986; Bergerat, 1987; Philip, 1987] have expanded these findings by demonstrating temporal variations in the observed long-wavelength spatially coherent stress patterns. This work has provided strong evidence for the occurrence of large-scale rotations in the paleo-stress fields and also showed [see Philip, 1987] that the state of stress can vary enough to produce quite different deformations on relatively short time scales of approximately 5 Ma. At the same time, numerical modeling [Richardson et al., 1979; Wortel and Cloetingh, 1981, 1983; Cloetingh and Wortel, 1985, 1986] has yielded better understanding of the causes of the observed variations in stress levels and stress directions in the various lithospheric plates. These studies have demonstrated a causal relationship between the processes at plate boundaries and the deformation in the plate interiors [e.g., Johnson and Bally, 1986].

Most students of sedimentary basins agree that intraplate stresses play a crucial role during basin formation. The formation of sedimentary basins by lithospheric stretching, for example, requires tensional stress levels of the order of at least a few kbars [Cloetingh and Nieuwland, 1984; Houseman and England, 1986]. On the other hand, however, the effect of intraplate stresses on the subsequent evolution of sedimentary basins has been largely ignored in geodynamic modeling. However, recent work by Cloetingh et al. [1985], Cloetingh [1986] and Karner [1986] has demonstrated that temporal fluctuations in intraplate stress levels may have important consequences on basin stratigraphy and provide a tectonic explanation for short-term sea-level variations inferred from the stratigraphic record [Vail et al., 1977; Haq et al., 1987]. Vail and co-workers traditionally have interpreted their cyclic variations in onlap/offlap patterns in terms of a glacio-eustatic origin. This preference was primarily based on the inferred global change

[1] Authors now at Department of Sedimentary Geology, Institute of Earth Sciences, Free University, P.O. Box 7161, 1007 MC Amsterdam, The Netherlands.

of the relative sea-level variations and the lack of a tectonic mechanism to explain Vail et al.'s third-order cycles. Although previous authors [e.g., Bally, 1982; Watts, 1982] have argued for a tectonic cause for apparent sea-level variations, they were unable to identify a tectonic mechanism operating on a time scale appropriate to explain the observed short-term changes of sea level [Pitman and Golovchenko, 1983]. A problem with the glacio-eustatic interpretation, however, is the lack of evidence in the geological and geochemical records for significant Mesozoic and Cenozoic glacial events prior to mid Tertiary time [Pitman and Golovchenko, 1983]. Hence, plate dynamics, and associated changes in stress levels in the plate's interiors offer a tectonic framework for quantitative dynamic stratigraphy.

Conversely, we have explored [Cloetingh, 1986; Lambeck et al., 1987] the use of the stratigraphic record as a new source of information on paleo-stress fields. These studies were carried out using a simplified elastic rheology for the lithosphere. Here we model the effect of intraplate stresses on lithospheric deflection for a more realistic depth-dependent rheology of the lithosphere [Goetze and Evans, 1979]. This modeling demonstrates that significantly lower stress levels are required to simulate the observed onlap and offlap patterns at the flanks of sedimentary basins. These findings bear on the scale of the underlying plate reorganizations versus a more local origin of the stress changes. More precise estimates of the magnitudes of the fluctuations in stress level are important for a quantitative assessment of the relative contributions of eustasy and stress-induced subsidence to the apparent sea-level record. Note that mechanisms for (thermally induced) long-term changes in sea level [e.g., Kominz, 1984; Heller and Angevine, 1985] fall beyond the scope of the present paper.

Intraplate Stresses and Elastic Deflection of the Lithosphere

In classical studies Smoluchowski [1909], Vening Meinesz [see Heiskanen and Vening Meinesz, 1958] and Gunn [1944] have investigated the flexural response of the lithosphere to applied horizontal forces. The flexural response of a uniform elastic lithosphere at a position x to an applied horizontal force F and a vertical load $q(x)$ is given by:

$$D\frac{d^4w}{dx^4} - F\frac{d^2w}{dx^2} + (\rho_m - \rho_i)gw = q(x)$$

where w is the displacement of the lithosphere, and D is the flexural rigidity ($D = E\,T^3 / 12\,(1-\nu^2)$), with E the Young's modulus, T the plate thickness, and ν the Poissons's ratio. The axial load F is equivalent to the product of the intraplate stress σ_N and the plate thickness T. ρ_m and ρ_i are respectively the densities of mantle material and the infill of the lithospheric depression, usually water or sediment, and g is the gravitational acceleration. The solution to this classical equation is easily obtained for some simple loading cases [e.g., Turcotte and Schubert, 1982].

Early studies showed that, in the absence of vertical loads, horizontal forces alone are quite inefficient at producing vertical deflections of the lithosphere. For compressional forces below the buckling limit the induced vertical deflections of the lithosphere are close to zero. These results and the lack of evidence for the existence of such horizontal forces, at that time, caused their possible effects to be ignored.

The situation is quite different in the presence of already existing vertical loads on the lithosphere. For example, in sedimentary basins constituting significant vertical loads, relatively low levels of intraplate stresses suffice to modulate the deflection, and hence the preexisting basin configuration [Cloetingh et al., 1985]. Crustal stretching or lithospheric flexure in response to vertical loading are, by their nature, inherently associated with such a preexisting deflection of the lithosphere.

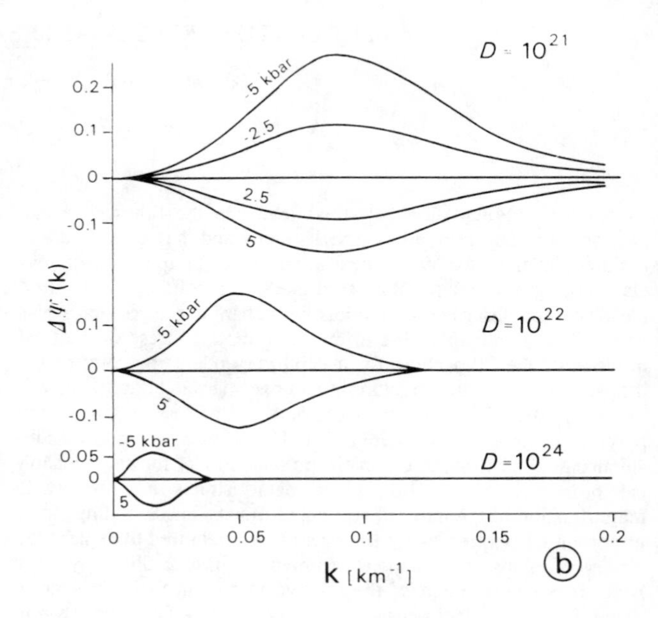

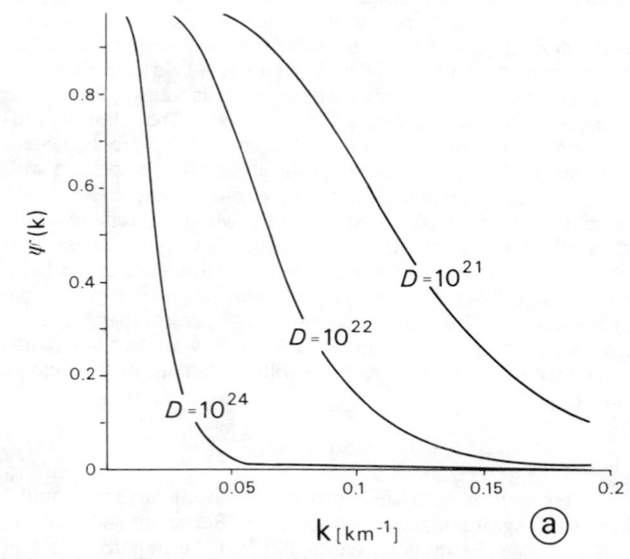

Fig. 1. *(a)* Flexural response functions $\Psi(k)$, in the absence of an intraplate stress field, for thin elastic plates with various flexural rigidities D in Nm, plotted as a function of wavenumber k. *(b)* The effect of intraplate stresses σ_N (tension is positive) on the response function. $\Delta\Psi_i(k) = \Psi_i(k) - \Psi(k)$, with $\Psi_i(k)$ the response function for an applied intraplate stress field. Results are given for the same flexural rigidities for various intraplate stresses σ_N in kbars. [After Stephenson and Lambeck, 1985].

Analytical solution. In analytical solutions of the equation describing the flexural behavior of thin elastic plates, the loading response of the plate is traditionally decomposed into its harmonic components by transforming the equation to the Fourier domain [Stephenson and Lambeck, 1985].

$$\Psi_i(k) = [1+\frac{D(2\pi k)^4 - F(2\pi k)^2}{\rho_m g}]^{-1}$$

If $F=0$, then $\Psi_i(k) = \Psi(k)$, the flexural response function of the plate in the absence of intraplate stress. The wave number at which intraplate stresses most affect the flexural response of the lithosphere is almost completely determined by the plate's flexural rigidity [Stephenson and Lambeck, 1985]. These authors showed that the presence of intraplate stresses has a small but perceptible effect on this wave number, but exerts a controlling influence on the magnitude of the response for a given flexural rigidity. These features are illustrated in Figure 1, which shows the effect of intraplate stress fields of a magnitude of a few kbars on the flexural response of an elastic lithosphere.

Numerical solution. The analytic formulation of specific simplified problems, as given above, shows explicitly how the solution depends on various parameters. Numerical modeling techniques, however, have the advantage of allowing more realistic geometries and variations in parameters to be handled, adding flexibility to the analysis. Cloetingh et al. [1985] considered the case of an elastic lithosphere evolving through time in response to changing thermal condition and loading with a wedge of sediments [Turcotte and Ahern, 1977]. They showed that vertical deflections of the lithosphere of up to a hundred meters may be induced by the action of lithospheric stress fields with a magnitude of up to a few kbars (Figure 2). When horizontal compression occurs, the peripheral bulge flanking the basin is magnified, while simultaneously migrating in a seaward direction, such that the basin flanks are uplifted. As a result an offlap develops, and an apparent fall in sea level results, which may expose the sediments and produce an unconformity. Simultaneously, the basin center undergoes deepening (Figure 2b), resulting in a steeper basin slope. For a horizontal tensional intraplate stress field, the flanks of the basin subside such that the shoreline migrates landward, producing an apparent rise in sea level, so that renewed deposition, with a corresponding facies change, is possible. In this case the center of the basin is shallowed (Figure 2b), and the basin slope is reduced.

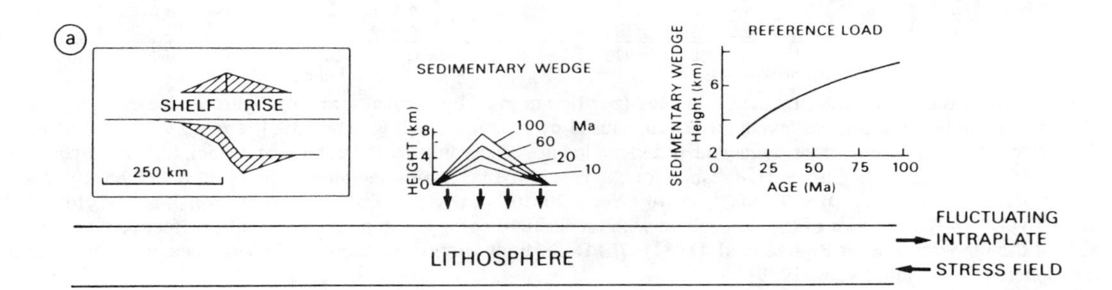

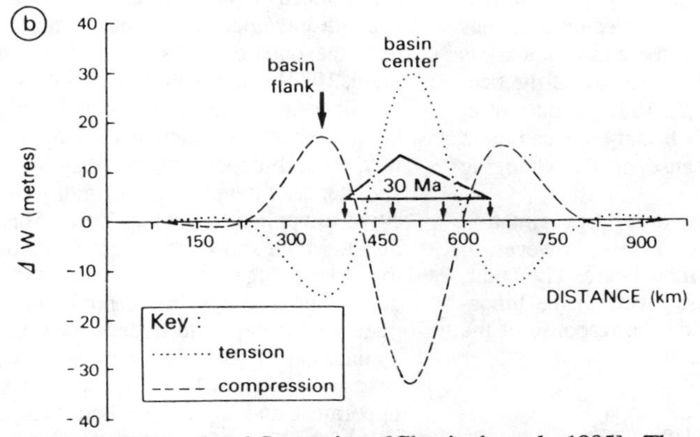

Fig. 2. *(a)* Model for apparent sea-level fluctuations [Cloetingh et al., 1985]. The vertical displacements of the lithosphere evolve through time because of thermal contraction, stiffening of the lithosphere, and loading with a sediment wedge. Inset shows position of this wedge on the outer shelf,slope and rise of a passive margin. Sedimentation is assumed to be sufficiently rapid to equal approximately the subsidence rate [Turcotte and Ahern, 1977]. *(b)* Effect of variations in intraplate stress fields on the deflection of 30 Ma old oceanic lithosphere. Differential subsidence or uplift (meters) relative to the deflection in the absence of an intraplate stress field is given for intraplate stress fields of 1 kbar compression (an in-plane force of -2.17×10^{12} Nm^{-1} (dashed)), and 1 kbar tension, (an in-plane force of 2.17×10^{12} Nm^{-1} (dotted)). Sign convention: uplift is positive, subsidence is negative. Note the opposite effects at the flanks of the basin and at the basin center. [After Cloetingh et al., 1985].

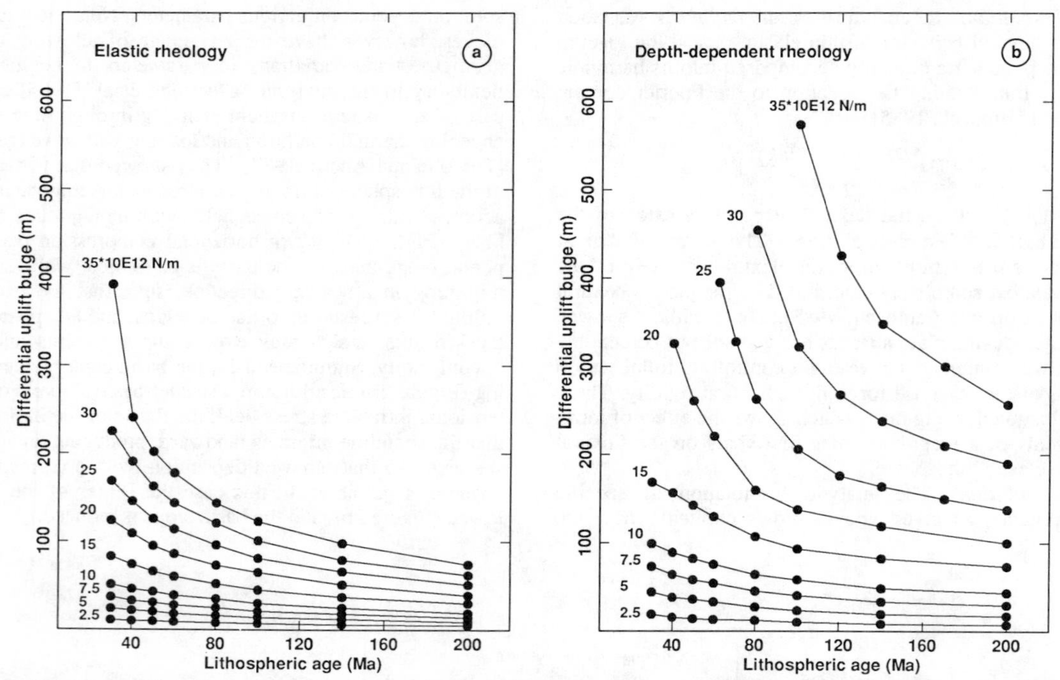

Fig. 3. Differential uplift $|\Delta W|$ (m) at basin edge (position marked by vertical arrow in Figure 2) due to superposition of variations in intraplate stress levels on flexure caused by sediment loading. The $|\Delta W|$ are plotted as a function of the age of the underlying lithosphere and sediment loading according to the reference model [After Turcotte and Ahern, 1977] of Figure 2. Curves give results for changes in the level of intraplate compression ΔF of $2.5\times10^{12}\ Nm^{-1}$, $5\times10^{12}\ Nm^{-1}$, $7.5\times10^{12}\ Nm^{-1}$, $10\times10^{12}\ Nm^{-1}$, $15\times10^{12}\ Nm^{-1}$, $20\times10^{12}\ Nm^{-1}$, $25\times10^{12}\ Nm^{-1}$, $30\times10^{12}\ Nm^{-1}$, and $35\times10^{12}\ Nm^{-1}$, respectively. *(a)* Deflections of the lithosphere with an elastic rheology with an increase of the effective elastic thickness of the lithosphere after Bodine et al. [1981]. *(b)* Deflections adopting a depth-dependent oceanic rheology of the lithosphere [Goetze and Evans, 1979].

Analysis of the flexural response of oceanic lithosphere to tectonic processes [Bodine et al., 1981] and seismotectonic studies [Wiens and Stein, 1983] show an increase in the elastic thickness of the oceanic lithosphere with age. Thus the response of the oceanic lithosphere to sediment loads [Watts et al., 1982], and to intraplate stress fields, is time dependent not only because the sediment load accumulates with time but also because of the changing mechanical properties of the lithosphere.

The elastic model is quite useful as a first order approximation of the mechanical behavior of the lithosphere. It fails, however, to take into account the finite strength of the lithosphere. The latter, by its nature, provides upper limits to stress levels in the lithospheric plates, and has a profound influence on the response of the lithosphere to applied tectonic loads.

Depth-dependent Rheology and Stress-induced Lithospheric Deflection

A more realistic model of the mechanical properties of the lithosphere is based on the extrapolation of rock-mechanics data from laboratory experiments to geological conditions [Goetze and Evans, 1979]. These workers developed a depth-dependent rheology of the lithosphere combining Byerlee's law [Byerlee, 1978] in the brittle domain with temperature-dependent constitutive equations describing the deformation in the ductile regime. Although extrapolated over several orders of magnitudes, the resulting strength envelopes have been demonstrated to be quite consistent with the outcome of studies of intraplate seismicity [Wiens and Stein, 1983] and bending of the lithosphere [McAdoo et al., 1985].

Since an elastic plate is assumed to accommodate arbitrarily high stresses, it implicitly has infinite strength. In contrast, a plate with a depth-dependent rheology has a finite strength at any depth and, hence, a lower flexural rigidity. It will, therefore, be more sensitive to applied horizontal loads. This feature is illustrated in Figure 3, which shows the magnification of the vertical deflection at the basin edge resulting from the incorporation of a depth-dependent rheology in the modeling. Due to its finite strength, a plate with a depth-dependent rheology will begin to fail upon the application of tectonic loads in the lithosphere. This failure process will begin where lithospheric strengths are lowest, at the uppermost and lowermost boundaries of the mechanically strong part of the lithosphere. As a result, the mechanically strong part of the lithosphere will be thinned to a level which ultimately depends on the ratio of the integrated strength of the lithosphere to the applied stress. This results in a reduction of the flexural wavelength of the plate when a tectonic load is applied. This effect applies to both compressional and tensional stresses, and adds a complexity not present in the case of an elastic rheology, where compression and tension have an opposite effect on the flexural shape of the basin (Figure 2b). Rheological weakening of

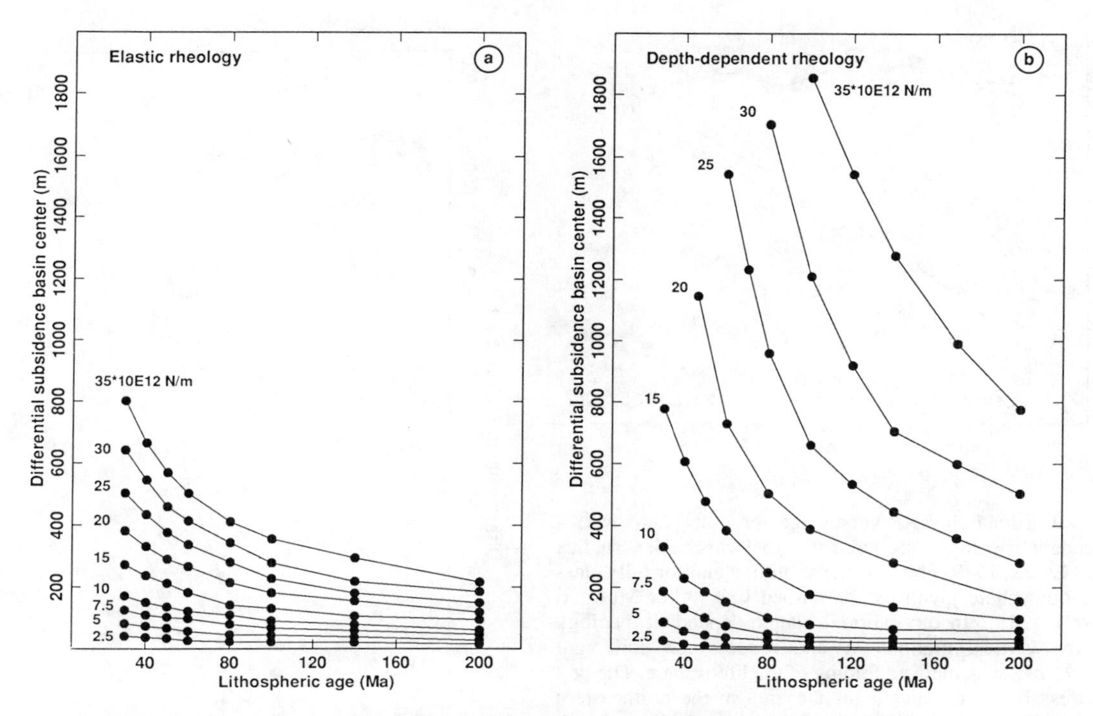

Fig. 4. Differential subsidence $|\Delta W|$ (meters) at basin center due to superposition of variations in in-plane compression on flexure caused by sediment loading. *(a)* Deflection for elastic rheology. *(b)* Deflection for depth-dependent oceanic rheology. Figure conventions as in Figure 3.

the lithosphere can also occur due to flexural stresses induced by the sediment loading itself [Cloetingh et al., 1982]. On the other hand, sediments which fill in the depression add to lithospheric strength by increasing the normal stress on the rocks at depth, an effect that is to a large extent compensated by weakening of the lithosphere caused by thermal blanketing induced by the sediment loading. We have, therefore, ignored the effects of the sediment load on lithospheric strength.

The sign of the vertical motions induced by in-plane forces is the same for a depth-dependent rheology as for an elastic model. For the depth-dependent rheology, the stresses weaken the lithosphere and thus affect the wavelength of the deflection. Since the flexural bulge is commonly located below the shelf at passive margins [Watts and Thorne, 1984], its vertical displacement due to varying stress levels will dominate the stratigraphic response and apparent sea-level record induced by intraplate stresses. Figures 3a and 3b demonstrate that in-plane forces have a greater effect on the height of the bulge for a depth-dependent rheology than for an elastic lithosphere. Figures 3 and 4 show how the vertical motion of the peripheral bulge and the basin center depends on the age of the underlying lithosphere. The differential uplift or subsidence, the difference in deflection for a change in in-plane force, is calculated for the basin flank (Figure 3) and center (Figure 4) as a function of in-plane force. Figures 3 and 4 show, that compared to the elastic plate model, a depth-dependent rheology of the lithosphere substantially enhances the ability of intraplate stresses to induce vertical deflections at the basin edge. Due to the age-dependence of the lithospheric strength, the magnification is particularly pronounced for young lithosphere.

For large compressional intraplate stresses, close to lithospheric strength, the lithosphere will buckle. The region in the northeastern Indian Ocean south of the Bay of Bengal provides a well studied example of this process. The deformation occurs by broad basement undulations, with wavelengths of roughly 200 km and amplitudes up to 3 km, and numerous high-angle reverse faults [Weissel et al., 1980]. The strike of the undulations and reverse faults is approximately east-west, in agreement with the present-day north-south orientation of the stress field in the area [Bergman and Solomon, 1985]. The basement undulations coincide with undulations in gravity and geoid anomalies [McAdoo and Sandwell, 1985; Zuber, 1987]. For elastic plate models, unrealistically large stresses of several tens of kbars (equivalent to in-plane forces of the order of $2.5{\times}10^{14}$ Nm^{-1}) are required to induce the observed folding of the 60-80 Ma old oceanic lithosphere [Weissel et al., 1980]. McAdoo and Sandwell [1985] studied this folding and noted the importance of incorporating the depth-dependent rheology [Goetze and Evans, 1979] in models of the response of the lithosphere to large intraplate stresses. A depth-dependent rheology dramatically lowers the compressional stress level required to induce the observed folding to approximately 5-6 kbar, or equivalently in-plane forces of the order of $2{\times}10^{13}$ Nm^{-1} (Figure 5). This figure shows that McAdoo and Sandwell's [1985] results are in excellent agreement with estimates from numerical modeling of intraplate stresses [Cloetingh and Wortel, 1985, 1986; see Figure 5]. In these models the exceptionally high level of the intraplate stresses in the northeastern Indian Ocean results from the special dynamic situation of the present Indian plate. The calculated stresses depend on the plate geometry and the boundary

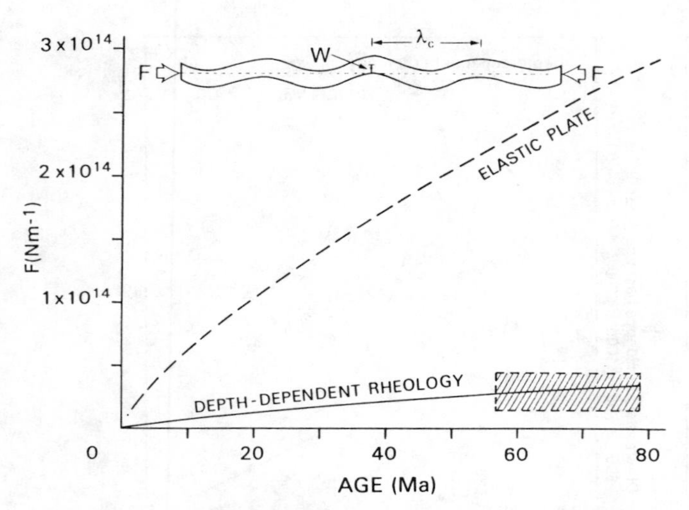

Fig. 5. Buckling load $F\,(Nm^{-1})$ versus age for lithosphere with a depth-dependent rheology inferred from rock-mechanics studies [Goetze and Evans, 1979] given by the solid line and for fully elastic oceanic lithosphere given by the dashed line [After McAdoo and Sandwell, 1985]. Incorporation of depth-dependent rheology magnifies the vertical motions W and reduces the horizontal wavelength λ_c of stress-induced folding of the lithosphere. The box indicates stress levels calculated for the area in the northeastern Indian Ocean [Cloetingh and Wortel, 1985] where folding of oceanic lithosphere under the influence of compressional stresses has been observed [McAdoo and Sandwell, 1985].

forces applied, which are assumed to vary with the age of the subducted lithosphere. In this case, the large age contrasts, the Himalayan collision and the specific geometry of the plate give rise to high stresses. In plates not involved in continental collision, rifting, or plate reorganizations, stresses will, in general be lower, with values as low as a few hundred bars (equivalent to in-plane force of the order of $2{\times}10^{12}\ Nm^{-1}$) calculated for the present Nazca plate [Wortel and Cloetingh, 1983].

Syn-depositional folding of oceanic lithosphere might have some interesting analogues in the continents [e.g., Hoffman et al., 1988; van Wamel, 1987]. Rheological models indicate that continental lithosphere can be substantially weaker than oceanic lithosphere [Kirby, 1985; Carter and Tsen, 1987]. This primarily reflects mineralogic differences between oceanic and continental lithosphere and is thought to have a profound influence on the nature of the rifting [Vink et al., 1984; Steckler and Ten Brink, 1986; Shudofsky et al., 1987]. Similarly, we expect that a given change in intraplate stress will induce larger vertical motions in continental lithosphere than in oceanic lithosphere, with important implications for vertical motions at intracratonic and foreland basins.

Lithospheric folding will occur only if high compressional stress levels are induced in the lithosphere, by special dynamic situations, such as collision processes like those presently occurring between the Indian and Eurasian plates. Figure 6 shows the transition between vertical motions of the order of a hundred meters, which are reflected in the apparent sea-level record, and the more dramatic vertical motions with magnitudes of the order of a few kilometers which result from the accumulation of stress to levels close to the strength of the lithosphere.

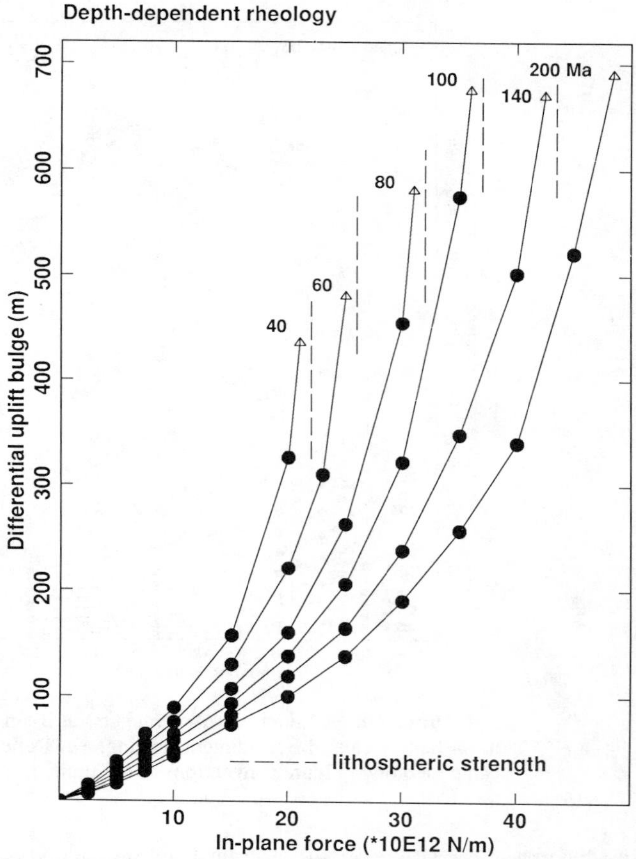

Fig. 6. Uplift of the lithosphere $|\Delta W|$ (km) at the position of the flexural bulge flanking the basin induced by in-plane compressional forces up to the lithospheric strength for various lithospheric ages. Curves have been calculated for an oceanic depth-dependent lithospheric rheology. The dashed vertical lines indicate the strength of lithosphere in compression for ages of 40, 60, 80, 100, 140 and 200 Ma. Arrows indicate the transition to displacements of the lithosphere reaching magnitudes of several kilometers, which occurs when in-plane compression approaches the integrated strength of the lithosphere.

Intraplate Stress and Basin Stratigraphy

Figure 7 schematically illustrates the relative movement between sea level and the lithosphere at the flank of a flexural basin immediately landward of the principal sediment load predicted by numerical calculations [Cloetingh et al., 1985] using the elastic plate model. The synthetic stratigraphy at the basin edge is schematically shown for three situations. In one, long-term flexural widening of the basin results from cooling [Watts, 1982] in the absence of an intraplate stress field (Figure 8a). Figures 8b and 8c show the same situation with a superimposed transition to 500 bar compression (Figure 8b) or tension (Figure 8c) at 50 Ma. As noted by Watts [1982], the thermally induced flexural widening of the basin (Figure 8a) provides an adequate explanation for long-term phases of coastal onlap. However by its long-term nature, it fails to produce the punctuated character of the stratigraphy of sedimentary basins. Figures 8b,c demonstrate that the incorporation

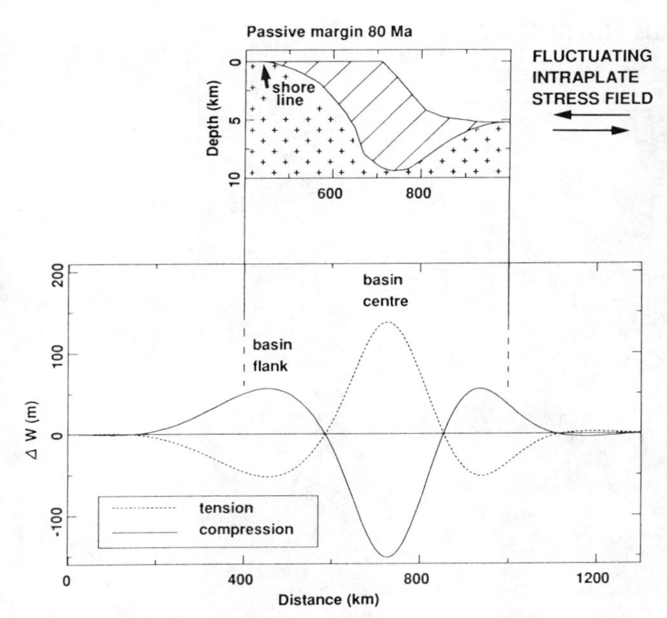

Fig. 7. Flexural deflections at a sedimentary basin induced by changes in intraplate stress field. Top: an 80 Ma old passive margin initiated by stretching. The wedge of sediments flexurally loads an elastic plate. The thickness of the plate varies horizontally due to lateral changes in the temperature structure of the lithosphere. Bottom: Differential subsidence or uplift (meters) induced by a change to 1 kbar compression (solid line) and 1 kbar tension (dashed line).

of intraplate stresses in elastic models of basin evolution can in principle predict a succession of alternating rapid onlaps and offlaps observed along the flanks of basins such as the U.S. Atlantic margin [Sleep and Snell, 1976; Figure 9]. We use the U.S. Atlantic margin for a numerical simulation of the stratigraphy for several reasons. The margin stratigraphy has been extensively documented [e.g., Poag and Schlee, 1984] and has previously been the subject of detailed quantitative modeling [Sleep and Snell, 1976; Watts and Thorne, 1984; Steckler et al., 1988]. Sleep and Snell [1976] proposed a visco-elastic model of the lithosphere to account for the observed late-stage narrowing of the North-Carolina margin. Watts and Thorne [1984] and Steckler et al. [1988] employed a two-layer stretching model adopting an elastic rheology of the lithosphere and zero intraplate stresses. They assumed eustatic long-term and short-term sea-level fluctuations throughout the basin evolution. Our modeling approach resembles the one taken by Watts and Thorne [1984] and Steckler et al. [1988] in adopting a two-layer stretching model for basin initiation but also incorporates the effects of finite and multiple stretching phases and intraplate stresses in the stratigraphic modeling. We use a finite-difference approach for the thermal calculations [Verwer, 1977].

Although of limited impact for the late-stage development of the basin, the incorporation of finite stretching rates severely affects syn-rift and early post-rift subsidence and sedimentation [Jarvis and McKenzie, 1980; Cochran, 1983]. There is general agreement that the initial rifting phase began in the Late Carnian (approximately 225 Ma), whereas sea floor spreading began at about 180 Ma [Manspeizer, 1985; Ziegler, 1988]. Thus the initiation of subsidence is associated with a long period of rifting and stretching. Jurassic deposition has been restricted to the deeper part of the margin now located under the outer shelf. This may be explained by post-Jurassic widening of the basin due to a second stretching phase, or by subcrustal attenuation under the inner shelf part of the basin inhibiting its subsidence. That this thermal anomaly should have been very large is evident from the long duration

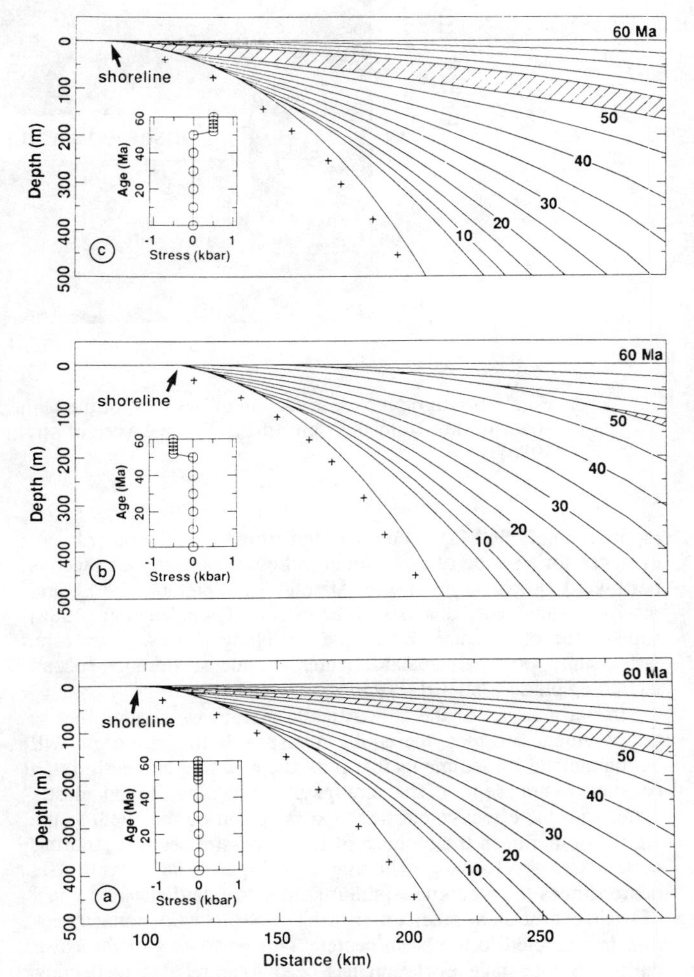

Fig. 8. Synthetic stratigraphy for a 60 Ma old passive margin, that was initiated by lithospheric stretching followed by thermal subsidence and flexural infilling of the resulting depression. Hachuring indicates the position of a sedimentary package bounded by isochrons of 50 Ma and 52 Ma after basin formation. *(a)* Continuous onlap associated with long-term cooling of the lithosphere in the absence of intraplate stress fields. *(b)* A transition to 500 bar in-plane compression at 50 Ma induces uplift of the peripheral bulge, narrowing of the basin and a phase of rapid offlap, which is followed by a long-term phase of gradual onlap due to thermal subsidence. *(c)* A transition to 500 bar in-plane tension at 50 Ma induces downwarp of the peripheral bulge, widening of the basin and a phase of rapid basement onlap.

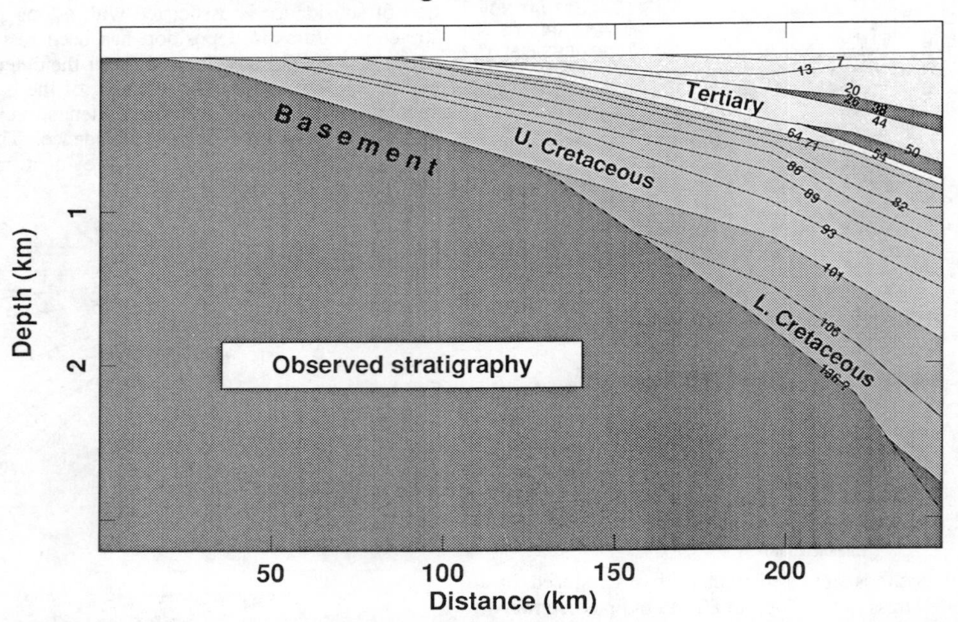

Fig. 9. Stratigraphic cross section of the shelf of the United States Atlantic margin at Cape Hatteras. The shelf break is about 40 km from the right of the figure. Ages of stratigraphic boundaries are given in Ma [After Sleep and Snell, 1976].

(approximately 36 Ma) of cooling after rifting. On the other hand, evidence for a period of extensional tectonics and Early Cretaceous northward propagation of the Atlantic rift [Ziegler, 1988] and results of subsidence analysis of the margin [Greenlee et al., 1988] support the occurrence of multiple stretching phases. Therefore, we assume both subcrustal attenuation and a (minor) second stretching phase from 131-119 Ma.

In our analysis of the U.S. Atlantic margin we assume that as the basement subsides, the equilibrium profile of the margin will be maintained by sediments that infill the resulting depression to a constant water depth. The stratigraphy modeled for an elastic plate, with the effective elastic thickness given by the depth to the $400^o C$ isotherm, in the absence of intraplate stresses and ignoring eustatic sea-level changes is shown in Figure 10a. Figure 10a demonstrates the well known failure of the standard elastic models of basin evolution to predict narrowing of basins with younger sediments restricted to the basin center. This narrowing of the basin during its late-stage evolution has been interpreted as reflecting either the response of the basin to a phase of visco-elastic relaxation [Sleep and Snell, 1976], or to a long-term eustatic sea-level fall [Watts and Thorne, 1984]. The total thickness of the Cenozoic sediments provides an independent constraint for the magnitude of the proposed long-term lowering in sea level. Our modeling yields an upper estimate of approximately 100 meters for the long-term post-Late Cretaceous fall in sea level. We therefore incorporated a long-term sea-level curve with a highstand of 100 meters during the end of the Cretaceous, a curve equivalent to the minimum curve by Kominz [1984]. The resulting stratigraphic model (Figure 10b) demonstrates that although the incorporation of long-term changes in sea level enhances the Cenozoic narrowing of the margin, a long-term post-Late Cretaceous decline in sea level alone cannot cause both the documented basin narrowing and the total thickness of sediments accumulated during this time. We propose that much of the observed non-depositional or erosional character of the shelf surface is caused by stress-induced uplift of the basin flank. Similarly, short-term changes in intraplate stress levels can produce the Early Eocene and Oligocene onlap/offlap phases. Figure 10c shows the best fit to the observed stratigraphy for a two-layered stretching model and an elastic rheology, incorporating long-term sea level changes after Kominz [1984] and a fluctuating intraplate stress level in the stratigraphic modeling. It appears that the margin stratigraphy can be simulated by relaxation of overall tensional Mesozoic intraplate stress fields and a transition to compressional stress, whose level increases with time during the Tertiary.

During rifting phases, eventually followed by continental break-up, the tensional stresses will be reduced. Rifting in the Southern Atlantic, for example, rather than instantaneously, occurred as discrete rifting phases, with stepwise relaxation of tensional stresses. This process might explain the enigmatic high-frequency sea-level fluctuations in Cretaceous times that do not correlate with accelerations in plate spreading or increases in ridge lengths [Schlanger, 1986]. Similarly, the correlation of short-term sea-level fluctuations at both sides of the Atlantic may reflect rifting-related accumulation and relaxation of tensional stresses. In this model, the accumulation of tensional stresses induces periods of apparent rise in sea level. These are followed by periods of sea-level lowering, in general of shorter duration, associated with the rapid relaxation of tensional stresses. Hence, in the period just prior to rifting, sea levels should rise and then fall during the continental break-up phase. Such a break-up unconformity is commonly observed in the stratigraphic record of passive margins, as

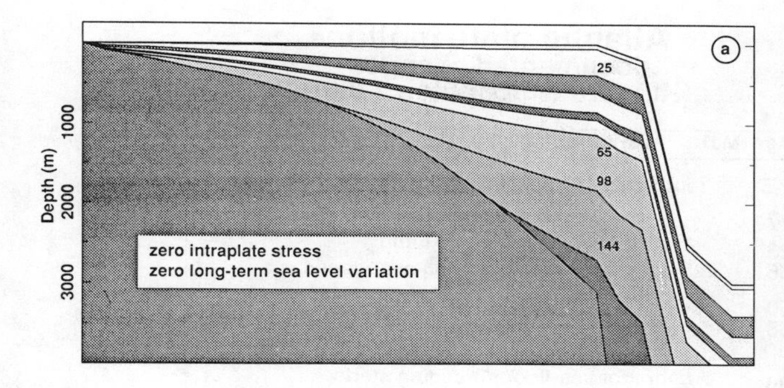

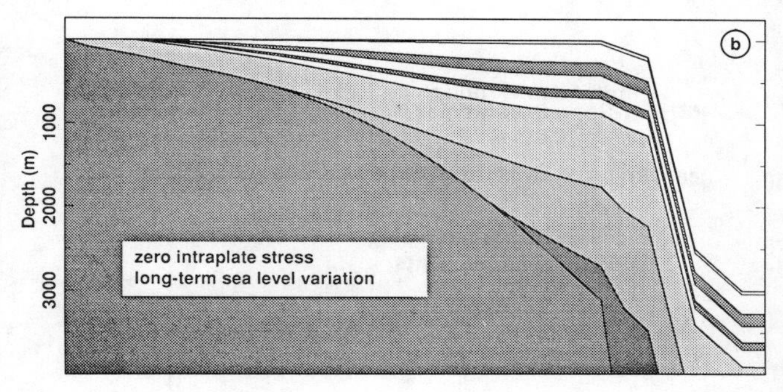

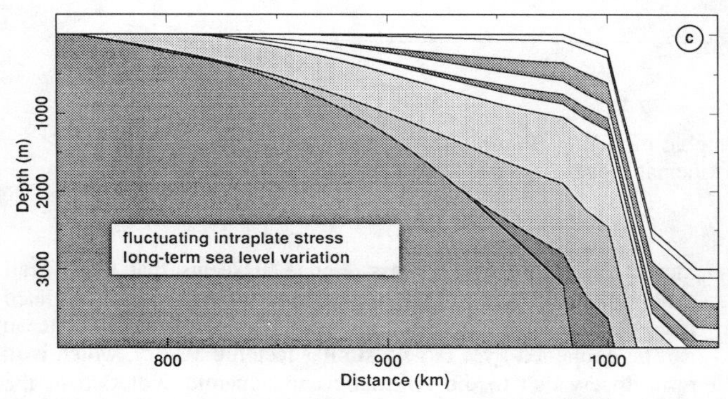

Fig. 10. U.S. Atlantic margin stratigraphy modeled for elastic rheology of the lithosphere. *(a)* Modeled stratigraphy in the absence of intraplate stresses, assuming zero changes in long-term sea level. *(b)* Modeled stratigraphy in the absence of intraplate stresses, but adopting long-term changes in sea level after Kominz [1984]. *(c)* Modeled stratigraphy showing the combined effect of long-term changes in sea level and a fluctuating intraplate stress field.

illustrated by the major lowering in sea level coinciding with the onset of the opening of the South Atlantic [Haq et al., 1987]. These break-up unconformities have not been successfully explained by prior geodynamic models of sedimentary basins, not including the effects of intraplate stresses.

The position of coastal onlap reflects the position where rate of subsidence equals rate of sea-level fall. During application of stress the rate of subsidence is temporarily changed. Consequently, the equilibrium point of the coastal onlap is shifted in position. The thermally induced rate of long-term subsidence strongly decreases with the age of the basin [Turcotte and Ahern, 1977]. Hence, the production of offlaps during late stages of passive margin evolution requires much lower rates of change of sea level than in earlier stages of basin evolution [Thorne and Watts, 1984]. If these offlaps result from fluctuations in intraplate stress levels, the rate of stress changes required diminish with age during the post-rift evolution of the basin. This is of particular relevance for an assessment of the relative contributions of tectonics and eustasy to Cenozoic unconformities. For example, Cenozoic unconformities developed at old passive margins in association with short-term basin narrowing could be produced by relatively mild changes in intraplate stress levels. Such late-stage narrowing of Phanerozoic platform basins and passive margins is frequently observed [Sleep and Snell, 1976], without clear evidence for active tectonism.

Hence, the incorporation of intraplate stresses in elastic models of basin evolution can, in principle, predict a succession of onlaps and offlaps such as observed along the flanks of the U.S. Atlantic margin and the Tertiary North Sea Basin [Kooi et al., 1988]. Such a stratigraphy can be interpreted as a natural consequence of the mechanical widening and narrowing of basins by fluctuations in intraplate stress levels superimposed on the long-term broadening of the basin with cooling since its formation.

Figure 11 shows the paleo-stress field inferred from the modeling of the stratigraphy of the U.S. Atlantic margin. The stress levels given for the elastic models for basin stratigraphy provide upper limits. As discussed in the previous section, the incorporation of depth-dependent rheology in the modeling will lower the predicted stress levels. Similarly, the resolution of the inferred paleo-stress pattern can be enhanced by incorporating more sophisticated sedimentation models and high-resolution paleobathymetry in the analysis [Kooi et al., 1988]. The long-term trend of the paleo-stress pattern is that of a change from overall tension during Cretaceous times to a stress regime of accumulating compression during Tertiary times. Superimposed on this long-term trend are more abrupt changes. Both the character and timing of these changes are largely consistent with independent data on the kinematic evolution of the Central Atlantic [Klitgord and Schouten, 1986]. From 175 Ma- 59 Ma rifting in the Atlantic evolves from the initiation of the Gulf of Mexico - Central Atlantic - Ligurian Tethys spreading (175 Ma), by a number of discrete steps (170, 150, 132, 119, 80, 67 Ma), to the start of spreading in the Northern Atlantic (59 Ma). The paleo-stress curve inferred from the stratigraphy suggests that, in particular, the rifting events around 180 Ma have been associated with a major relaxation of tensional stresses, followed by renewed accumulation of tension. It is interesting to compare the paleo-stress curve for the Tertiary with the tectonic history of the Atlantic. The predicted phase of relaxation of tensional stresses and the transition to a more neutral stress regime around 50 Ma coincides with the termination of the dramatic phase of Thulean volcanism in the northern Atlantic and the break-up in the Greenland-Rockall and Norwegian-Greenland sea [Ziegler, 1988; see also Tucholke and Mountain, 1986]. Similarly, the predicted transition to a more compressional stress regime coincides with the timing of the Caribbean orogeny, the Pyrenean orogeny and the cessation of spreading in the Labrador Sea [Klitgord and Schouten, 1986]. As noted by Issler and Beaumont [1987], sea-floor spreading ended in the Labrador Sea simultaneously with widespread shelf shallowing, tectonism and coastal erosion, features consistent with an increase in the level of compressional stress. The change in the compressional stress level at the time of the mid-Oligocene regression, coincides with a major reorganization in the Central Atlantic: the African plate boundary jump.

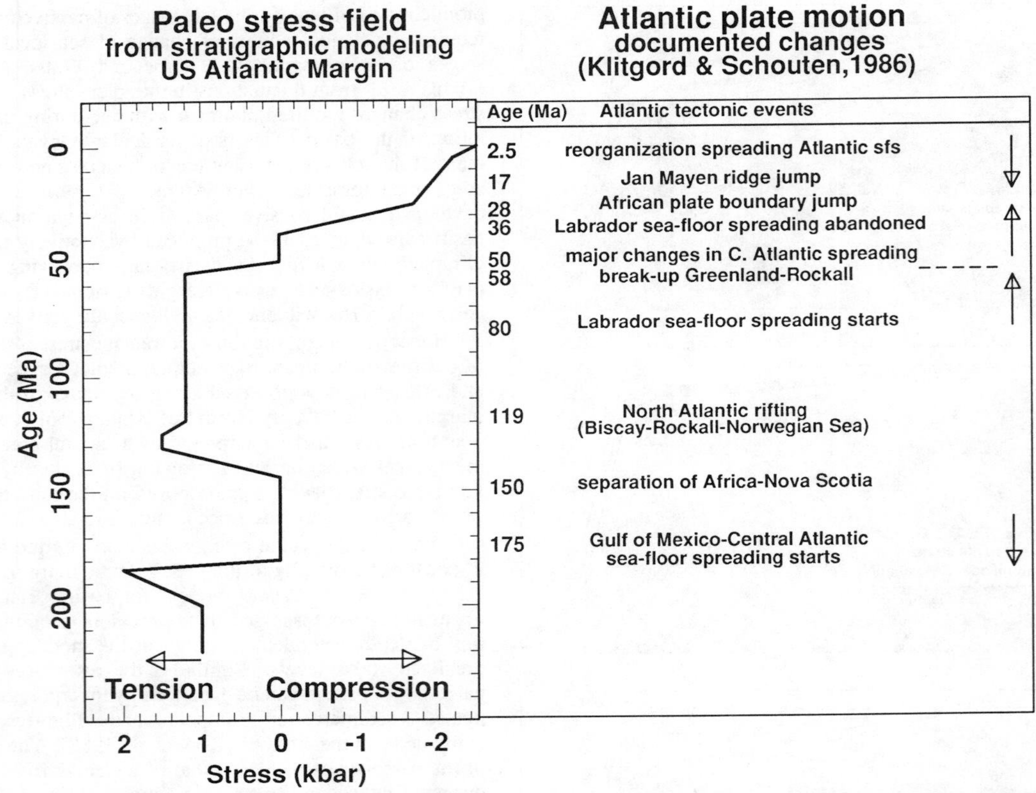

Fig. 11. The paleo-stress curve inferred from the stratigraphic modeling (Figure 10c) of the U.S. Atlantic margin. Tension is positive, compression is negative. Timing of kinematic events in the North/Central Atlantic is given at right [After Klitgord and Schouten, 1986].

Hence, the inferred transition in the horizontal stress field seems to be qualitatively consistent with that expected from the plate kinematics during the same time period. The transition from Mesozoic overall tension to a more compressional regime during Cenozoic times also agrees with a recent stratigraphic study [Hubbard, 1988] on the timing and nature of the sequence boundaries in the Arctic and the Atlantic. This author demonstrated that the majority of Mesozoic sequence boundaries in three rifted continental margin basins are associated with rifting phases, while Cenozoic unconformities in the Arctic correspond to the timing of compressional phases. In this context, it is also interesting to note that a paleo-stress curve derived from the apparent sea-level record of the North Sea area, assuming that the sea levels are controlled by the effects of intraplate stresses, also shows a change from a long-term tensional regime during Jurassic-Cretaceous times to a more compressional stress regime during Tertiary times [Lambeck et al., 1987]. These findings have been recently confirmed by detailed stratigraphic modeling, incorporating the role of intraplate stresses, of the North Sea Central Graben area [Kooi et al., 1988]. The inferred paleo-stress curve is consistent with the transition from rift-wrench tectonics during Mesozoic times to compressional tectonics during Tertiary times in Northwestern Europe [Ziegler, 1982, 1987]. Also, in other respects, the paleo-stress curve appears to mirror the tectonic evolution of Northwestern Europe: rifting episodes correspond to relaxation of tensional paleo-stresses, and Alpine orogenic phases were found to correspond to episodes of increased compressional stress. Hence, it seems that the overall synchroneity in the timing of sea-level changes and associated unconformities for margins at opposite sides of the Atlantic Ocean can be explained by a largely similar tectonic history, which is in turn closely tied to the kinematic and dynamic evolution of the Atlantic/Tethys domain.

Although we have so far concentrated on the relationship between tectonics and stratigraphy for rifted basins, the effect of intraplate stress fields is important to a wider range of sedimentary environments. Other settings where lithosphere is flexed downward under the influence of sedimentary loads occur in foreland basins [Beaumont, 1981; Quinlan and Beaumont, 1984; Tankard, 1986]. These authors interpreted the development of unconformities in the Appalachian foreland basin in terms of uplift of the peripheral bulge caused by viscoelastic relaxation of the lithosphere. However, the presence of tensional or compressional intraplate stresses, the latter being more natural in this tectonic setting, can amplify or reduce the height of the peripheral bulge by an equivalent amount and thus greatly influence the stratigraphic record.

Discrimination of Tectonics and Eustasy

The Exxon curves [Vail et al., 1977; Haq et al., 1987], though based on data from basins throughout the world, are heavily weighted in favor of the Northern Atlantic and the North Sea areas. Therefore, the inferred global cycles may strongly reflect the

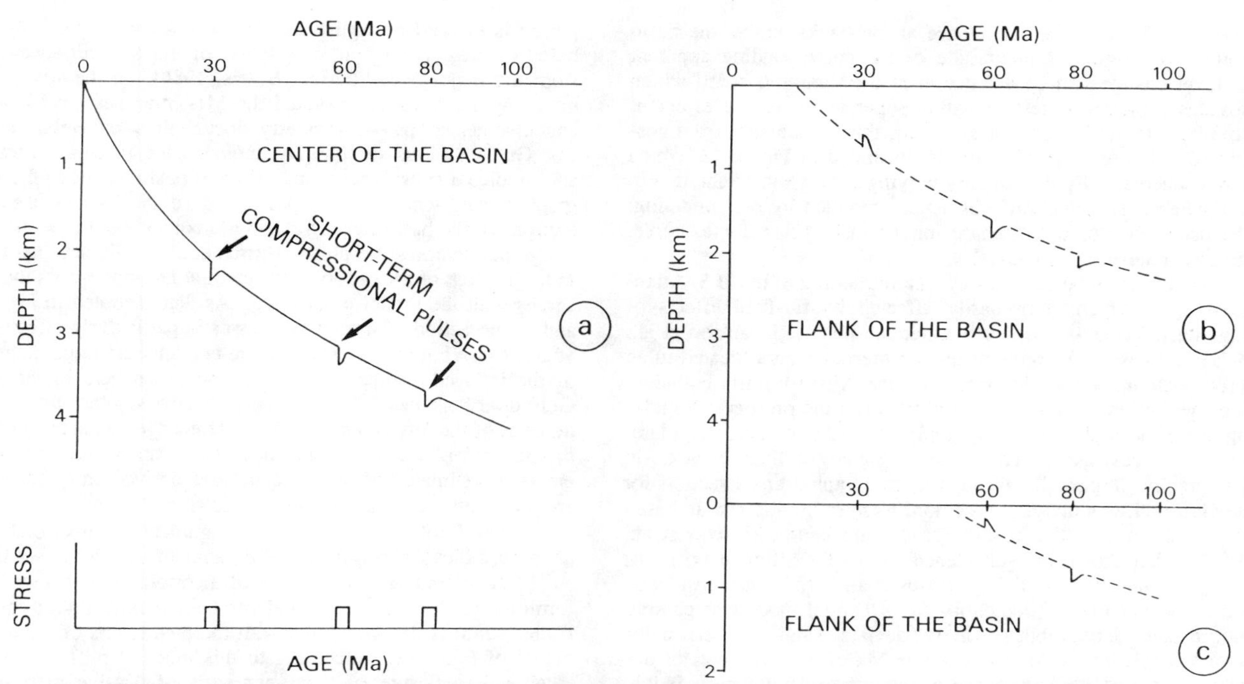

Fig. 12. Effect of intraplate stresses on subsidence curves at three different positions a,b, and c in the basin. *(a)* Effect of compression on subsidence predicted for a well in the basin center. *(b)* and *(c)* show the effect of compressional stress on subsidence for locations at the flanks of the basin, closer to the position of the flexural node. Note in these cases the different effects of compression at different time intervals, which are caused by widening of the basin during its long-term thermal evolution.

seismic stratigraphic record of basins in a tectonic setting dominated by rifting events -and associated changes in intraplate stress levels- in the northern and central Atlantic [e.g., Miall, 1986; Hallam, 1988]. The inferred global character of the Exxon curves has been subject to considerable debate. Parkinson and Summerhayes [1985] and Miall [1986] questioned the global character, pointing out that the synchroneity of the inferred sea-level changes is more widespread than global. Others [e.g., Kerr, 1984] regard the Exxon curves as a worldwide correlation tool applicable for dating mapped unconformities. Similarly, considerable debate has been going on on the causes of the short-term sea-level fluctuations. The question of global synchroneity is also an important one in this respect, as it has played a crucial role in arguments favoring a glacio-eustatic cause for the short-term sea-level changes versus a tectonic origin. Only the larger short-term fluctuations in sea level, with magnitudes in excess of 50 meters, require stress changes large enough to be related to major plate boundary reorganizations. This observation explains the strong correlation between the timing of plate reorganizations and rapid lowerings in sea level [Bally, 1982].

The regional character of intraplate stresses sheds new light on deviations [e.g., Hubbard et al., 1985; Hallam, 1988] from "global" sea-level cycles. Although such deviations are a natural feature of Cloetingh et al.'s [1985] model, the occurrence of short-term deviations does not preclude the presence of global events elsewhere in the stratigraphic record. These are to be expected when major plate reorganizations affect the intraplate stress fields simultaneously in more than one plate or when glacio-eustasy dominates. Examples of major plate reorganizations include those during Mid-Oligocene [Engebretson et al., 1985] and Early Cenozoic [Rona and Richardson, 1978; Schwann, 1985] times. In particular the mid-Oligocene is a time with a high level of tectonic activity in the northern and southern Atlantic, with the concomitant occurrence of a major Alpine folding phase [Ziegler, 1982] and, for example, uplift of the shelf of the African Atlantic margins [Lehner and De Ruiter, 1977]. Furthermore, differences in the rheological structure of the lithosphere, which influence its response to applied intraplate stresses (Figure 4), might also explain differences in the magnitudes of the inferred sea levels such as observed between the North Sea region and the Gippsland Basin off southeastern Australia [Vail et al., 1977].

Independent recent studies of the magnitude of the mid-Oligocene lowering point to a value much smaller than previously thought. The magnitude of this fall in sea level, which is by far the largest shown in the Vail et al. [1977] and Haq et al. [1987] curves, is now estimated to be between 50 [Miller and Fairbanks, 1985; Watts and Thorne, 1984] and 100 [Schlanger and Premoli-Silva, 1986] meters. Hence, a significant part of the short-term sea-level record inferred from seismic stratigraphy might have a characteristic magnitude of a few tens of meters, which can be explained by relatively modest stress fluctuations. The superposition of a glacio-eustatic event and a major tectonic reorganization might explain the exceptional magnitude of the Oligocene sea-level lowering.

The discrimination of regional events from eustatic signals in the sea-level record of individual basins is usually a subtle matter, especially if biostratigraphic correlation is imprecise [Hallam, 1988]. As noted earlier, intraplate stresses cause different and opposite effects on the subsidence at the flanks and in the basin center. The sign and magnitude of the corresponding apparent sea-level change will be a function of the sampling point, which provides a means of testing within separate basins the effect of intraplate stresses or of distinguishing this mechanism from eustatic contributions. This feature is illustrated in Figure 12, which shows schematically the laterally varying expression of changes in intraplate stress levels on subsidence predicted by our modeling. The occurrence of out-of-phase intrabasinal cycles, for example, can be explained by the model.

In contrast to the present-day tectonic setting of the U.S. Atlantic margin, which is primarily affected by far-field effects of ridge-push forces in the North-American plate [Richardson et al., 1979], the passive margins of the Mediterranean are located in an active tectonic setting dominated by the Africa-Eurasia collision. The Mediterranean margins, therefore, offer the prospect of studying the near-field effect of intraplate stresses induced by plate-tectonic forces, associated with the ongoing collision, on basin stratigraphy. Figure 13 displays a stratigraphic cross section for the Gulf de Lions margin in the northwestern Mediterranean based on recent work by the Institut Francais du Petrole [Burrus et al., 1987]. Also shown are subsidence curves for different positions along the margin. As noted by Burrus et al. [1987], the subsidence curves closely follow predictions from thermal models of passive margin subsidence, but strongly deviate from the thermally predicted subsidence for the last five Ma of the evolution of the basin. Rapid excess subsidence of the order of 500 meters is initiated at the basin center, while uplift of the order of a few hundred meters is induced at the shelf (Figure 13). The thick offlap sequences and time-equivalent unconformities in Figure 13 correspond to the Messinian salinity crisis [Bessis, 1986]. This period is marked by a distinct drop in sea level commonly attributed to dessication due to isolation of the Mediterranean basin from the major ocean basins [Bessis, 1986]. In the stratigraphic modeling we have incorporated the Messinian sea-level lowering and changes in paleobathymetry documented by Bessis [1986]. The Gulf de Lions stratigraphy, modeled for an elastic rheology, and predicted subsidence along different positions along the stratigraphic cross section are shown in Figure 14. The rapid vertical motions of the basin starting at 7-5 Ma coincide with the timing of a regional compressive phase [Burrus et al., 1987; see also Figure 15]. The sign of the observed differential motions across the basin agrees with the model predictions. As demonstrated in Figures 3 and 4, the action of intraplate stresses is particularly effective for young passive margins. This feature explains the large magnitude of the induced vertical motions of the lithosphere in the young Gulf de Lions basin. These findings also suggest that vertical motions of the lithosphere caused by late-stage compression during the post-rift phases of extensional basins can produce substantial errors in estimates of crustal extension derived from subsidence analysis with standard stretching models.

Hallam [1988] has shown that a significant number of Jurassic unconformities are confined to the flanks of the North Sea basins. At the same time, the occurrence of a correlation between unconformities at the basin edge and the basin center [Wise and Van Hinte, 1986] is not in conflict with the predictions of the tectonic model of Figure 2. According to this model, uplift of the basin edge with exposure of the inner shelf of passive margins and steepening of the basin slope can be caused by the action of intraplate compressional stresses or, equivalently, by relaxation of a tensional stress regime. As pointed out by Miller et al. [1987], the frequently observed correlation between unconformities on the

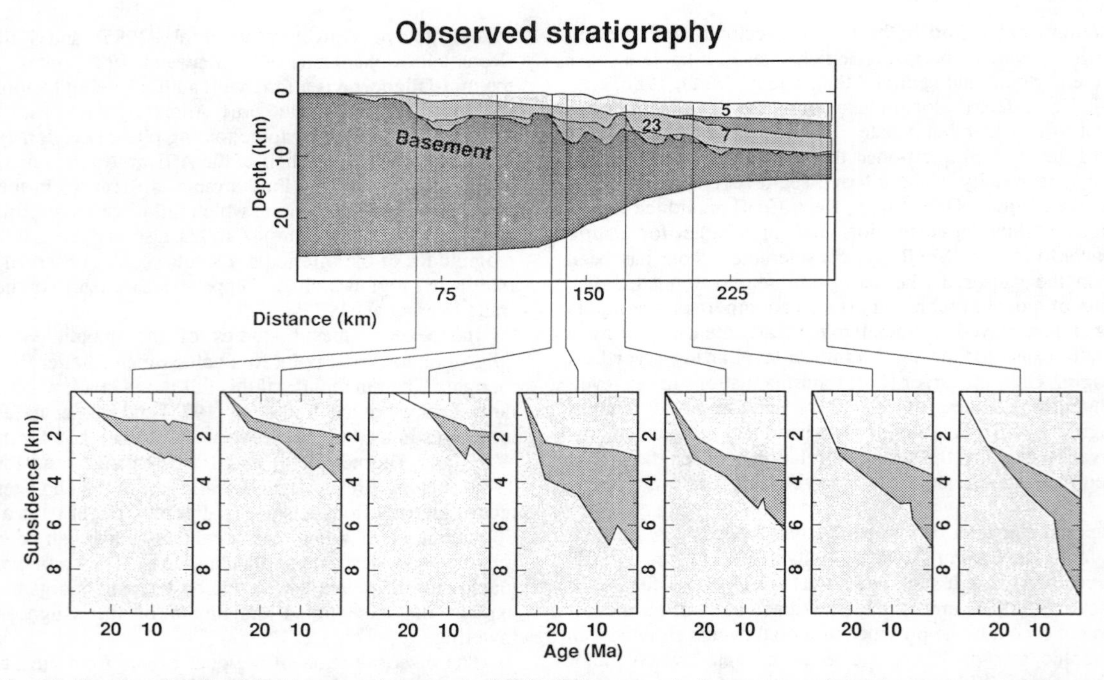

Fig. 13. Stratigraphy Gulf de Lions passive margin (NW Mediterranean). Lower part of the figure shows subsidence curves for different positions along stratigraphic cross section. Shading indicates unloading correction. [After Burrus et al., 1987].

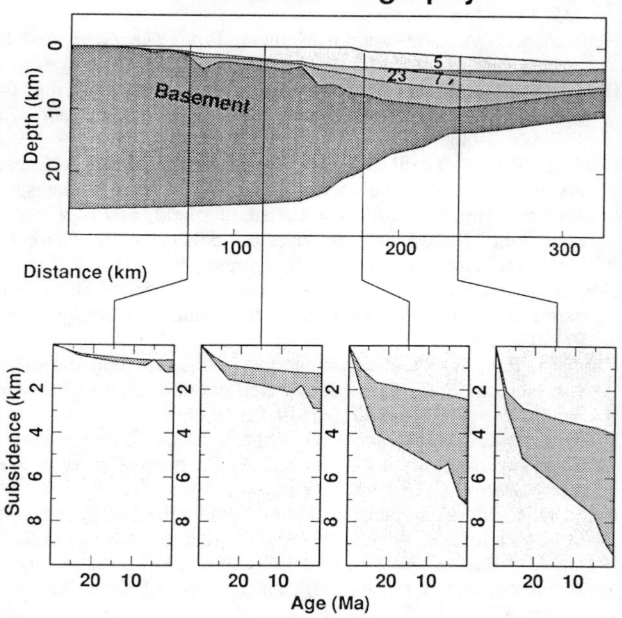

Fig. 14. Modeled stratigraphy Gulf de Lions passive margin for an elastic rheology of the lithosphere, adopting a strong compressive phase starting at the 7-5 Ma time interval. Curves showing predicted subsidence are given in the lower part of the Figure for positions along the modeled stratigraphic cross section.

shelf and in deeper parts of continental margin basins might simply result from subaereal exposure of the shelves. These authors argued that "the material eroded from the exposed shelves could have increased sediment supply to the actually restricted submarine shelf, stimulating increased slope failure and submarine erosion".

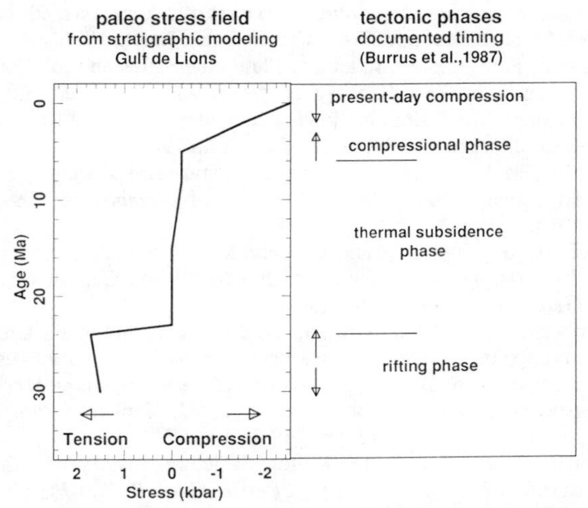

Fig. 15. The paleo-stress field inferred from the stratigraphic modeling (Figure 14) with timing of tectonic phases [After Burrus et al., 1987].

It seems that the essential factor controlling the inter-basin correlation of unconformities is the ratio of surface gradient to differences in water depth across the basin. Hence, care should be taken in selectively interpreting the occurrence of intrabasinal correlations solely in terms of eustatic changes in sea level.

The stratigraphic modeling of the U.S. Atlantic margin described in this paper and similar modeling for the North Sea Basin [Kooi et al., 1988; see also Lambeck et al., 1987], strongly suggests that tectonics might be the controlling factor underlying the apparent sea-level record even during periods with a non-icefree world.

Conclusions

Numerical modeling demonstrates that the incorporation of intraplate stresses in quantitative models of basin evolution can predict a succession of onlap and offlap patterns such as observed along the flanks of the U.S. Atlantic margin. Such a stratigraphy can be viewed as the natural consequence of the short-term narrowing of basins by moderate fluctuations in intraplate stress levels, superimposed on the long-term broadening of the basin with cooling since its formation. The effect of intraplate stresses on vertical motions is magnified when the lithosphere is treated having a depth-dependent, rather than elastic, rheology. For intraplate stress levels close to the lithospheric strength, vertical motions of the lithosphere of the order of a few kilometers are induced. A paleostress field inferred from the U.S. Atlantic margin stratigraphy is characterized by a transition from overall tension during Mesozoic times to a regime of more compressional character during Cenozoic times. These findings are in agreement with the outcome of a similar analysis of North Sea Basin stratigraphy and strongly suggest that the short-term apparent sea-level record of the basins at both sides of the Northern and Central Atlantic reflects the tectonic evolution of the Atlantic and global tectonic effects. Differential subsidence across passive margins provides a criterion to discriminate eustatics from tectonism. Stress-induced subsidence and uplift explains the observed record of vertical motions in for example the Gulf de Lions basin.

Acknowledgments. Partial support for this work was provided by NATO grant 0148/87. Thanks are due to Kurt Lambeck and Herb McQueen for various contributions. Julian Thorne, Randell Stephenson, Rinus Wortel, Charles Angevine, Seth Stein and Peter Ziegler are thanked for useful discussions. Peter Ziegler and Norm Sleep provided thoughtful reviews.

References

Bally, A.W., Musings over sedimentary basin evolution, *Philos. Trans Roy. Soc. Lond., A305*, 325-338, 1982.

Beaumont, C., The evolution of sedimentary basins on viscoelastic lithosphere: Theory and examples, *Geophys. J. R. Astron. Soc., 55*, 471-498, 1978.

Beaumont, C., Foreland basins, *Geophys. J. R. Astron. Soc., 65*, 291-329, 1981.

Beaumont, C., C.E. Keen, and R. Boutilier, On the evolution of rifted continental margins: comparison of models and observations for the Nova Scotia margin, *Geophys. J. R. Astron. Soc., 70*, 667-715, 1982.

Beaumont, C., and A.J. Tankard (editors), *Sedimentary Basins and Basin-Forming Mechanisms*, Can. Soc. Petrol. Geol. Memoir, *12*, 527 pp., 1987.

Bell, J.S., and D.I. Gough, Northeast-southwest compressive stress in Alberta: Evidence from oil wells, *Earth Planet. Sci. Lett., 45,* 475-482, 1979.

Bergerat, F., Stress fields in the European platform at the time of Africa-Eurasia collision, *Tectonics, 6,* 99-132, 1987.

Bergman, E.A., Intraplate earthquakes and the state of stress in oceanic lithosphere, *Tectonophysics, 132,* 1-35, 1986.

Bergman, E.A., and S.C. Solomon, Earthquake source mechanisms from body-waveform inversion and intraplate tectonics in the northern Indian Ocean, *Phys. Earth Planet. Int., 40,* 1-23, 1985.

Bessis, F., Some remarks on the study of subsidence of sedimentary basins, application to the Gulf de Lions margin (Western Mediterranean), *Mar. Petrol. Geol., 3,* 37-63, 1986.

Bodine, J.H., M.S. Steckler, and A.B. Watts, Observations of flexure and the rheology of oceanic lithosphere, *J. Geophys. Res., 86,* 3695-3707, 1981.

Burrus, J., F. Bessis, and B. Doligez, Heat flow, subsidence and crustal structure of the Gulf of Lions (NW Mediterranean): a quantitative discussion of the classical passive margin model, in *Sedimentary Basins and Basin-Forming Mechanisms,* edited by C. Beaumont and A.J. Tankard, Can. Soc. Petrol. Geol. Memoir, *12,* pp. 1-15, 1987.

Byerlee, J.D., Friction of rocks, *Pageoph, 116,* 615-626, 1978

Carter, N.L., and M.C. Tsen, Flow properties of continental lithosphere, *Tectonophysics, 136,* 27-63, 1987.

Cloetingh, S., Intraplate stresses: A new tectonic mechanism for fluctuations of relative sea level, *Geology, 14,* 617-621, 1986.

Cloetingh, S., H. McQueen, and K. Lambeck, On a tectonic mechanism for regional sealevel variations, *Earth Planet. Sci. Lett., 75,* 157-166, 1985.

Cloetingh, S., and F. Nieuwland, On the mechanics of lithospheric stretching and doming: a finite element analysis, *Geologie Mijnbouw, 63,* 315-322, 1984.

Cloetingh, S., and R. Wortel, Regional stress field of the Indian plate, *Geophys. Res. Lett., 12,* 77-80, 1985.

Cloetingh, S., and R. Wortel, Stress in the Indo-Australian plate, *Tectonophysics, 132,* 49-67, 1986.

Cloetingh, S.A.P.L., M.J.R. Wortel, and N.J. Vlaar, Evolution of passive continental margins and initiation of subduction zones, *Nature, 297,* 139-142, 1982.

Cochran, J.R., Effects of finite rifting times on the development of sedimentary basins, *Earth Planet. Sci. Lett., 66,* 289-302.

Engebretson, D.C., A. Cox, and R.G. Gordon, Relative motions between oceanic and continental plates in the Pacific Basin, *Geol. Soc. Am. Spec. Pap., 206,* 56 pp., 1985.

Goetze, C., and B. Evans, Stress and temperature in the bending lithosphere as constrained by experimental rock mechanics, *Geophys. J. R. Astron. Soc., 59,* 463-478, 1979.

Greenlee, S.M., F.W. Schroeder, and P.R. Vail, Seismic stratigraphy and geohistory of Tertiary strata from the continental shelf off New Jersey: Calculation of eustatic fluctuations from stratigraphic data, in *The Geology of North America, I-2, The Atlantic Continental Margin: U.S.,* edited by R.E. Sheridan and J.A. Grow, pp. 399-416, Geol. Soc. Am., 1988.

Gunn, R., A quantitative study of the lithosphere and gravity anomalies along the Atlantic Coast, *J. Franklin Inst., 237,* 139-154, 1944.

Hallam, A., A reevaluation of Jurassic eustasy in the light of new data and the revised Exxon curve, *Soc. Econ. Palaeont. Miner. Spec. Publ., 42,* in press.

Haq, B., J. Hardenbol, and P.R. Vail, Chronology of fluctuating sea level since the Triassic (250 million years to present), *Science, 235,* 1156-1167, 1987.

Heiskanen, W.A., and Vening Meinesz, F.A., *The Earth and its gravity field,* New York, Mc Graw-Hill, 470 pp.

Heller, P.L., and C.L. Angevine, Sea level cycles during the growth of Atlantic type oceans, *Earth Planet. Sci. Lett., 75,* 417-426, 1985.

Hoffman, P.F., R. Tirrul, J.E. King, M.R. St-Onge, and S.B. Lucas, Axial projections and modes of crustal thickening, eastern Wopmay orogen, northwest Canadian shield, in *Processes in continental lithospheric deformation,* edited by S.P. Clark Jr., Geol. Soc. Am. Spec. Pap., 218, in press.

Houseman, G., and P. England, A dynamical model of lithosphere extension and sedimentary basin formation, *J. Geophys. Res., 91,* 719-729, 1986.

Hubbard, R.J., Age and significance of sequence boundaries on Jurassic and Early Cretaceous rifted continental margins, *Am. Assoc. Petrol. Geol. Bull., 72,* 49-72, 1988.

Hubbard, R.J., J. Pape, and D.G. Roberts, Depositional sequence mapping to illustrate the evolution of a passive margin, *Am. Assoc. Petrol. Geol. Mem., 39,* 93-115, 1985.

Issler, D.R., and C. Beaumont, Thermal and subsidence history of the Labrador and West Greenland continental margin, in *Sedimentary Basins and Basin-Forming Mechanisms,* edited by C. Beaumont and A.J. Tankard, Can. Soc. Petrol. Geol. Memoir, *12,* pp. 45-69, 1987.

Jarvis, G.T., and D.P. McKenzie, Sedimentary basin formation with finite extension rates, *Earth Planet. Sci. Lett., 48,* 42-52, 1980.

Johnson, B., and A.W. Bally (editors), Intraplate deformation: characteristics, processes and causes, *Tectonophysics, 132,* 1-278, 1986.

Karner, G.D., Effects of lithospheric in-plane stress on sedimentary basin stratigraphy, *Tectonics, 5,* 573-588, 1986.

Kerr, A.R., Vail's sea-level curves are'nt going away, *Science, 226,* 677-678, 1984.

Kirby, S.H., Rock mechanics observations pertinent to the rheology of the continental lithosphere and the location of strain along shear zones, *Tectonophysics, 119,* 1-27, 1985.

Klein, R.J., and M.V. Barr, Regional state of stress in western Europe, in *Rock stress and rock stress measurements,* edited by O. Stephensson, pp. 33-44, Centek Publ., Lulea, 1986.

Klitgord, K.D., and H. Schouten, Plate kinematics and the Central Atlantic, in *The Geology of North America, vol. M., The Western North Atlantic Region,* edited by P.R. Vogt and B.E. Tucholke, pp. 351-378, Geol. Soc. Am., 1986.

Kominz, M.A., Oceanic ridge volumes and sea-level change - an error analysis, *Am. Assoc. Petrol. Geol. Memoir, 36,* 109-126, 1984.

Kooi, H., S. Cloetingh, and G. Remmelts, Intraplate stresses and the stratigraphic evolution of the North Sea Central Graben, *Geologie Mijnbouw,* in press.

Lambeck, K., S. Cloetingh, and H. McQueen, Intraplate stresses and apparent changes in sea level: the basins of north-western Europe, in *Sedimentary Basins and Basin-Forming Mechanisms,* edited by C. Beaumont and A.J. Tankard, Can. Soc. Petrol. Geol. Mem., *12,* pp. 259-268, 1987.

Lehner, P., and P.A.C. De Ruiter, Structural history of Atlantic margin of Africa, *Am. Assoc. Petrol. Geol. Bull., 61,* 961-981, 1977.

Letouzey, J., Cenozoic paleo-stress pattern in the Alpine foreland

and structural interpretation in a platform basin, *Tectonophysics, 132,* 215-231, 1986.

Manspeizer, W., Early Mesozoic history of the Atlantic passive margin, in *Geological Evolution of the United States Atlantic margin,* edited by C.W. Poag, pp. 1-23, Nostrand Reinhold, New York, 1985.

McAdoo, D.C., C.F. Martin, and S. Poulouse, Seasat observation of flexure: evidence for a strong lithosphere, *Tectonophysics, 116,* 209-222, 1985.

McAdoo, D.C., and D.T. Sandwell, Folding of oceanic lithosphere, *J. Geophys. Res., 90,* 8563-8569, 1985.

McKenzie, D.P., Some remarks on the development of sedimentary basins, *Earth Planet. Sci. Lett., 40,* 25-32, 1978.

Miall, A.D., Eustatic sea level changes interpreted from seismic stratigraphy: a critique of the methodology with particular reference to the North Sea Jurassic record, *Am Assoc. Petrol. Geol. Bull., 70,* 131-137, 1986.

Miller, K.G., and R.G. Fairbanks, Oligocene-Miocene global carbon and abyssal circulation changes, *Am. Geophys. Un. Geophys. Monogr., 32,* 469-486, 1985.

Miller, K.G., R.G. Fairbanks, and G.S. Mountain, Tertiary oxygen isotope synthesis, sea-level history, and continental margin erosion, *Paleoceanography, 2,* 1-19, 1987.

Parkinson, N., and C. Summerhayes, Synchronous global sequence boundaries, *Am. Assoc. Petrol. Geol. Bull., 69,* 685-687, 1985.

Philip, H., Plio-Quaternary evolution of the stress field in Mediterranean zones of subduction and collision, *Ann. Geophys., 5B,* 301-320, 1987.

Pitman, W.C. III, and X. Golovchenko, The effect of sea level change on the shelf edge and slope of passive margins, *Soc. Econ. Paleont. Miner. Spec. Publ., 33,* 41-58, 1983.

Poag, C. W., and J.S. Schlee, Depositional sequences and stratigraphic gaps on submerged Unites States Atlantic margin, *Am. Assoc. Petrol. Geol. Memoir, 36,* 165-182, 1984.

Price, R.A., Large-scale gravitational flow of supracrustal rocks, southern Canadian Rockies, in: *Gravity and Tectonics,* edited by K.A. deJong and R. Scholten, pp. 491-502, Wiley, New York, 1973.

Quinlan, G.M., and C. Beaumont, Appalachian thrusting, lithospheric flexure and the Paleozoic stratigraphy of the eastern interior of North America, *Can. J. Earth Sci., 21,* 973-996, 1984.

Richardson, R.M., S.C. Solomon, and N.H. Sleep, Tectonic stress in the plates, *Rev. Geophys. Space Phys., 17,* 981-1019, 1979.

Rona, P.A., and E.S. Richardson, Early Cenozoic plate reorganization, *Earth Planet. Sci. Lett., 40,* 1-11.

Royden, L., and C.E. Keen, Rifting processes and thermal evolution of the continental margin of eastern Canada determined from subsidence curves, *Earth Planet. Sci. Lett., 51,* 343-361, 1980.

Schlanger, S.O., High-frequency sea-level oscilations in Cretaceous time: an emerging geophysical problem, *Am. Geophys. Un. Geodyn. Series, 15,* 61-74, 1986.

Schlanger, S.O., and I. Premoli-Silva, Oligocene sealevel falls recorded in mid-Pacific atoll and archipelagic apron settings, *Geology, 14,* 392-395, 1986.

Schwan, W., The worldwide active middle/Late Eocene geodynamic episode with peaks at 45 and 37 My b.p. and implications and problems of orogeny and sea-floor spreading, *Tectonophysics, 115,* 197-234, 1985.

Shudofsky, G.S., S. Cloetingh, S. Stein, and R. Wortel, Unusually deep earthquakes in East Africa: constraints on the thermomechanical evolution of a continental rift system, *Geophys. Res. Lett., 14,* 741-744, 1987.

Sleep, N.H., Thermal effects of the formation of Atlantic continental margins by continental break up, *Geophys. J. R. Astron. Soc., 24,* 325-350, 1971.

Sleep, N.H., and N.S. Snell, Thermal contraction and flexure of mid-continent and Atlantic marginal basins, *Geophys. J. R. Astron. Soc., 45,* 125-154, 1976.

Smoluchowski, M., Uber eine gewisse Stabilitatsproblem der Elasticitatslehre und dessem Beziehung zur Entstehung Faltengebirge, *Bull. Int. Ac. Sci. Cracovie, 2,* 3-20, 1909.

Steckler, M.S., and U.S. Ten Brink, Lithospheric strength variations as a control on new plate boundaries: examples from the northern Red Sea region, *Earth Planet. Sci. Lett., 79,* 120-132, 1986.

Steckler, M.S., A.B. Watts, and J.R. Thorne, Subsidence and basin modeling at the U.S. Atlantic passive margin, in *The Geology of North America, v. I-2, The Atlantic Continental Margin: U.S.,* edited by R.E. Sheridan and J.A. Grow, pp. 399-416, Geol. Soc. Am., 1988.

Stephenson, R., and K. Lambeck, Isostatic response of the lithosphere with in-plane stress: Application to central Australia, *J. Geophys. Res., 90,* 8581-8588, 1985.

Tankard, A.J., Depositional response to foreland deformation in the Carboniferous of eastern Kentucky, *Am. Assoc. Petrol. Geol. Bull., 70,* 853-868, 1986.

Thorne, J., and Watts, A.B., Seismic reflectors and unconformities at passive continental margins, *Nature, 311,* 365-368, 1984.

Tucholke, B.E., and G.S. Mountain, Tertiary paleoceanography of the western north Atlantic Ocean, in *The Geology of North America, vol. M., The Western North Atlantic Region,* edited by P.R. Vogt and B.E. Tucholke, pp. 631-650, Geol. Soc. Am., 1986.

Turcotte, D.L., and J.L. Ahern, On the thermal and subsidence history of sedimentary basins, *J. Geophys. Res., 82,* 3762-3766, 1977.

Turcotte, D.L., and G. Schubert, *Geodynamics,* 450 pp., Wiley, New York, 1982.

Vail, P.R., R.M. Mitchum Jr., and S. Thompson III, Global cycles of relative changes of sea level, *Am. Assoc. Petrol. Geol. Memoir, 26,* 83-97, 1977.

Verwer, J.G., A class of stabilized three-step Runge-Kutta methods for the numerical integration of parabolic equations, *J. Comp. Appl. Math., 3,* 155-166, 1977.

Vink, G.E., W.J. Morgan, and W.L. Zhao, Preferential rifting of continents: a source of displaced terranes, *J. Geophys. Res., 89,* 10072-10076, 1984.

Wamel, W.A. van, On the tectonics of the Ligurian Apennines (northern Italy), *Tectonophysics, 142,* 87-98, 1987.

Watts, A.B.,Tectonic subsidence, flexure and global changes of sea level, *Nature, 297,* 469-474, 1982.

Watts, A.B., G.D. Karner, and M.S. Steckler, Lithospheric flexure and the evolution of sedimentary basins, *Philos. Trans. R. Soc. Lond., A305,* 249-281, 1982.

Watts, A.B., and J.Thorne, Tectonics, global changes in sea level and their relationship to stratigraphical sequences at the U.S. Atlantic continental margin, *Mar. Petr. Geol., 1,* 319-339, 1984.

Weissel, J.K., Anderson, R.N., and Geller, C.A., Deformation of the Indo-Australian plate, *Nature, 287,* 284-291, 1980.

Wiens, D.A., and Stein, S., Age dependence of oceanic intraplate seismicity and implications for lithospheric evolution, *J. Geophys. Res., 88*, 6455-6468, 1983.

Wiens, D.A., and Stein, S., Intraplate seismicity and stresses in young oceanic lithosphere, *J. Geophys. Res., 89*, 11442-11464, 1984.

Wise, S.W., J.E. Van Hinte, et al., Mesozoic-Cenozoic depositional environment revealed by Deep Sea Drilling Project Leg 93 drilling on the continental rise of the eastern Unites States: *Geol. Soc. Spec. Publ., 21*, 35-66, 1986.

Wortel, R., and S. Cloetingh, On the origin of the Cocos-Nazca spreading center, *Geology, 9*, 425-430, 1981.

Wortel, R., and S. Cloetingh, A mechanism for fragmentation of oceanic plates, *Am. Assoc. Petrol. Geol. Memoir, 34*, 793-801, 1983.

Ziegler, P.A., *Geological Atlas of Western and Central Europe*, 130 pp., Shell Int. Petrol. Mij (The Hague)/ Elsevier (Amsterdam), 1982.

Ziegler, P.A., Intraplate compressional deformations in the Alpine foreland - a geodynamic model, *Tectonophysics, 137*, 389-420, 1987.

Ziegler, P.A., Evolution of the Arctic-North Atlantic and the western Tethys, *Am. Assoc. Petrol. Geol. Memoir, 43*, 1988.

Ziegler, P.A., Post-Hercynian plate reorganization in the Tethys and Arctic North Atlantic, in *Triassic-Jurassic rifting in North-America-Africa*, edited by W. Manspeizer, Elsevier, Amsterdam, in press.

Zoback, M.D., Wellbore break-out and in-situ stress, *J. Geophys. Res., 90*, 5523-5530, 1985.

Zuber, M.T., Compression of oceanic lithosphere: an analysis of intraplate deformation in the Central Indian Ocean Basin, *J. Geophys. Res., 92*, 4817-4825, 1987.

EFFECTS OF ASTHENOSPHERE MELTING, REGIONAL THERMOISOSTASY, AND SEDIMENT LOADING ON THE THERMOMECHANICAL SUBSIDENCE OF EXTENSIONAL SEDIMENTARY BASINS

R. A. Stephenson[1], S. M. Nakiboglu[2], and M. A. Kelly[1]

[1]Geological Survey of Canada, 3303-33rd St. N.W., Calgary, Alberta T2L 2A7, Canada
[2]Department of Civil Engineering, King Saud University, P.O. Box 800, Riyadh 11421, Saudi Arabia

Abstract. The general characteristics of the evolution of many sedimentary basins, especially those found at rifted or sheared continental margins, are conventionally explained by models of crustal subsidence driven by the thermal contraction of an anomalously hot lithosphere. A rigorous three-dimensional thermomechanical model of sedimentary basin formation is examined that comprises a sedimentary basin overlying a thinned, rheologically layered (elastic-viscoelastic) lithosphere in turn overlying an inviscid asthenosphere. Active sediment deposition, at a specified rate, occurs within the overlying basin and the thermal state of the sedimentary basin is fully coupled with that of the lithosphere. Temperatures in the model are governed by the effects of vertical and horizontal thermal conduction such that the lithosphere-asthenosphere boundary is defined as a (partial) melt isotherm or phase change boundary which migrates vertically depending on the transient thermal state. Vertical deformations of the lithosphere result from the purely mechanical effects of sediment loading as well as from changes in the ambient temperature field. The temperature anomalies contribute to these deformations not only by setting up body forces but also by creating thermal in-plane forces and associated bending moments.

Conventional local isostatic and simple elastic plate models of sedimentary loading on a thermally perturbed lithosphere may be inadequate because of the important interactions of thermal and mechanical forces. Several thermomechanical processes inherent to the present model have the tendency to reduce or retard the thermal subsidence mechanism. These include the regional isostatic compensation of the thermal body forces, given regional compensation of the sediment load, and the implicit accompanying effects of the thermal bending moments; lithospheric thickening (or thinning) due to the basal phase change which is accompanied by the introduction (or removal) of latent heat; the displacement of hot asthenospheric material by sedimentary loading; and the insulating effect of the thickening, low thermal conductivity, sedimentary sequence atop the cooling, thickening thermal lithosphere.

Introduction

In this paper we reappraise the character and evolution of large-scale crustal subsidence resulting from the purely conductive cooling of a thinned and heated lithosphere. There is almost universal agreement that this kind of mechanism is responsible for the post-rift, first-order development of many of the earth's continental margin and intra-cratonic sedimentary basins. McKenzie [1978] presented a simple and concise quantitative formulation of the transient lithosphere temperatures and resulting crustal heat flux and subsidence patterns predicted by such a mechanism and his paper forms the benchmark for numerous subsequent studies of the regional development of specific sedimentary basins [e.g., Watts and Steckler, 1981; Bond and Kominz, 1984; Nunn et al., 1984]. McKenzie's formulation, in which lithospheric thinning results from pure shear extension, is based on the assumption that the resulting anomalous temperature field decays by transient conduction of heat in the vertical direction and subsidence occurs according to local isostasy. The effects of a thermally coupled asthenosphere and of a sedimentary layer of low conductivity are neglected. The impetus for undertaking the present investigations is to assess these and some other important physical processes that are implicit to McKenzie's concept of a sedimentary basin forming atop a cooling lithosphere and that may significantly alter the numerics of his model. The low-order effects of the processes that we consider here arise from lithosphere attributes that would typically be considered in models of other lithosphere geodynamic processes of similar scale but which hitherto have not been considered in quantitatively modeling the evolution of post-rift thermally-subsiding extensional sedimentary basins.

We present here a rigorous thermomechanical model of sedimentary basin evolution comprising a linear elastic-viscoelastic layered lithosphere coupled with both the sediment layer and underlying asthenosphere. The model allows us to consider, in particular, (1) the explicit effects on the heat budget of a cooling lithosphere of a partial liquidus-solidus boundary coinciding with the lithosphere- asthenosphere boundary, as is assumed to be the case in the development of oceanic lithosphere [e.g. Oldenburg, 1975], as well as the implicit thermal consequences of assuming a finite thickness for continental lithosphere [e.g., Morgan and Sass, 1984] on the cooling characteristics of thinned lithosphere. Kono and Amano [1978] considered these factors in modeling the thickening of continental lithosphere with age and supported their conclusions with seismological and other observations from North America; (2) the thermoelastic consequences of rheological layering in the lithosphere, including the generation of bending moments resulting in uplift of a cooling lithosphere capped by a relatively strong, elastic layer [Nakiboglu and Lambeck, 1985]. Beaumont et al. [1982] assumed that the cooler part of the rheologically layered lithosphere is elastic, and that the mechanical loads of sediment and water are therefore compensated regionally. However, in their models thermal subsidence was calculated assuming local isostasy and thermal bending moments were therefore neglected. Bills [1983] pointed out the possible role of thermal bending moments during vertical deformations of the lithosphere and considered that thermal flexural effects may significantly affect sedimentary basin formation; (3) the thermal equilibrium changes in the lithosphere resulting from the presence of the overlying, relatively lower thermal conductivity, sedimentary layer such as Beaumont et al. [1982] incorporated into their thermal subsidence model of the Nova Scotian passive margin sedimentary basin. Also considered is the influence on all of the above of horizontal heat conduction in the lithosphere during its post-extensional passive thermal subsidence phase, especially at rifted continental margins, with the concomitant large horizontal thermal gradients established there, shown by Cochran [1983] to significantly affect the character of syn-rift vertical deformations occurring at rifted margins.

Thermomechanical Model

Thermal Model

The thermal model governing equations for the two-dimensional conduction of heat in the sedimentary layer and underlying rifted lithosphere are

$$\frac{\partial T_s}{\partial t} = \kappa_s \left(\frac{\partial^2}{\partial x^2} + \frac{\partial^2}{\partial z^2}\right) T_s \tag{1a}$$

and

$$\frac{\partial T}{\partial t} = \kappa \left(\frac{\partial^2}{\partial x^2} + \frac{\partial^2}{\partial z^2}\right) T + \frac{q(x,y,z)}{\rho C_p}, \tag{1b}$$

where ρ, C_p, k, $\kappa = k/\rho C_p$ and T are respectively the density, specific heat capacity, thermal conductivity, diffusivity, and temperature distribution within the lithosphere; the subscript s denotes quantities pertaining to the sedimentary layer; $q(x,y,z)$ is the radiogenic heat source at point (x,y,z); and t is time. The radiogenic heat generation in the sediment layer is neglected in equation (1a) but this need not be the case. The origin of the coordinate system is at the surface of the crust with the z-axis pointing downwards and $x = 0$ arbitrarily chosen as the left-hand edge of the model (Figure 1). As the lithosphere moves down due to subsidence the coordinate system moves with it. In equations (1) the thermal properties within the layers are assumed to be independent of position, time, and temperature.

The boundary condition at the top of the sedimentary layer is $T_s(x,y,z,t) = 0°C$; $z = -h_s(x,y,z)$, where $h_s(x,y,t)$ is the sediment thickness at location x,y at time t, which is assumed to be given. The boundary conditions at the zero and distal edges of the sedimentary layer are imposed by letting $h_s(x,y,t) = 0$ for $x = 0$ and $x = x_{max}$. The continuity condition at the interface of the sediment layer and the lithosphere is $k_s \, \partial T_s(x,y,z,t)/\partial z = k \, \partial T(x,y,z,t)/\partial z$; $z = 0$.

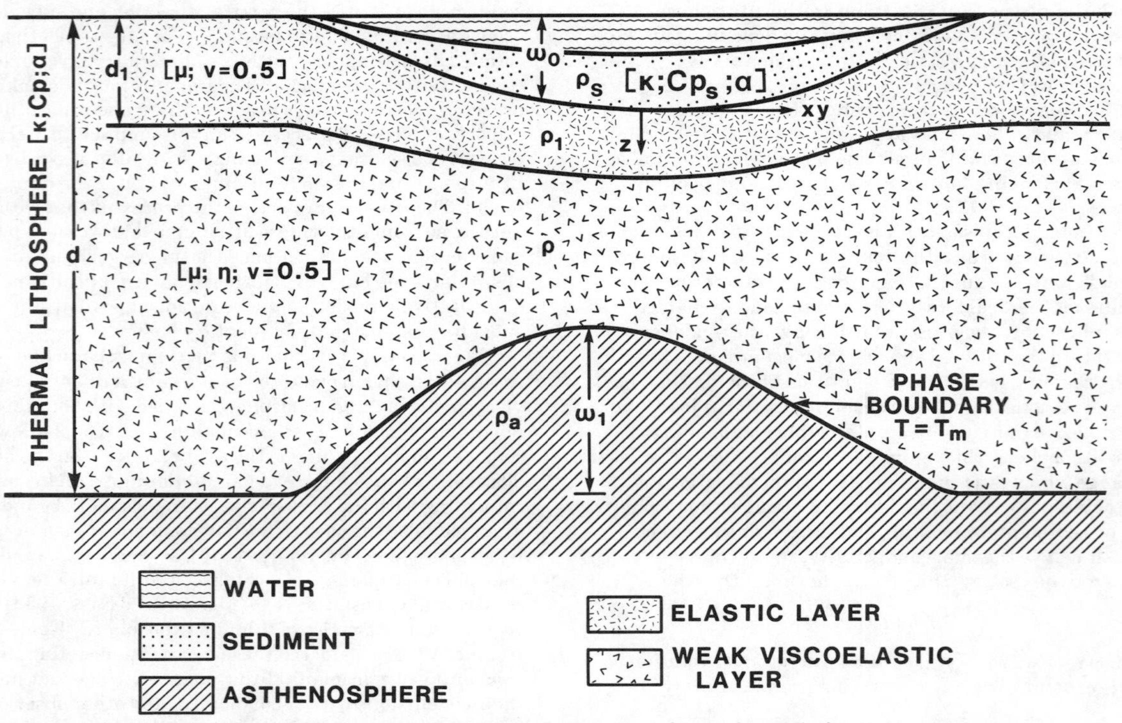

Fig. 1. Schematic representation of the coupled thermomechanical model of sedimentary basin subsidence; symbols and parameters are listed in Table 1.

The temperature at the surface of the lithosphere is equal to the temperature at the base of the sediment layer or is zero if $h_s = 0$. The temperatures at the vertical boundaries of the lithosphere are adjusted according to nearby temperatures as the model evolves.

At the base of the rifted lithosphere, taken to be a sharp phase boundary, the boundary condition is [Ozisik, 1980, p. 404]

$$\rho L \frac{\partial s}{\partial t} = \left[1 + (\frac{\partial s}{\partial x})^2 + (\frac{\partial s}{\partial y})^2 \right] \left(k \frac{\partial T}{\partial z} - k_a \frac{\partial T_a}{\partial z} - F \right), \quad (2)$$

where L is the latent heat, $z = s(x,y,t)$ is the depth of the phase boundary, F is the rate of convective heat flow from asthenosphere into lithosphere, and subscript a indicates asthenosphere. An analogous equation can be written for the unperturbed lithosphere outside the sedimentary basin and, combined with (2), gives

$$\rho L \frac{\partial}{\partial t}(s - s^o) \simeq \left[1 + (\frac{\partial s}{\partial x})^2 + (\frac{\partial s}{\partial y})^2 \right] k \frac{\partial}{\partial z}(T - T^o), \quad (3)$$

where the superscribed nought refers to the unperturbed region; it follows that $\partial(s\text{-}s^o)/\partial t$ is the relative velocity of the phase boundary with respect to the unperturbed base of the lithosphere. The thermal coupling of the lithosphere and underlying asthenosphere is fully accounted for in the preceding conservation equation but the equation is written entirely in terms of lithospheric temperatures within and outside the basin. If the unperturbed lithosphere is sufficiently old with finite thickness d and steady state temperature gradient $\partial T^o/\partial z = T_m/d$, equation (3) reduces to

$$\rho L \frac{\partial s}{\partial t} = \left[1 + (\frac{\partial s}{\partial x})^2 + (\frac{\partial s}{\partial y})^2 \right] k \left(\frac{\partial T}{\partial z} - \frac{T_m}{d} \right). \quad (4)$$

The initial condition in the lithosphere is deduced from the assumption of steady state temperature distribution at the end of the rifting process [e.g., McKenzie, 1978; Royden and Keen, 1970]; referring to Figure 1, in which ω_0 and ω_1 are the displacements of the upper and lower lithosphere boundaries respectively:

$$T = \frac{T_m}{d - \omega_0(x,y) - \omega_1(x,y)} z \; ; \quad t = 0. \quad (5)$$

Equations (1)-(5) were discretized and solved numerically using an explicit method of finite differences [cf. Hastaoglu, 1986].

Mechanical Model

The mechanical responses of a rheologically layered lithosphere to conductive cooling, and to loading of an overlying water and sediment layer, were developed in Nakiboglu and Lambeck [1985] and Lambeck and Nakiboglu [1981] respectively and combined and modified slightly to deduce the appropriate regional compensation model for sedimentary basin applications in Stephenson et al. [1987]. The upper elastic layer of the two-layered lithosphere model is the cooler, more competent part of the lithosphere with a thickness equal to its long-term effective elastic thickness. The base of the underlying viscoelastic layer is defined by the thermal thickness of the lithosphere. It is assumed that the two layers are perfectly coupled such that the vertical and horizontal deformations are continuous across the interface. The effective viscosity of the viscoelastic layer is expected to range with depth from asthenosphere viscosities of $\sim 10^{20}$ Pa s [Nakiboglu and Lambeck, 1983] to an effective long-term value of $\sim 10^{23}$ Pa s. The implied relaxation time constant is therefore $\tau \leq 4\text{x}10^4$ yr and is much smaller than both the thermal conduction or Fourier constant, $\sim 10^7$ yr, and the evolution period of sedimentary basins, $\sim 10^8$ yr. The governing equations for the asymptotic behavior of an elastic-viscoelastic incompressible lithosphere at times sufficiently large compared with τ are [Nakiboglu and Lambeck, 1985]

$$D\nabla^4\omega_t + (\rho_a - \rho_w) g w_t = 3\alpha\mu d_1^2 \nabla^2 (T_{01} - 2T_{11}) - 3\alpha g(\rho_1 d_1 T_{01} + \rho d T_{02}), \quad (6a)$$

for deformations ω_t induced by thermal forces, and

$$D\nabla^4\omega_m + (\rho_a - \rho_w) g w_m = (\rho_s - \rho_w) h_s g + w_t \rho_w g, \quad (6b)$$

for deformations ω_m arising from sediment and water load. In equations (6) the rigidity of the lithosphere μ is taken to be the same for both layers, d and d_1, the respective thicknesses of the thermal lithosphere and elastic layer; $D = \mu d_1^3/[6(1\text{-}\nu)]$ is the flexural rigidity of the elastic layer where ν is Poisson's ratio; α is the linear thermal expansion coefficient; ρ_1, ρ, ρ_a and ρ_w are the densities of crust, lower lithosphere, asthenosphere, and water; g is the acceleration of gravity; ∇^2 is the two-dimensional Laplace operator written on horizontal coordinates;

$$T_{i1} = \frac{1}{d_1} \int_0^{d_1} z^i T(z) dz \quad (i = 0,1)$$

are the thermal moments in the elastic layer; and

$$T_{02} = \frac{1}{d} \int_{d_1}^{d} T(z)\, dz$$

is the zero-order thermal moment in the lower lithosphere where temperatures $T(z)$ for successive timesteps are deduced from the numerical solution of the thermal model (equations 1-5). It is assumed that the rifted lithosphere is initially submerged in water, and that the accumulating sediment load replaces the water load. The total thermomechanical subsidence is $\omega = \omega_t + \omega_m$. As seen in equation (6a), the anomalous temperature field induces body forces proportional to T_{01} and T_{02} as well as thermal bending moments which increase with T_{11}, μ, and d_1. Both of these effects are compensated regionally as are the surface loads.

The Green's function for the differential equations (6a) and (6b) and a semianalytical procedure for convolving it over cylindrical loads is given by Nakiboglu and Lambeck [1985] and was used by Stephenson et al. [1987]. The Green's function appropriate for line loads normal to a two-dimensional section across a passive continental margin is

$$G(|\mathbf{x} - \mathbf{x}'|) = \frac{\ell^3}{2\sqrt{2}} exp(-u)[cos(u) + sin(u)], \quad (7)$$

where the flexural distance is

$$l = \left[\frac{D}{(\rho_a - \rho_w) g} \right]^{\frac{1}{4}}$$

and

$$u = \frac{|\mathbf{x} - \mathbf{x}'|}{\sqrt{2} l}$$

with **x** and **x′** being the vectors of the deformation and loading positions respectively. Convolving the Green's function with the right hand side of equations (6) yields the solution as

$$\omega_t(x) = -\frac{9\alpha}{2\sqrt{2} d_1} \int_{x'} exp(-u)[cos(u) - sin(u)]$$

$$\times [T_{01}(x') - 2T_{11}(x')]\, dx' \tag{8a}$$

$$- \frac{3\alpha}{2\sqrt{2} (\rho_a - \rho_w) l} \int_{x'} exp(-u)[cos(u) - sin(u)]$$

$$\times [\rho_1 d_1 T_{01}(x') + \rho d T_{02}(x')]\, dx'$$

and

$$\omega_m(x) = \frac{1}{2\sqrt{2}(\rho_a - \rho_w) l} \int_{x'} exp(-u)[cos(u) + sin(u)] \tag{8b}$$

$$\times [(\rho_s - \rho_w) h_s(x') + \rho_w \omega_t(x')]\, dx'.$$

The first integral in equation (8a) represents the deformation arising from bending moments set in the elastic layer by the anomalous temperature field. The second term gives the regional isostatic deformation caused by the changes in the weight of the crustal and subcrustal lithosphere arising from thermal contraction. The mechanical deformation ω_m in equation (8b) arises from the additional loads of sediment replacing water and of water infilling thermal subsidence. Equations (8) are evaluated across the continental margin by discretizing the mechanical and thermal loads in a manner consistent with the finite difference discretization of the thermal model.

The basic input variable to the model calculations is β, the lithosphere stretching parameter as defined by McKenzie [1978], from which initial values of ω_0 and ω_1 are determined assuming local isostasy at the cessation of lithosphere rifting.

Model Results

Asthenosphere Partial Melt

Figure 2 shows thermal subsidence and lithosphere thickness as functions of the square root of time for several values of β for the present model. No sediments are included so that the evolving basins are water-filled only and the basins are taken to be in a state of local isostasy. Model parameters are those listed in Table 1. The degree of partial melt in the asthenosphere beneath the stretched lithosphere is taken to be 5% so that the value of L used in the computations is 5% of the value listed in Table 1, following the method of Oldenburg [1975]. It should be remembered that the model implicitly accounts for the introduction of heat equally to the base of the

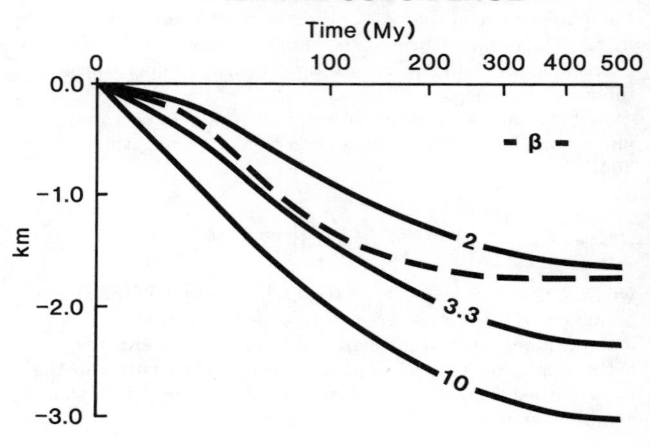

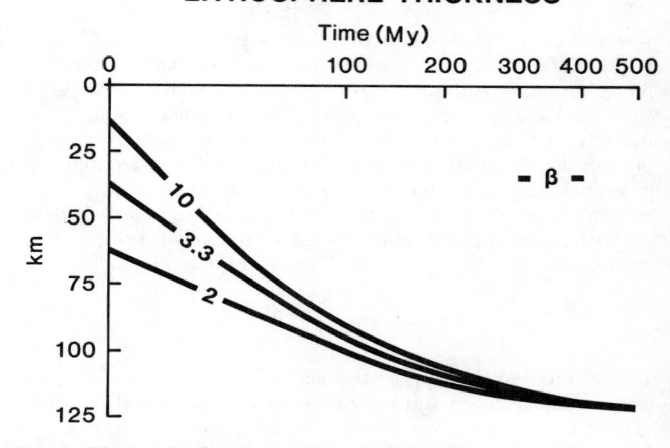

Fig. 2. Thermal subsidence (top) and lithosphere thickness (bottom) for various βs as functions of $t^{1/2}$, the former compared to equivalent β=2 McKenzie [1978] model (dashed curve). No sedimentation; local isostasy.

stretched and to the base of the unperturbed lithosphere by mantle conduction and convection (equation 3) such that the thickness of the unperturbed lithosphere remains constant. In other words, the thickness of the unperturbed lithosphere is assumed to be the finite thickness of the continental lithosphere. The slope of the linear equation relating lithosphere thickness and the square root of time (for times less than about 50 Myr) is approximately 8.6 km $Myr^{-1/2}$ for the β=10 curve in the lower half of Figure 2. This compares to the conventionally adopted value of 9.4-9.6 km $Myr^{-1/2}$ for oceanic lithosphere [Parker and Oldenburg, 1973; Oldenburg, 1975]. The present model reproduces this value for a β of approximately 12.

Superimposed on the top half of Figure 2 (dashed curve) is the equivalent thermal subsidence predicted by the model of McKenzie [1978] for β=2 calculated using the present physical parameters where applicable. Because McKenzie's model does not incorporate into its heat budget the processes noted above,

TABLE 1: Thermal and mechanical parameters used in model calculations.

Parameter	Definition	Value
ρ_1	Mean crustal density	$2.7\text{x}10^3$ kg m^{-3}
ρ	Lower lithosphere density	$3.2\text{x}10^3$ kg m^{-3}
ρ_a	Asthenosphere density	$3.2\text{x}10^3$ kg m^{-3}
ρ_s	Sediment density	$2.5\text{x}10^3$ kg m^{-3}
ρ_w	Water density	$1.0\text{x}10^3$kg m^{-3}
k	Thermal conductivity of lithosphere	3.1 Wm^{-1} K^{-1}
k_s	Thermal conductivity of sediments	2.0 Wm^{-1} K^{-1}
C_p	Specific heat of lithosphere	1210 J kg^{-1} K^{-1}
C_{p_s}	Specific heat of sediment	900 J kg^{-1} K^{-1}
T_m	Melting point temperature of lithosphere	1200°C
g	Radiogenic heat source in crust	0.0/0.4 μWm^{-3}
α	Linear thermal expansion coefficient	$1.03\text{x}10^{-5}$ K^{-1}
L	Latent heat of melting of lithosphere	$4.2\text{x}10^5$ J kg^{-1}
d	Thickness of thermal lithosphere	125 km
μ	Effective rigidity of lithosphere	$0.39\text{x}10^5$ MPa
d_1	Effective elastic thickness of lithosphere	0/20/40 km
ν	Poisson's ratio of lithosphere	0.5
η	Newtonian viscosity of lower lithosphere	$\leq 10^{23}$ Pa s
τ	Relaxation constant of lower lithosphere	$\leq 4 \text{ x } 10^4$ yr

namely, the latent heat of crystallization derived from the solidifying asthenosphere and convective and conductive heat sources from the asthenosphere, it predicts a significantly earlier thermal equilibration of the stretched lithosphere.

Figure 3 demonstrates the effect of variations in the degree of partial melt in the asthenosphere on the rate at which the lithosphere thickens beneath, in this case, a water-filled thermally subsiding basin with $\beta = 2$ and in a state of local isostasy. Because the basins are water-filled (cf. the following section on sediment loading effects), the equilibrium thickness eventually attained by the lithosphere (not shown in the figure) will not depend on the degree of partial melt and will be the same for all three curves in Figure 3 ($d = 125$ km). Similarly, the rate of thermal subsidence is affected by the degree of partial melt but the asymptotic amount of thermal subsidence is not.

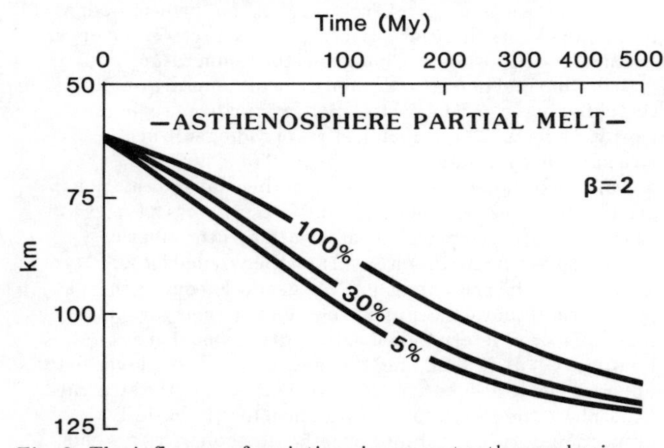

Fig. 3. The influence of variations in percent asthenospheric partial melt on lithosphere thickness as a function of $t^{1/2}$. $\beta = 2$; no sedimentation; local isostasy.

Regional Thermoisostasy

Figure 4 illustrates the thermal effects of a cooling, thinned lithosphere beneath a circular basin. The elastic layer thickness of the lithosphere is taken to be either 20 km or 40 km. Sediment accumulation is assumed to proceed at a constant rate of 0.2 mm yr^{-1} until the basin is completely filled. The thermal body forces are predominantly generated by the cooling of the lower lithosphere and they induce subsidence. The thermal body force subsidence is reduced appreciably if the elastic layer is thick (i.e. 40 km; solid lines), as the thermal load is more regionally compensated isostatically. On the other hand, the thermal bending moments, because of the combined effects of cooling of deeper and warming of shallow regions,

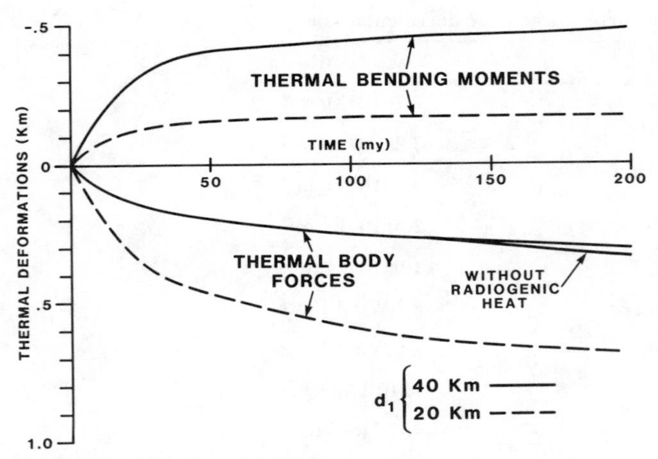

Fig. 4. The influence of regional thermoisostasy on deformations induced by thermal bending moments and thermal body forces at the centre of a 100 km radius circular sedimentary basin on a lithosphere with an elastic layer d_1 of 40 km (solid curve) and 20 km (dashed curve), the former also shown excluding the effects of radiogenic heating in the upper crust. Sedimentation rate is 0.2 mm yr^{-1}; $\beta = 2$.

generate a thermal uplift that increases in magnitude proportionately with elastic thickness. In Figure 4 the combined effects of the thermal bending moments and thermal body forces, for the particular circular basin under consideration, are seen to yield no net thermal subsidence if the elastic layer is 40 km thick. Therefore, neglecting the contribution of bending moments leads to overestimation of thermal subsidence, perhaps especially so during the later part of a basin's evolution history when the lithosphere is cooler and stronger.

Also shown on Figure 4 are the effects on the heat budget of the cooling lithosphere of radiogenic heat sources in the upper 15 km of the lithosphere. Changes to the temperature distribution and hence to thermal deformations are negligible compared to those from partial melting and from regional isostatic effects. The reducing effect of radiogenic heat sources on cooling accumulates only to about 20 m in 200 Myr.

Figure 5 shows the distribution of thermal deformations, arising from both body and in-plane forces, across a typical passive continental margin that formed by lithospheric stretching (with maximum $\beta = 4$) and then cooled for 200 Myr according to the present model. The general geometry of the passive continental margin was chosen to be comparable to those offshore eastern Canada studied and modeled by Beaumont et al. [1982]. The thickness of the elastic layer in the unperturbed lithosphere is taken to be 20 km. In the stretched lithosphere the elastic layer is assumed to thin initially, proportionately to the thinning of the lithosphere as a whole, and then to thicken in the same manner as the whole lithosphere cools and thickens. The base of the elastic layer therefore approximately coincides with the 200°C isotherm, probably a lower limit [Beaumont et al., 1982]. No sediments were loaded and other parameters are those listed in Table 1. In the abyssal part of the subsiding ocean basin, the effect of the thermal bending moments is one of uplift and therefore reduces net thermal subsidence, as observed in the circular basin shown in Figure 4. However, on the continental slope and shelf, the effect of the thermal bending moments is to accentuate thermal subsidence. In this region, deformations resulting from thermal bending moments are dominated by the peripheral, hence subsiding, effects of the rapid, substantially greater degree of lithospheric cooling occurring in the adjacent, deep ocean basin ($\beta = 4$ in this case) rather than the immediate, hence uplifting, effects of the directly underlying, less attenuated lithosphere. In particular, the in-plane thermo-elastic forces are solely responsible for the net thermal subsidence occurring over several tens of kilometres of continental shelf overlying unthinned lithosphere. This region would otherwise be in a state of upwarp due to the peripheral effects of the thermal body forces generated by the cooling margin and adjacent ocean basin. Net thermal subsidence in this region could permit

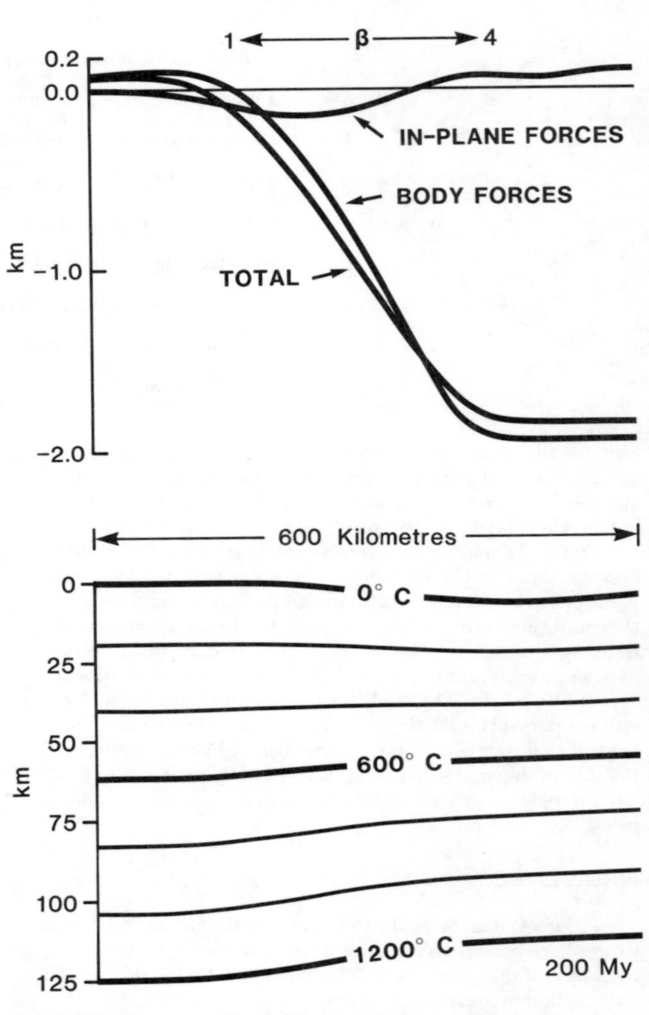

Fig. 5. The decomposition of thermal deformations (top) and thermal structure of the lithosphere (bottom) at a rifted, passive continental margin after 200 Myr of cooling. Initial β varied linearly 1-4. No sedimentation; regional isostasy ($d_1 = 20$ km).

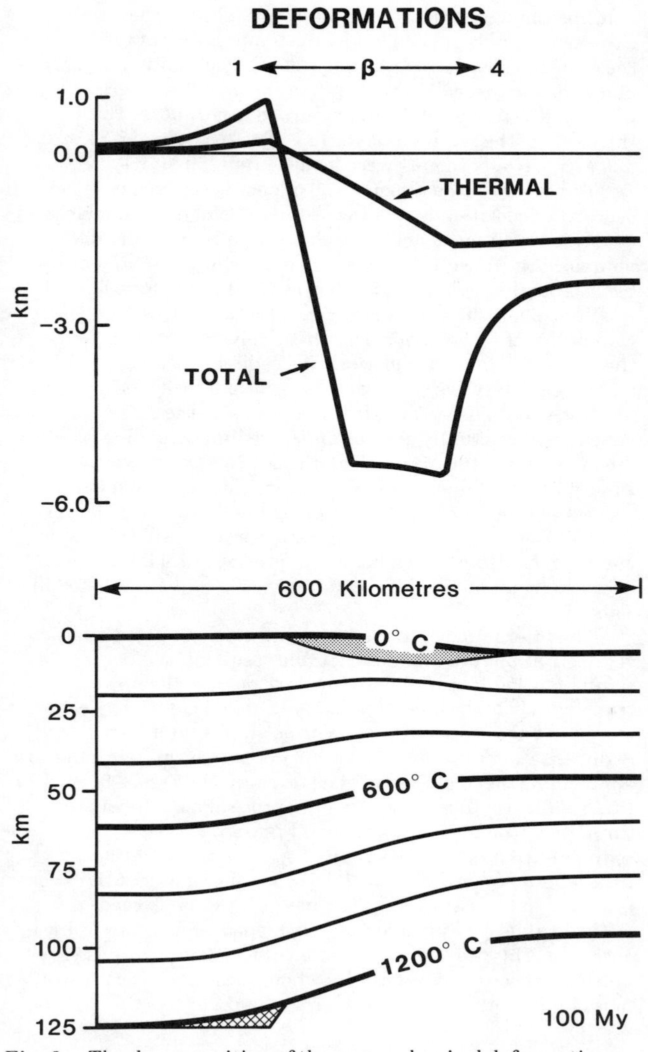

Fig. 6a. The decomposition of thermomechanical deformations (top) and thermal structure of the lithosphere (bottom) at a rifted, passive continental margin with sedimentation after 100 Myr of cooling. Initial β varied linearly 1-4. The shaded area represents the volume of lithosphere "melted" since cooling began. Maximum sedimentation rate is 0.075 mm yr^{-1}; local isostasy.

additional sediment loading cratonward of the edge of the thinned lithosphere which, in turn, due to the regional isostatic compensation of the sediment load, could extend the influence of platformal sedimentary basins on unthinned continental crust adjacent to passive continental margins.

Furthermore, regional thermoisostasy significantly limits the consequences of horizontal heat conduction in the lithosphere in the near margin part of the continental lithosphere, an area where they would otherwise have the tendency to produce thermal uplift (Figures 6). These effects across the continental margin are concentrated where the second derivative of the depth of the base of the lithosphere (hence isotherms) across the margin is greatest. They contribute to slightly greater thermal subsidence in the vicinity of the continental rise because the lithosphere there will tend to thicken more rapidly. Conversely, at the edge of the unthinned continent, horizontal heat transfer has the effect of vertically "melting" several kilometres of the adjacent lower lithosphere during the first 100 Myr or so of post-rift thermal evolution, as shown in the lower halves of Figures 6. In so doing it expands by several scores of kilometres the zone of attenuated continental lithosphere and therefore, if these temperature changes are isostatically compensated locally by the lithosphere (Figure 6a), may have the tendency to produce a flanking thermal uplift that peaks in magnitude at about 100 Myr (for the case with 5%

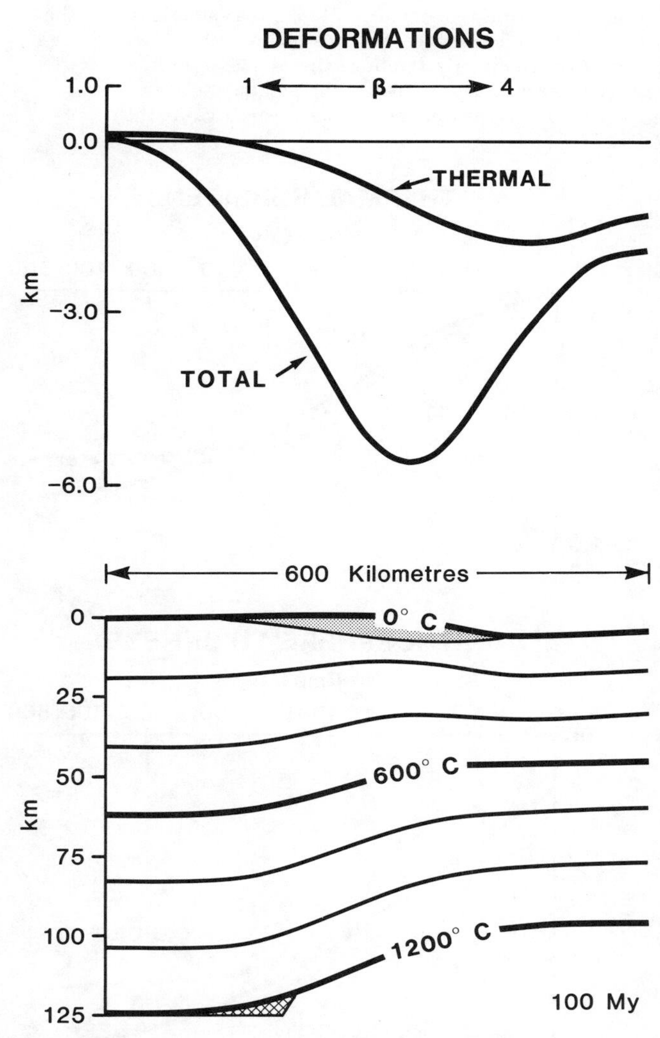

Fig. 6b. The decomposition of thermomechanical deformations (top) and thermal structure of the lithosphere (bottom) at a rifted, passive continental margin with sedimentation after 100 Myr of cooling. Initial β varied linearly 1-4. The shaded area represents the volume of lithosphere "melted" since cooling began. Maximum sedimentation rate is 0.075 mm yr^{-1}; regional isostasy ($d_1 = 20$ km).

partial melt). For the regional thermoisostasy case (Figure 6b), the flanking thermal uplift is much reduced and is significantly offset toward the centre of the continent. In either case, the newly melted zone subsequently solidifies and the flanking thermal uplift decays, possibly extending platformal basin formation even further cratonward.

Sediment Loading

The reducing effects of sediment accumulation on lithospheric cooling and subsidence are illustrated in Figure 7. A sediment layer hinders cooling and mantle solidification at the base of the lithosphere. The asymptotic thickness reached by a cooling lithosphere covered by sediments is appreciably less than its initial unperturbed thickness and is less than the asymptotic thickness reached by a cooling lithosphere not covered by sediments. It follows that the total thermal subsidence is less for a basin with a sediment cover, proportional to the thickness of the cover, because the

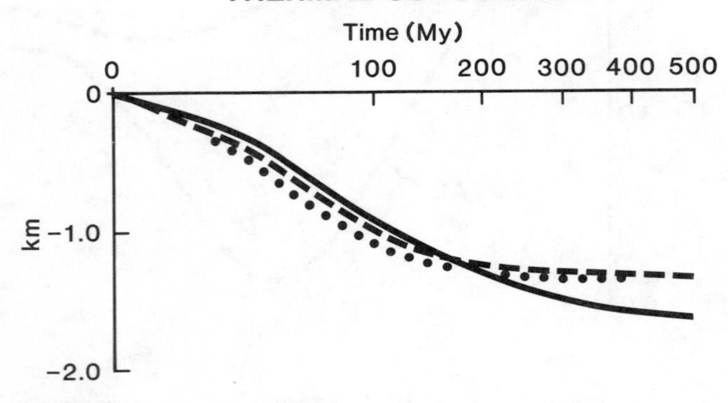

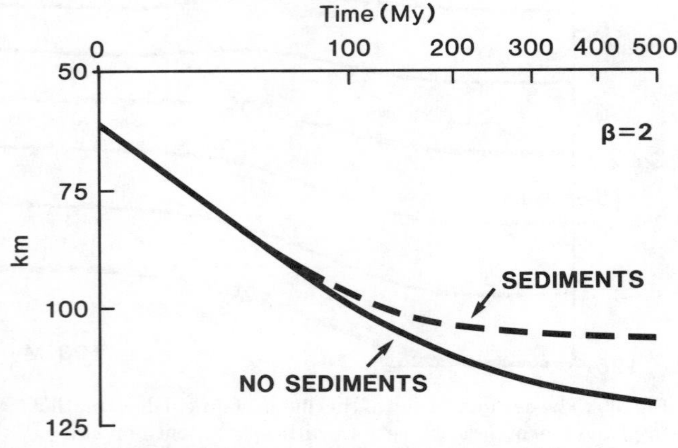

Fig. 7. Thermal subsidence (top) and lithosphere thickness (bottom) as functions of $t^{1/2}$ with sedimentation (dashed curves) and with no sedimentation (solid curves), the former compared to the temporally decaying exponential curve 1.4[exp(-t/62.75)-1] (dotted); $\beta=2$; local isostasy.

equilibrium temperatures ultimately reached in the lithosphere are higher. That is, the temperature at the top of the cooled lithosphere, at the base of the sedimentary layer, is non-zero and therefore greater than the initial temperature presumed at the top of the unperturbed lithosphere. The thickness of the sediment layer in Figure 7 attains 13.3 km at 500 Myr. (No sediments were loaded after 250 Myr.)

The other reason thermal subsidence is reduced because of sediment deposition is that the sediment load itself displaces an equivalent volume of hot lithosphere (that is, an equivalent volume of the upwelled asthenosphere below the sedimentary basin) into the region below the presumed asymptotic base of the lithosphere. The cooling and contraction of this material cannot therefore be invoked as ultimately contributing to the thermal subsidence of the overlying sedimentary basin. This effectively advective cooling of the lithosphere, as sediments are loaded onto its surface, is the reason why the rate of thermal subsidence is actually greater for the sediment loading case during the first 100 Myr or so in Figure 7. Also shown on Figure 7 (dotted line) is an exponentially decaying curve (with respect to time) such as those used by Bond et al. [1987], derived from McKenzie [1978], in modeling observed basin subsidence curves. The curve shown has an amplitude of 1.4 km and a decay constant of 62.8 Myr, equal to $d^2/\pi^2\kappa$ for values listed in Table 1.

The interaction of the sediment insulation effect with the sedimentation rate produces notable results. If the sedimentation rate is high in comparison with the upward transfer of heat in the sediment layer, then heat energy is stored and the sediment cover acts as an insulator. Sedimentation rates of ~0.05 mm yr^{-1} are too small to generate appreciable thermal inertia (dashed curves in Figure 8) and lithospheric cooling and subsidence proceed monotonously throughout the basin history as in Figure 7. If the sedimentation rate is very rapid, e.g. ~0.2 mm yr^{-1}, the lithosphere is overcooled initially when the sediment thickness is small. In this case, as the sediment thickness exceeds a critical value, cooling is completely stopped and a slow uplift is observed. This sedimentation rate value, 0.2 mm yr^{-1}, is very much an upper limit, especially when persisting over several tens of millions of years as shown in Figure 8.

Discussion

The purpose of the present work is to assess the effects of several hitherto little examined thermomechanical processes during the post-rift, passive thermal subsidence phase of sedimentary basin formation. In this, we ignore the geodynamic processes responsible for the rifting phase itself and adopt the conventional assumption that the rifting occurs in a short time compared to sedimentation history [e.g. McKenzie, 1978; Beaumont et al., 1982]. The main corollary of this assumption is that the rifting and subsidence phases of evolution are decoupled; that is, the effects of sedimentation, during a short period of rifting, on subsequent subsidence and temperature fields are assumed to be negligible. Some of the implications of these assumptions have been discussed by Beaumont et al. [1982] and Cochran [1983].

Also conventionally assumed in post-rift thermal subsidence models, and adopted here, is that the temperature field in the lithosphere has reached a steady state by the end of rifting. Hence, the rifting time, although short compared to the thermal time constant of the lithosphere, is taken to be long

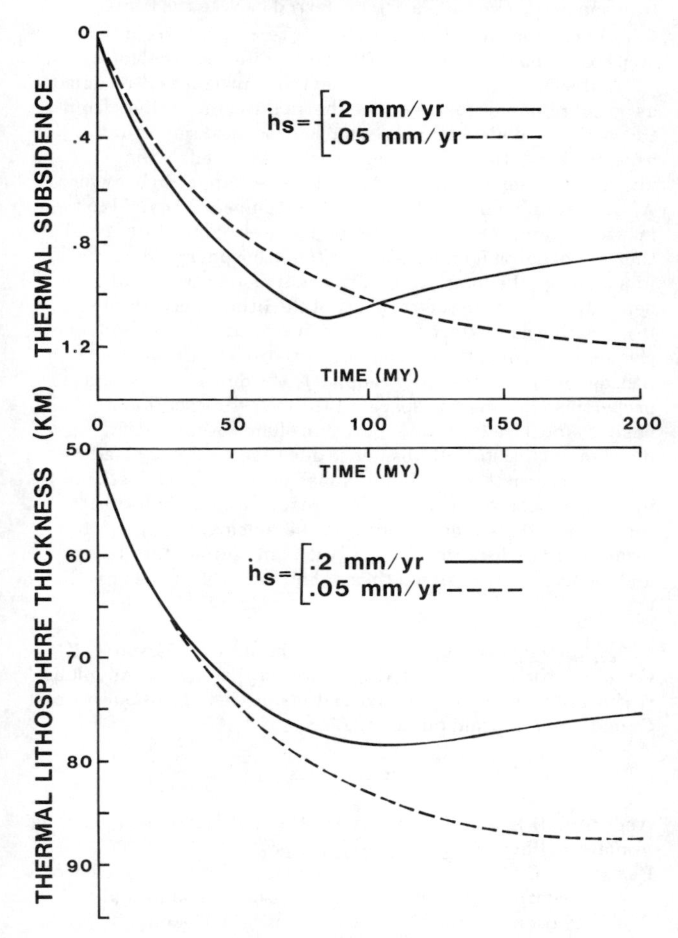

Fig. 8. Thermal subsidence (top) and lithosphere thickness (bottom) as functions of $t^{1/2}$ with sedimentation rates of 0.05 mm yr^{-1} (dashed curves) and 0.2 mm yr^{-1} (solid curves). $\beta = 2$; local isostasy.

enough (i.e. > 7 Myr) that short-wavelength variations in temperature have attenuated to a negligible level [e.g. Royden and Keen, 1980]. Furthermore, the lithosphere is assumed to be in local isostasy during rifting. This allows a single parameter, typically the stretching factor β, to completely characterize the rifted lithosphere if the stretching is assumed to occur uniformly with depth.

Much has been published recently on the basin subsidence manifestations of nondepth-dependent lithosphere stretching [e.g. Royden and Keen, 1980] and/or "pure" shear (as in McKenzie's formulation) versus "simple" shear [i.e. Wernicke, 1985] lithosphere stretching models [e.g. Mudford, 1988]. Nondepth-dependent and simple shear models can in some cases be equivalent. The most important and diagnostic elements of these processes appear to take place during the syn-rift development of sedimentary basins. Our investigations are restricted to the post-rift phase and are not intrinsically related to the mode of lithospheric thinning, whether by a stretching or some other mechanism, which initiated the driving thermal disequilibrium. Although we assume a pure shear, uniform stretching initiation of the lithosphere thermal anomaly our conclusions should apply equally as well to passive subsidence generated by these and other kinds of initiating mechanisms.

It is not inappropriate to model the thermal evolution of the stretched lithosphere beneath a continental margin sedimentary basin as that of a cooling thermal boundary layer because it is fully coupled to the oceanic lithosphere and the oceanic lithosphere is by convention modelled as a cooling thermal boundary layer [e.g. Parker and Oldenburg, 1973; Parsons and Sclater, 1977]. The present model provides a unified approach to the thermal evolution of oceanic lithosphere and to the thermal evolution of sedimentary basins. It suggests that a lithosphere stretching factor of about 12 is equivalent to oceanic crust generation, predicting a rate of lithosphere thickening similar to that observed for the oceans [e.g. Oldenburg, 1975].

The main implications of the cooling thermal boundary layer model for the formation of thermally subsiding sedimentary basins relate to the thermal consequences of the latent heat of crystallization of the partially molten asthenosphere beneath thinned lithosphere and of other heat sources from the asthenosphere to the base of the lithosphere. Geophysical evidence for significant volumes of partially melted material in the asthenosphere beneath mid-ocean ridges and beneath relatively young oceanic lithosphere is ample [e.g. Anderson and Sammis, 1970; Solomon, 1973]. Recent seismological studies have suggested that rifting of continental crust is accompanied, at least in places, by considerable generation of partial melt with subsequent magmatic crustal underplating [White et al., 1987; Keen and de Voogd, 1988]. The physical processes responsible for syn-rift generation of the partial melt phase, for example adiabatic depressurization [e.g. Foucher et al., 1982; McKenzie, 1984], perhaps augmented by the effects of local convective circulation in the mantle [Mutter et al., 1988], are assumed to have terminated by the onset of post-rift thermal subsidence so that only the heating effects of crystallization of the existing partial melt have been included in the calculations. Similarly, the potential buoyancy effects of a compositionally less dense melt [e.g. Brown and Beaumont, 1988] have also have excluded from consideration. Mantle processes resulting in a finite thickness for continental lithosphere [e.g. Karner et al., 1983; Morgan and Sass, 1984] require that heat is transported to the base of the lithosphere [e.g. Oldenburg, 1975], modeled here simply as an unspecified basal lithospheric heat flux, due to an asthenospheric temperature gradient or to convection in the asthenosphere.

These thermal processes, which may be intrinsic to the development of extensional sedimentary basins, in possibly varying degrees from basin to basin, have the tendency to reduce and/or retard the thermal subsidence mechanism. The heat released by a phase change at the base of the rifted lithosphere will have a delaying effect on its cooling. The commonly exponential form of geological subsidence patterns, as recorded by sedimentary basin stratigraphy, is justifiably interpreted in terms of post-extensional thermal conductive cooling, as postulated by McKenzie, but the quantification of such things as the onset of thermal subsidence (or, equivalently, the cessation of lithospheric rifting) and the degree of crustal and/or subcrustal extension prior to thermal subsidence (and concomitant heat input to the overlying sedimentary basin) measured by the direct comparison of back-stripped observed subsidence curves to McKenzie's [1978] model could lead to spurious conclusions.

The results suggest that the intrinsic cooling time constant for thinned lithosphere underlain by partially-molten asthenosphere may be significantly greater than the typically adopted figure of about 65 Myr (based on the conductive diffusion of heat from a slab) but that sediment loading effects, coupled to the thermal state of the lithosphere, cause the thermal subsidence curve to affect a form having the shorter decay constant. It follows that sedimentation rates within a sedimentary basin may be an important factor when inferring thermal - or tectonic - subsidence patterns and that some extensional basins potentially may subside thermally for longer periods of time than the upper limit of ~200 Myr typically assigned to them.

The quantitative methodology presented here efficiently accommodates the calculation of flexural deformations resulting from distributed surface loads and from the distributed thermal loads. If the upper part of the lithosphere behaves elastically in the time frame of sedimentary basin evolution [e.g. Karner et al., 1983; Quinlan and Beaumont, 1984] such that the former are flexurally compensated then the distributed thermal loads must also be flexurally compensated. Incorporation of deformations arising from thermal bending moments in a rheologically layered lithosphere and the regional compensation of these loads and those arising from thermal body forces are realistic attributes which can also strongly affect the computation of β. It has been shown that thermoelastic bending moments generated by the cooling of highly stretched continental lithosphere at the distal part of the continental margin - or by the cooling of the adjacent oceanic lithosphere - may generate subsidence in the proximal part of the margin where negligible stretching has taken place. Subsidence in this platformal part of the continental margin, generated by these means, are seen to enhance flexural subsidence arising from the adjacent continental shelf/rise sedimentary load.

Summary and Conclusions

A rigorous thermomechanical model of sedimentary basin evolution comprising a linear elastic-viscoelastic layered lithosphere coupled with both the sediment layer and underlying asthenosphere has been examined. The elastic upper layer in the model is defined by the long-term effective elastic thickness of the lithosphere. The lower warmer part of the lithosphere is relatively weak, and thermomechanical stresses relax there and migrate upwards in a short time compared with the time scale for thermal conduction. The base of the thermal lithosphere is taken to be a sharp phase boundary at melting point temperature which is moving due to not only thermal and mechanical deformations but also solidification of the upwelling asthenosphere. The thermal problem is solved by the method of finite difference considering the sedimentary layer, lithosphere, and the asthenosphere as a single thermal system with appropriate continuity and boundary conditions across the moving interfaces between the three layers. Radiogenic heat sources are assumed to be distributed uniformly within the top 15 km of the crust. The thermal and mechanical fields are coupled. The phase change and thermal moments and forces are calculated by numerical methods and are introduced, together with variable sediment and water loads, into a semianalytical convolution method for determining deformations. The deformations and sediment and lithosphere thicknesses are transferred back to the finite difference program. The procedure is repeated at fixed time steps based on the stability of the finite difference calculations.

Lithospheric response to thermal perturbances is modeled as one of regional isostasy. The thermal isostasy differs from the isothermal mechanical response in several important aspects. First, the density perturbations caused by the anomalous temperature field give rise to changes in body forces. Also, nonuniformity of the anomalous temperature field in the horizontal and vertical directions creates in-plane forces and thermal bending moments within the lithosphere. Most importantly, the magnitude of thermal moments and forces depends strongly on the rheology of the lithosphere, in contrast to an isothermal external load which is entirely independent of rheology. In turn, the thermal evolution of the thinned lithosphere is intrinsically coupled to the development and properties - in time and space - of the overlying sedimentary basin so much so that the separation of an isothermal mechanical (isostatic) subsidence due to sedimentary loading from the thermal - or tectonic - subsidence using back-stripping techniques, could lead to spurious results and perhaps to a fundamental misunderstanding of the thermal history of the lithosphere underlying thermally driven sedimentary basins, and, hence, to the regional thermal histories of the basins themselves.

Acknowledgments. The authors thank Alan Jessop (GSC, Calgary), Kurt Lambeck (ANU, Canberra), and an anonymous reviewer for their constructive criticism. Geological Survey of Canada contribution number 22088.

References

Anderson, D.L. and C. Sammis, Partial melting in the upper mantle, Phys. Earth Planet. Interiors, 3, 41-50, 1970.

Beaumont, C., C.E. Keen, and R. Boutilier, On the evolution of rifted continental margins: comparison of models and observations for the Nova Scotian margin, Geophys. J. R. Astron. Soc., 70, 667-715, 1982.

Bills, B.G., Thermoelastic bending of the lithosphere: implications for basin subsidence, Geophys. J.R. Astron. Soc., 75, 169-200, 1983.

Bond, G.C. and M.A.Kominz, Construction of tectonic subsidence curves for the early Paleozoic miogecline, southern Canadian Rocky Mountains: implications for subsidence mechanisms, age of breakup, and crustal thinning, Geol. Soc. Am. Bull., 95, 155-173, 1984.

Bond, G.C, P.A. Nickeson, and M.A. Kominz, Evidence for breakup of a late Proterozoic supercontinent and its effects on evolution of early Paleozoic passive margin basins (abstract), XIX General Assembly, IUGG, Abstracts, 1, 184, 1987.

Brown, K.C. and C. Beaumont, The effect of partial melting on syn-rift subsidence during pure extension of the lithosphere (abstract),Geol. Ass. Can.,Prog. with Abstracts, 13, A14, 1988.

Cochran, J.R., Effects of finite rifting times on the development of sedimentary basins. Earth Planet. Sci. Lett., 66, 289-302, 1983.

Foucher, J.P, X. LePichon, and J.C. Sibuet, The ocean continent transition in the uniform stretching model: role of partial melting in the mantle, Philos. Trans. R. Soc. London, Ser. A, 305, 27-43, 1982.

Hastaoglu, M.A., Numerical solution to moving boundary problems: application to melting and solidification, Int. J. Heat Mass Transfer, 29, 495-499, 1986.

Karner, G.D., M.S. Steckler, and J.A. Thorne, Long-term thermomechanical properties of continental lithosphere, Nature, 304, 250-253, 1983.

Keen, C.E. and B. de Voogd, The continent-ocean boundary at the rifted margin off eastern Canada: new results from deep seismic reflection studies,Tectonics, 7, 107-124, 1988.

Kono, Y. and M. Amano, Thickening model of the continental lithosphere, Geophys. J. R. Astron. Soc., 54, 405-416, 1978.

Lambeck, K. and S.M. Nakiboglu, Seamont loading and stress in the ocean lithosphere 2: viscoelastic and elastic-viscoelastic models, J. Geophys. Res., 86, 6961-6984, 1981.

McKenzie, D., Some remarks on the development of sedimentary basins, Earth. Planet. Sci. Let., 40, 25-32, 1978.

McKenzie, D., The generation and compaction of partially molten rock, J. Petrol., 25, 713-765, 1984.

Morgan, P. and J.H. Sass, Thermal regime of the continental lithosphere, J. Geodyn., 1, 143-166, 1984.

Mudford, B.S., A quantitative analysis of lithosphere subsidence due to thinning by simple shear, Can. J. Earth. Sci., 1988.

Mutter, J.C., W.R. Buck, and C.M. Zehnder, Convective partial melting. 1. A model for the formation of thick basaltic sequences during the initiation of spreading, J. Geophys. Res., 93, 1031-1048, 1988.

Nakiboglu, S.M., and K. Lambeck, A re-evaluation of the isostatic rebound of Lake Bonneville, J. Geophys. Res., 88, 10439-10447, 1983.

Nakiboglu, S.M., and K. Lambeck, Comments on thermal isostasy, J. Geodyn., 2, 51-65, 1985.

Nunn, A.J., A.D. Scardina, and R.H. Pilger, Thermal evolution of the North-Central Gulf Coast, Tectonics, 3, 723-740, 1984.

Oldenburg, D.W., A physical model for the creation of the lithosphere, Geophys. J. R. Astron. Soc., 43, 425-451, 1975.

Ozisik, M.N., Heat Conduction, 687 pp., John-Wiley and Sons, New York, 1980.

Parker, R.L. and D.W. Oldenburg, Thermal model of ocean ridges, Nature Phys. Sci., 242, 137-139, 1973.

Parsons, B. and J.G. Sclater, An analysis of the variation of ocean floor bathymetry and heat flow with age, J. Geophys. Res., 82, 803-827, 1977.

Quinlan, G.M. and C. Beaumont, Appalachian thrusting, lithospheric flexture, and the Paleozoic stratigraphy of the Eastern Interior of North America, Can. J. Earth Sci., 21, 973-996, 1984.

Royden, L. and C.E. Keen, Rifting process and thermal evolution of the continental margin of eastern Canada determined from subsidence curves, Earth Planet. Sci. Lett., 51, 343-361, 1980.

Solomon, S.C., Shear wave attenuation and melting beneath the Mid-Atlantic Ridge, J. Geophys. Res., 78, 6044-6059, 1973.

Stephenson, R.A., A.F. Embry, S.M. Nakiboglu, and M.A. Hastaoglu, Rift-initiated Permian to Early Cretaceous subsidence of the Sverdrup Basin, in Sedimentary Basins and Basin-forming Mechanisms, Memoir 12, edited by C. Beaumont and A.J. Tankard, pp. 213-231, Can. Soc. Petr. Geol., Calgary, 1987.

Watts, A.B. and M.S. Steckler, Subsidence and tectonics of Atlantic-type continental margins, Proc. 26th IGC, Oceanologica Acta, 143-153, 1981.

Wernicke, B., Uniform-sense normal simple shear of the continental lithosphere, Can. J. Earth Sci., 22, 108-125, 1985.

White, R.W., G.D. Spence, S.R. Fowler, D.P. McKenzie, G.K. Westbrook, and A.N. Bowen, Magmatism at rifted continental margins, Nature, 330, 439-444, 1987.

RELATIONSHIP OF EUSTATIC OSCILLATIONS TO REGRESSIONS AND TRANSGRESSIONS ON PASSIVE CONTINENTAL MARGINS

Charles L. Angevine

Department of Geology and Geophysics, University of Wyoming, Laramie, Wyoming 82071

Abstract. Third-order (1-10 my) sea-level oscillations may cause cycles of transgression and regression on passive continental margins. However, transgressions need not be synchronous with highstands nor regressions with lowstands. Theoretical modelling shows that a natural time lag exists between a harmonic eustatic fluctuation and the resultant shoreline oscillation. The magnitude of the time lag will range from zero up to one-fourth the period of the eustatic cycle, depending on the passive margin's geometry and subsidence rate. Because the time lag can vary from one margin to another, there is no reason to expect regression-related unconformities to be globally synchronous. The amplitude of the shoreline oscillation depends not only on the amplitude of the sea level change, but also on the time lag. Transgressions and regressions are most extensive when the time lag is small. Application of the model to the Oligocene section of the U.S. Atlantic margin indicates that a major mid-Oligocene regression may have coincided with a glacioeustatic lowstand.

Introduction

One of the more long-lived debates in the earth sciences concerns the origin of sedimentary sequences observed on continental margins [Seuss, 1906; Barrell, 1918; Sloss, 1962; Vella, 1965; Vail et al., 1977b; Pitman, 1978; Watts, 1982]. As defined by Vail et al. [1977a], these sequences are packages of conformable sediments, representing a time span of 1 to 10 my, that are bounded by unconformities or horizons that can be correlated with unconformities. Strata within individual sequences show continuous onlap onto the continental margin, whereas sequence boundaries mark abrupt seaward shifts in coastal onlap. Originally, Vail et al. [1977a] argued that onlapping sediments record a marine transgression, whereas sequence boundaries signal an abrupt regression. Subsequently, more detailed studies [Vail et al., 1984] of a small number of sequences showed that onlapping sediments in the upper parts are nonmarine, indicating that regressions are gradual rather than abrupt. Based on the apparent global synchroneity of many sequence boundaries, Vail et al. [1977b] and Haq et al. [1987] propose that transgressions and regressions are caused by oscillatory eustatic variations. Some workers [Hallam, 1984; Miall, 1986] question whether the resolution of biostratigraphic correlations is sufficient to show that sequence boundaries are truly synchronous between basins.

A fundamental difficulty in understanding the origin of sedimentary sequences is that transgressions and regressions can be caused by changes in the rate of tectonic subsidence or sedimentation, as well as eustatic variations [Sloss, 1962; Curray, 1964]. Hubbard et al. [1985], for instance, argue that, on the Newfoundland and Beaufort Sea margins of Canada, many sequence boundaries are associated with delta-lobe switching and non-deposition rather than sea-level fluctuations. Pitman [1978] and Pitman and Golovchenko [1983] demonstrate that, on subsiding passive margins, a decreasing sea-level history, with abrupt changes in the rate of fall, can produce an onlap history similar to that described by Vail et al. [1977a]. Variations in tectonic subsidence rates may ensure that sedimentary sequences differ in age between margins [Thorne and Watts, 1984; Parkinson and Summerhayes; 1985]. Watts [1982], Watts et al. [1982], and Watts and Thorne [1984] show that a significant fraction of coastal onlap may be explained by the increase in lithosphere rigidity that occurs as margins cool following rifting. Recent work by Cloetingh et al. [1985], Cloetingh [1986], and Karner [1986] shows that modest fluctuations in intraplate stress levels can cause vertical deflections of the lithosphere, thereby providing a tectonic explanation for cycles of transgression and regression. This mechanism may explain why many sequence boundaries seem to coincide with major plate reorganizations [Bally, 1981; Watts, 1982; Hallam, 1984; Hubbard et al., 1985; Hubbard, 1988].

Studies of oxygen isotopes in foraminifer from deep-sea sediments suggest that glacioeustatic variations have occurred during much of late Tertiary time [Matthews and Poore, 1980; Keigwin and Keller, 1984; Miller et al., 1985; Miller et al., 1987; Moore et al., 1987]. Figure 1 shows one recent attempt [by Miller et al., 1985] to correlate the glacioeustatic record with coastal onlap curves. Individual lowstands in the sea-level record are constrained by $\delta^{18}O$ measurements, but other parts of the curve are based on the assumption that short-term glacioeustatic variations cause symmetric rises and falls. The long-term sea-level fall is based on a reconstruction by Kominz [1984]. Onlap curve is based on seismic stratigraphy and well data. Miller et al. [1985] correlate the mid-Oligocene unconformity with the maximum rate of sea-level fall. The new sea-level curve from Exxon [Haq et al., 1987] seems to make a similar correlation. However, erosion during regressions, possible delays in the onset of sedimentation during the subsequent transgression, and uncertainties in biostratigraphic correlations all conspire to make precise dating of unconformities difficult. Miller et al. [1985] estimate an uncertainty of 1 to 2 my in the ages of the Oligocene unconformities. Such a large uncertainty makes it equally possible to correlate the mid-Oligocene unconformity with the glacioeustatic lowstand. It is important to note that not all Tertiary unconformities are associated with $\delta^{18}O$ shifts [Miller et al., 1987].

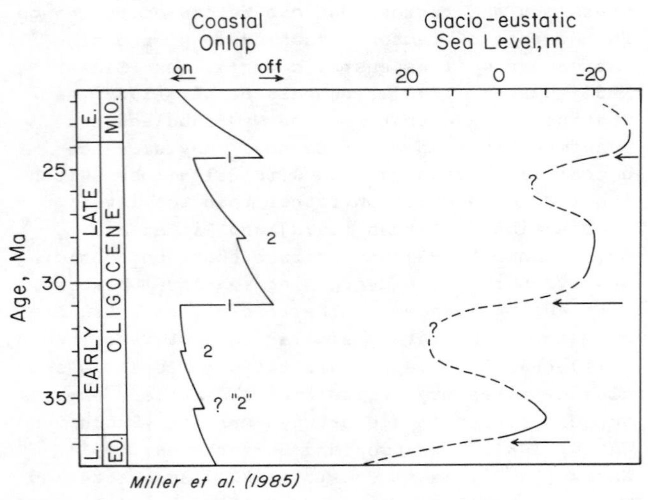

Fig. 1. Proposed correlation [after Miller et al., 1985] between Atlantic passive-margin stratigraphy and glacioeustatic oscillations during Oligocene time. The eustatic history is based on $\delta^{18}O$ data and presumes uniform rises and falls. Major sequence-bounding unconformities (indicated by 1; 2 indicates a minor event) are portrayed as being synchronous with maximum rates of sea-level lowering (shown by arrows).

It is not my intent to argue for, or against, third-order eustatic fluctuations. Resolution of that issue must await a better understanding of $\delta^{18}O$ variations in marine sediments [cf. Chappell and Shackleton, 1986; Moore et al., 1987]. Rather, the purpose of this paper is to investigate how shoreline positions will vary in response to third-order harmonic variations in sea level. I will show that, on passive margins, there is a natural time lag between shoreline movement and the causal eustatic oscillation. The time lag can range from zero up to one-fourth the period of the sea-level oscillation, depending on the margin's width and surface slope, and the maximum subsidence rate on the shelf. Thus, sequence boundaries associated with a particular sea-level fall may differ in age by as much as 2.5 my. On older passive margins, which tend to be broader and subside more slowly than younger margins, maximum regressions will be nearly synchronous with lowstands in the short-term eustatic oscillation. The tendency for sedimentation rates to lag behind sea-level changes will reinforce this near synchroneity.

Sedimentation Model

Pitman's [1978] model for clastic sedimentation on subsiding continental margins provides an ideal basis for investigating the interplay between tectonic subsidence and eustasy. Subsidence is modeled as a rigid rotation about a landward hinge line with a maximum subsidence rate of R_{SS} at the shelf edge (Figure 2). Sedimentation rates are assumed to be proportional to subsidence rates so that a graded slope (α) is maintained on the shelf and coastal plain. Under these conditions, the shoreline position (X_L) is governed by

$$\frac{dX_L}{dt} + \frac{R_{SS}}{D\,\alpha} X_L = \frac{1}{\alpha}(R_{SL} + A) \qquad (1)$$

where t is time, D the margin width, R_{SL} the rate of sea-level fall, and A the regional sedimentation rate. A tectonic time constant ($\tau_T = D\alpha/R_{SS}$, where D is the margin's width) controls the rate of shoreline migration. Pitman [1978] shows that if sea level falls at a constant rate (R_{SLO}) for a long period of time, compared to τ_T, then the

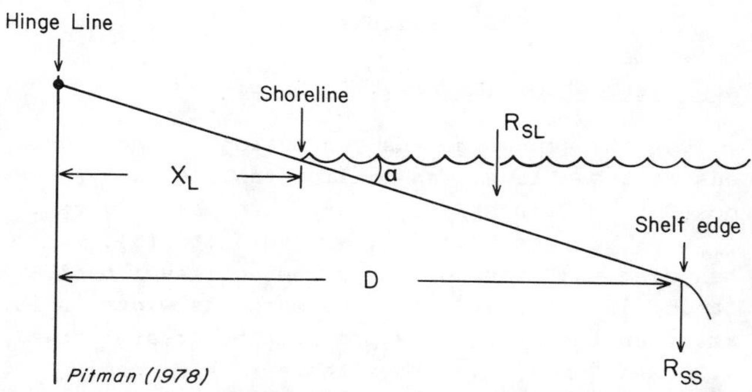

Fig. 2. Simplified model of a passive continental margin [after Pitman, 1978]. Subsidence rates increase linearly from zero at the hinge line to a maximum (R_{SS}) at the shelf edge. Erosion and sedimentation keep pace with subsidence and eustatic variations, maintaining a graded slope (α) on the coastal plain and continental shelf. R_{SL} is the rate of sea-level change (positive for falls). X_L is the shoreline location, measured from the hinge line, and D is the width of the margin.

shoreline will move to an equilibrium location (X_L^{EQ}) where the rate of tectonic subsidence is equal to the rate of sea-level lowering plus the regional sedimentation rate:

$$X_L^{EQ} = (R_{SLO} + A)\,\frac{D}{R_{SS}} \tag{2}$$

The strongest justification for using Pitman's [1978] sedimentation model is that it neatly reproduces the monotonous tapering-wedge geometry of passive margin sequences while yielding analytical, rather than numerical, results. Nevertheless, the model does have certain deficiencies: no allowance is made for the secular variations in subsidence rate and margin width that accompany lithospheric cooling, the flexural rigidity of the lithosphere is ignored, subsidence rates increase linearly across the margin, and sedimentation keeps pace with sea level and subsidence. Width and subsidence-rate variations can be ignored if the model is applied only to older passive margins where these changes occur slowly. Likewise, the omission of lithosphere flexure is of little consequence because sediment-loading on the continental shelf is balanced by erosion on the coastal plain. Flexure cannot be ignored if the shelf edge is rapidly prograding or retreating. The manner in which subsidence rates increase across any margin will depend on the configuration of rifting, the subsequent thermal evolution of the margin, and the lithosphere's rigidity. No two margins will be alike, so it makes sense, at least for the present, to consider the simplest possible case. Finally, on the time scale of third-order eustatic variations (1 to 10 my), sedimentation rates may not always be in equilibrium with sea level. This may affect the magnitude and timing of transgressions and regressions. I will present a sedimentation model, later on, that allows for departures from equilibrium.

Periodic Sea-Level Fluctuations

Equation (2) was derived under the assumption that sea level falls at a constant rate, consequently it cannot be used to understand how a shoreline responds to a continuously varying sea level, like the one shown in Figure 1. A reasonable approximation to the Oligocene glacioeustatic variation is shown in Figure 3. The sea-level history consists of a short-term component of magnitude Δh and period T, superposed on a long-term fall. The rate of sea-level change is

$$R_{SL} = R_{SLO} + \frac{2\pi\,\Delta h}{T}\cos\left(\frac{2\pi t}{T}\right) \tag{3}$$

Solving (1) and (3), I find that the shoreline location will vary according to

$$X_L(t) = X_L^{EQ} - \frac{\Delta h\,\xi_T}{\alpha\sqrt{(1+\xi_T^2)}}\sin\left[\frac{2\pi t}{T} + \tan^{-1}\left(\frac{1}{\xi_T}\right)\right] \tag{4}$$

where ξ_T ($= 2\pi\tau_T/T$) is a dimensionless tectonic parameter. The shoreline simply oscillates about the equilibrium position predicted by (2). Although the periods of the shoreline and sea-level oscillations are identical, there is a phase shift between the two. This phase shift translates into a time lag (τ) that ranges from 0 to T/4, depending on the value of ξ_T:

$$\tau = \frac{T}{2\pi}\tan^{-1}\left(\frac{1}{\xi_T}\right) \tag{5}$$

Figure 3 shows that regressions are synchronous with lowstands when ξ_T is large (>>1) and synchronous with the maximum rate of fall when ξ_T is small (<<1). The magnitude of the shoreline oscillation, a function of ξ_T also, varies from 0 to Δh/α . Regressions and transgressions are most extensive when ξ_T is large, but vanish as ξ_T decreases towards zero. This solution can be extended to more complicated sea-level histories using standard Fourier-series techniques.

Variable Time Lag Between Shoreline and Eustatic Oscillations

It is of interest to understand how much ξ_T can vary by, from one passive margin to another. As

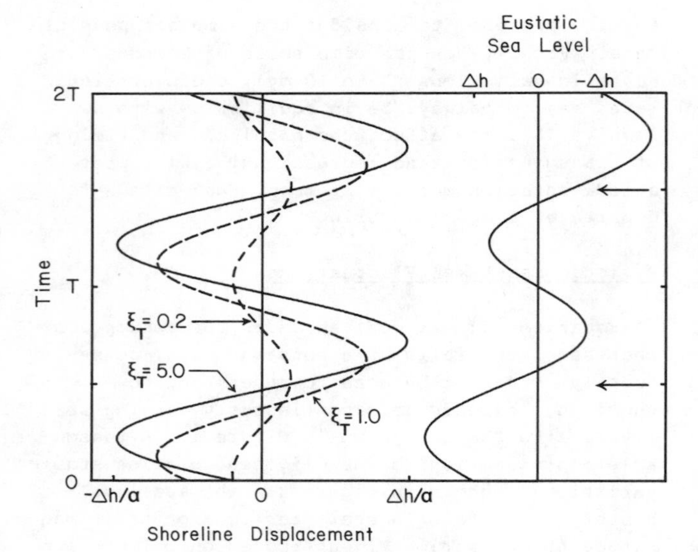

Fig. 3. Displacement of shoreline from equilibrium position, caused by a sinusoidal eustatic oscillation of period T and amplitude Δh (from Equation 4). The amplitude and phase of the shoreline oscillation vary with ξ_T ($= 2\pi\tau_T/T$), a dimensionless tectonic parameter that measures the influence of the tectonic time constant (τ_T). When ξ_T is small (<<1), the maximum regression occurs as sea level is falling most rapidly. For large values of ξ_T (>>1), the maximum regression is synchronous with the eustatic lowstand. From one margin to another, ages of sequence boundaries can vary by as much as one-fourth the period of the eustatic oscillation due to variations in subsidence rate and geometry.

was discussed earlier, the model's applicability is restricted to older margins where the width and subsidence rate are relatively constant. On older margins subsidence rates are low; a reasonable range is R_{SS} = 5 to 30 m/my [Pitman and Golovchenko, 1983]. The widths of margins, as measured from hinge line to shelf-edge, vary between 100 and 300 km. Although the choice of a margin slope is difficult (see discussion below), Pitman [1978] uses α = 1 to .1 m/km. Taking extreme values for each parameter, ξ_T can be as large as 56.5 or as small as 0.2, for third-order eustatic fluctuations (T = 1 to 10 my). Thus, transgressions or regressions on two different margins could differ in timing by as much as 2.5 my. Even if sedimentary sequences are caused by short-term eustatic fluctuations, there is no reason to expect sequence boundaries to be globally synchronous.

Discussion

Application to U.S. Atlantic Margin

For the mid-Oligocene regression to occur when eustatic sea level was falling fastest, as proposed by Miller et al. [1985; also, see Figure 1], ξ_T must be less than 0.2, according to (5). Assuming that the period of the sea-level oscillation is 6 my, and that the margin's width is 250 km, then the ratio of R_{SS}/α must be greater than 1.31 x 10^6 m/my, in order to meet this condition. Even with a relatively gentle slope of α = 0.1 m/km, the subsidence rate must be unreasonably large: R_{SS} = 131 m/my. I conclude that either the glacioeustatic fluctuation is not sinusoidal, or the sequence boundary is younger than indicated. A more realistic value for ξ_T can be found by assuming reasonable values for R_{SS} and α. Taking R_{SS} = 25 m/km and α = .5 m/km, then τ_T = 5 my and ξ_T = 5.2 , suggesting that the mid-Oligocene sequence boundary should be correlated to the eustatic lowstand. The actual time lag, calculated from (5), is only 180 ky. If the amplitude of the sea-level oscillation is 20 to 40 m (see Figure 1), then the shoreline could migrate 40 to 80 km seaward of its equilibrium position during the regression. Choosing a smaller slope for the shelf or a larger sea-level fluctuation would yield a more extensive regression.

Time Delays in Sedimentation

As was discussed earlier, the most unrealistic aspect of the model may be the assumption that sedimentation rates adjust instantaneously to variations in the rate of sea-level change. Studies of the U.S. continental margin [Curray, 1964; Swift, 1970; Pitman, 1978] suggest that much of the sediment carried to the margin by river systems is being deposited in river valleys and canyons that were flooded during the Holocene sea-level rise. Consequently, on some parts of the shelf, clastic sedimentation rates are low and relict sediments are exposed. A necessary consequence of low sedimentation rates is a gradual increase in water depth as tectonic subsidence continues.

Any lag in sedimentation rates (relative to a changing sea level) will tend to reinforce the near synchroneity of the shoreline and sea-level oscillations for third-order eustatic cycles. Under certain circumstances, the magnitudes of transgressions and regressions may even be amplified. This rather surprising result can be understood by considering the response of the shoreline to a sudden increase in the rate of sea-level fall. The shoreline will migrate toward the new equilibrium position predicted by (2), seaward of its previous location. Because sedimentation rates increase seaward of the shoreline, but lag behind the sea-level change, the sedimentation rate at the new equilibrium position will initially be too high, and the shoreline must migrate even farther seaward. As sedimentation rates equilibrate, the shoreline will migrate back to its predicted position. Amplification ought to be greatest when the sedimentary time lag is larger than the tectonic time constant.

In order to maintain a fixed slope, sedimentation (or erosion) rates at any location on the margin must equal the difference in tectonic subsidence rates between that location and the shoreline [Pitman, 1978]. The model can be made more realistic by introducing an explicit sedimentation time-constant (τ_S):

$$\tau_S \frac{\partial S}{\partial t} + S = \frac{R_{SS}}{D} (x - X_L) + A \qquad (6)$$

The term involving τ_s in (6) prevents sedimentation rates from adjusting instantaneously to changes in shoreline location. Assuming a sinusoidally-decreasing sea level (as in Figure 3), the magnitude of the shoreline oscillation becomes

$$B = \frac{\Delta h}{\alpha} \frac{\xi_T \{[\xi_T - \xi_S(1-\xi_S\xi_T)] + 1\}^{1/2}}{(1-\xi_S\xi_T)^2 + \xi_T^2} \qquad (7)$$

where ξ_S ($= 2\pi\tau_S/T$) is the dimensionless sedimentation parameter. The shoreline oscillation follows the eustatic oscillation with a phase of

$$\phi = \tan^{-1} \left[\frac{1}{\xi_T - \xi_S(1 - \xi_S\xi_T)}\right] \qquad (8)$$

When ξ_S goes to zero, (7) and (8) reduce to the amplitude and phase lag, respectively, given in (4). The sedimentation time constant is a measure of how rapidly sedimentation and erosion rates on the continental margin reach equilibrium following a transgression. Margins receiving higher influxes of terriginous sediment may reach equilibrium faster, simply because drowned estuaries are rapidly filled, allowing sediment to reach the continental shelf. The value of τ_S probably depends on a number of additional factors including, climate and lithology. A reasonable value for the sedimentation time constant might be τ_S = 1 my. However, it is expected that τ_S will vary from one area to another, even along a single margin.

Figures 4 and 5 show, respectively, the magnitude and phase-shift of the shoreline oscillation as functions of the sedimentation and tectonic parameters. Squares in the upper right-hand corner of each plot show where the parameters for third-order eustatic variations would lie if τ_T = 5 my and τ_S = 1 my. Reducing or increasing τ_S by an order of magnitude would shift the squares to the left or right, respectively, by one decade. Asterisks show the parameters appropriate to the Oligocene glacioeustatic fluctuation discussed earlier. The magnitude of the mid-Oligocene regression is increased by 10% whereas the time lag is reduced by 40% to 93 ky. If τ_S were an order of magnitude smaller, the sedimentation time lag would have a negligible effect on third-order shoreline oscillations.

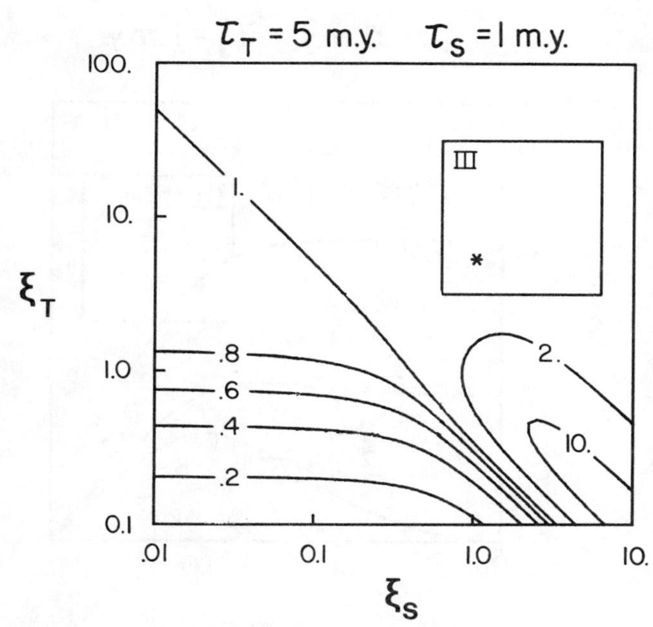

Fig. 4. Contours of the nondimensional amplitude of shoreline oscillation ($\alpha B/\Delta h$) as a function of tectonic (ξ_T) and sedimentation (ξ_S) parameters (from Equation 7). ξ_S ($= 2\pi\tau_S/T$) is a measure of how long geomorphic processes take to reestablish a graded shelf following a change of sea level (τ_S is the time constant for the return to grade). Unlike the previous case (Figure 3), where sedimentation kept pace with sea level (ξ_S = 0), here the amplitude of regressions and transgressions can be greater than $\Delta h/\alpha$. Square shows the parameter range for third-order eustatic oscillations (III: T=1-10 my), assuming τ_T = 5 my and τ_S = 1 my. Asterisk marks parameters appropriate for Oligocene sea-level variations on the U.S. Atlantic margin.

The greatest shortcoming of this model is that the slope of the shelf and coastal plain never

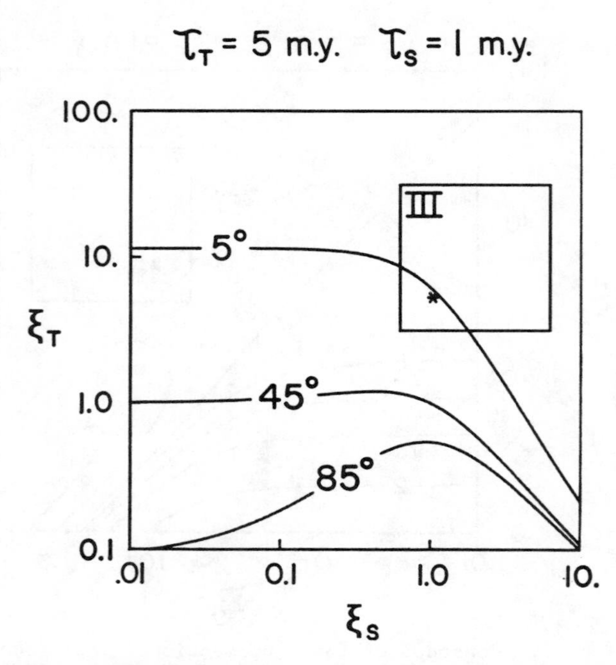

Fig. 5. Contours of the phase lag of the eustatic oscillation with respect to the shoreline oscillation, as a function of ξ_T and ξ_S (from Equation 8). Failure of sedimentation patterns to adjust instantaneously to changes in sea level reduces the time lag between shoreline and eustatic oscillations. See Figure 4 for more details.

changes, even though sedimentation rates lag behind sea level. However, it is doubtful that any sedimentation model parameterized by a single time constant will adequately characterize the adjustment of sedimentation patterns to a changing sea level. Unhappily, there seems to be a complete lack of theoretical models detailed enough to simulate sediment transport and deposition on continental shelves. The development of such models will greatly increase the usefulness of the elegant basin-evolution models that have appeared in recent years.

The effect of sea-level fluctuations on carbonate deposition has received a considerable amount of attention [Barrell, 1918; Sleep, 1976; Turcotte and Willemann, 1983; Cisne et al., 1984; Turcotte and Kenyon, 1984; Read et al., 1986; Schwarzacher, 1986]. These models differ from the present in that a flat platform develops on the continental margin rather than a graded slope. Nevertheless, water depths in carbonate systems tend to be out-of-phase with eustatic oscillations. The time lag can range from zero up to one-fourth the period of the eustatic cycle, depending on the period of the sea-level oscillation and the constant of proportionality between sedimentation rate and water depth. Unlike clastic-dominated shelves, the time lag for carbonate platforms is independent of subsidence rate and margin width.

Conclusions

Quantitative models of passive margin sedimentation are useful for understanding the timing and magnitude of the transgressions and regressions that result from a third-order, harmonic sea-level oscillation. In general, transgressions and regressions on passive margins are not synchronized with highstands and lowstands in the eustatic record. The time lag of the shoreline oscillation with respect to the eustatic oscillation, depends on the period of the eustatic oscillation, the slope of the coastal plain and continental shelf, the width of the margin, and the subsidence rate at the shelf-edge. The time lag is smallest when the period and subsidence rate are small, and the shelf slope and margin width are large. Shoreline and sea-level oscillations are approximately in-phase when the time lag is small. The maximum time lag corresponds to one-quarter of the period of the eustatic oscillation. With the maximum time lag, the shoreline oscillation is in-phase with the rate of sea-level change. For large time lags, regressions occur when sea level is falling fastest. The magnitudes of regressions and transgressions can vary from 0 to a maximum of $\Delta h/\alpha$, where Δh is the magnitude of the third-order eustatic oscillation and α is the shelf slope, depending on whether the time delay is large or small. Two sea-level variations, equal in amplitude but differing in period, can cause unequal regressions or transgressions.

If, as observations suggest, sedimentation cannot keep up with short-term sea-level changes, then time lags will be smaller and the magnitudes of regressions and transgressions will be larger. Better models of sediment transport and deposition must be developed in order to make realistic estimates of these effects.

Acknowledgements. I thank Michael A. Arthur and Paul L. Heller for helpful discussions during the course of this project. Acknowledgement is made to the Donors of the Petroleum Research Fund, administered by the American Chemical Society, for support of this research.

References

Bally, A.W., Basins and subsidence, Am. Geophys. Union Geodyn. Series, 1, 5-20, 1981.

Barrell, J., Rhythms and the measurement of geologic time, Geol. Soc. Am. Bulletin., 28, 745-904, 1917.

Chappell, J., and N.J. Shackleton, Oxygen isotopes and sea level: A reconciliation, Nature, 324, 137-140, 1987.

Cisne, J.L., R.F. Gildner, and B.D. Rabe, Epeiric sedimentation and sea level: synthetic ecostratigraphy, Lethaia, 17, 267-288, 1984.

Cloetingh, S., Intraplate stresses: a new tectonic mechanism for fluctuations of relative sea level, Geology, 14, 617-620, 1986.

Cloetingh, S., H. McQueen, and K. Lambeck, On a tectonic mechanism for regional sea level variations, Earth Planet. Sci. Lett., 75, 157-166, 1985.

Curray, J.R., Transgressions and regressions, in Papers in Marine Geology: Shepard Commemorative Volume, edited by R.L. Mills, pp. 175-203, MacMillan, New York, 1964.

Hallam, A., Pre-Quaternary sea-level changes, Ann. Rev. Earth Planet. Sci., 12, 205-243, 1984.

Haq, B.U., J. Hardenbol, and P.R. Vail, Chronology of fluctuating sea levels since the Triassic, Science, 235, 1156-1167, 1987.

Hubbard, R.J., Age and significance of sequence boundaries on Jurassic and Early Cretaceous rifted continental margins, Am. Assoc. Pet. Geol. Bull., 72, 49-72, 1988.

Hubbard, R.J., J. Pape, and D.G. Roberts, Depositional sequence mapping to illustrate the evolution of a passive continental margin, Am. Assoc. Pet. Geol. Memoir 39, 93-116, 1985.

Karner, G.D., Effects of lithospheric in-plane stresses on sedimentary basin stratigraphy, Tectonics, 5, 573-588, 1986.

Keigwin, L.D., and G. Keller, Middle Oligocene climatic change from equatorial Pacific DSDP Site 77, Geology, 12, 16-19, 1984.

Kominz, M.A., Oceanic ridge volumes and sea-level change: an error analysis, Am. Assoc. Pet. Geol. Memoir 36, 109-127, 1984.

Matthews, R.K., and R.Z. Poore, Tertiary $\delta^{18}O$ record and glacio-eustatic sea-level fluctuations, Geology, 8, 501-504, 1980.

Miall, A.D., Eustatic sea level changes interpreted from seismic stratigraphy: a critique of the methodology with particular reference to the North Sea Jurassic record, Am. Assoc. Pet. Geol. Bull., 70, 131-137, 1986.

Miller, K.G., R.G. Fairbanks, and G.S. Mountain, 1987, Tertiary oxygen isotope synthesis, sea level history, and continental margin erosion, Paleoceanography, 2, 1-19, 1987.

Miller, K.G., G.S. Mountain, and B.E. Tucholke, 1985, Oligocene glacioeustasy and erosion on the margins of the North Atlantic, Geology, 13, 10-13, 1985.

Moore, T.C., T.S. Loutit, and S.M. Greenlee, Estimating short-term changes in eustatic sea level, Paleoceanography, 2, 625-637, 1987.

Parkinson, N., and C. Summerhayes, Synchronous global sequence boundaries, Am. Assoc. Pet. Geol. Bull., 69, p. 685-687, 1985.

Pitman, W.C., Relationship between eustasy and stratigraphic sequences of passive margins, Geol. Soc. Am. Bull., 89, 1389-1403, 1978.

Pitman, W.C., and X. Golovchenko, The effect of sea-level change on the shelfedge and slope of passive margins, S.E.P.M. Special Publication 33, 41-58, 1983.

Read, J.F., J.P. Grotzinger, J.A. Bova, and W.F. Koerschner, Models for generation of carbonate cycles, Geology, 14, 107-110, 1986.

Schwarzacher, W., The effect of sea-level fluctuations in subsiding basins, Computers and Geosciences, 12, 225-227, 1986.

Seuss, E., The Face of the Earth, 2, 383 pp, Clarendon Press, Oxford. 1906.

Sleep, N.H., Platform subsidence mechanisms and "eustatic" sea-level changes, Tectonophysics, 36, 45-56, 1976.

Sloss, L.L., Stratigraphic models in exploration, J. Sed. Petrol., 32, p. 415-422, 1962.

Swift, D.J.P., Quaternary shelves and the return to grade, Marine Geology, 8, 5-30, 1970.

Thorne, J., and A.B. Watts, Seismic reflectors and unconformities at passive continental margins, Nature, 311, p. 365-368, 1984.

Turcotte, D.L., and P.M. Kenyon, Synthetic passive margin stratigraphy, Am. Assoc. Pet. Geol. Bull., 68, 768-775, 1984.

Turcotte, D.L. and R.J. Willemann, Synthetic cyclic stratigraphy, Earth Planet. Sci. Lett., 63, 89-96, 1983.

Vail, P.R., J. Hardenbol, and R.G. Todd, Jurassic unconformities, chronostratigraphy, and sea-level changes from seismic stratigraphy and biostratigraphy, A.A.P.G. Memoir 36, p.129-144,1984.

Vail, P.R., R.M. Mitchum, and S. Thompson, Relative changes of sea level from coastal onlap, A.A.P.G. Memoir 26, 63-81, 1977a.

Vail, P.R., R.M. Mitchum, and S. Thompson, Global cycles of relative changes of sea level, A.A.P.G. Memoir 26, 83-97, 1977b.

Vella, P., Sedimentary cycles, correlation, and stratigraphic classification, Trans. R. Soc. N.Z., 3, 1-9, 1965.

Watts, A.B., Tectonic subsidence, flexure, and global changes of sea level, Nature, 297, 469-474, 1982.

Watts, A.B., G.D. Karner, and M.S. Steckler, Lithospheric flexure and the evolution of sedimentary basins, Phil. Trans. R. Soc. London, Series A, 305, 249-281, 1982.

Watts, A.B., and J. Thorne, Tectonics, global changes in sea level and their relationship to stratigraphical sequences, Marine Pet. Geol., 1, 319-339, 1984.

CONTRASTING STYLES OF LITHOSPHERIC EXTENSION DETERMINED FROM CRUSTAL STUDIES ACROSS RIFT BASINS, EASTERN CANADA

C. E. Keen

Atlantic Geoscience Centre, Geological Survey of Canada, Bedford Institue of Oceanography, Dartmouth, Nova Scotia, Canada, B2Y 4A2

Abstract. Well constrained crustal cross-sections of the sedimentary basins on the rifted continental margin of the Grand Banks off eastern Canada have been derived by combining deep seismic reflection profiles, seismic refraction data, and borehole information. The cross-sections illustrate the crustal geometry resulting from extension during Mesozoic rifting and continental breakup. Furthermore, the syn- and post-rift subsidence history along the cross-sections was estimated , from which the amounts of crustal and mantle lithosphere stretching and thinning were inferred. The results show two contrasting regions which differ in the amount of lower lithospheric thinning. The Orphan Basin in the north is characterized by closely spaced, numerous faults in basement, which terminate at a mid-crustal decollement. Models of the subsidence history suggest that the sub-crustal lithosphere was thinned more than the crust during rifting. In contrast, half graben basins on the Grand Banks to the south are bounded by fewer, larger fault blocks, and the faults extend down to the lower crust or to the Moho, where they terminate in a decollement. The lower lithosphere does not appear to have thinned as much as the crust below these basins, although the thinning may extend well beyond the basin edges. These differences in the extensional geometry are related to thermal and mechanical properties of the extended lithosphere.

Introduction

About 1600 km of deep seismic reflection data has been collected across the rifted sedimentary basins and continental margins of the Grand Banks region, offshore Eastern Canada (Fig. 1). These data have been used to determine the geometry of extension during rifting, to constrain models of the evolution of the basins [Keen et al., 1987a,b] and to describe the deep structure of the continent-ocean boundary [Keen and de Voogd, 1987]. This paper describes how the seismic data, when combined with numerical models of basin subsidence, can delineate the relative roles played by the upper and lower parts of the lithosphere during lithospheric extension and rifting. The reader is referred to the above-mentioned papers for a complete description of the seismic data and of the subsidence modelling.

The thermal and mechanical response of the lithosphere to extension which occurs during rifting is a subject which has received much attention [see for example, McKenzie, 1978; Royden et al., 1980; Royden and Keen, 1980; Le Pichon and Sibuet, 1981; Hellinger and Sclater; 1983, Wernicke, 1985; Houseman and England, 1986; Buck, 1986; and Keen 1987a]. There are two main types of geometries suggested by these studies, and within each there are variants (Fig. 2).

Published in 1988 by the American Geophysical Union.

The first type involves necking of the lithosphere, so that extension produces thinning of both the upper and lower lithosphere over a given horizontal distance (Figs. 2a and b). The amount of thinning in the upper and lower lithosphere, given by β and δ, respectively, may be different. In these simple examples the Moho has been chosen as the level of detachment (Fig. 2b), so that β and δ represent crustal and sub-crustal thinning, respectively. Other levels of detachment may be more realistic choices. Different subsidence and thermal histories result from different distributions of β and δ. By comparison with the case $\beta = \delta$ (Fig 2a), $\beta > \delta$ will give higher ratios of syn-rift to post-rift subsidence, since the thermal anomaly, and hence the post-rift thermal cooling subsidence, caused by lower lithospheric thinning will be smaller. Conversely, $\beta < \delta$ will yield a lower ratio of syn-rift to post-rift subsidence than the $\beta = \delta$ case. This type of stretching can be conceived as necking of the lithosphere about a central point, producing in general a fairly symmetric rift. In order to conserve volume, the total amount of stretching in the upper lithosphere must be the same as that in the lower lithosphere [Hellinger and Sclater, 1983]. Exceptions probably occur if material was added or removed from the lithosphere during extension.

The second type of geometry involves offset of the lithosphere along a low angle detachment [Wernicke, 1985]. The low angle detachment or shear zone either extends through the entire lithosphere (Fig. 2c), or only through the upper lithosphere (Fig. 2d). In the latter case, the motion along the shear zone may be transfered to the lower lithosphere along a decollement, shown in Fig. 2d at the base of the crust, and extension in the lower lithosphere may be accommodated by ductile necking. Two different shapes of lower lithospheric necking are shown in Fig. 2d (curves 1 and 2, representing the base of the lithosphere). The distribution of lithospheric thinning given by β and δ will depend on the geometry of the shear zone (Fig. 2c). When only the upper lithosphere is offset along a shear zone, the mechanical response of the lower lithosphere to extension becomes important in deformation of that region (Fig. 2d). Unlike the necking models, there will be asymmetry in the subsidence and thermal histories of the basins. Given the geometry shown in Figs. 2c and d, the maximum crustal and lower lithospheric thinning are offset and the basins may experience more syn-rift subsidence and less post-rift subsidence than predicted by the simple necking model shown in Fig. 2a, given the same maximum β value. This

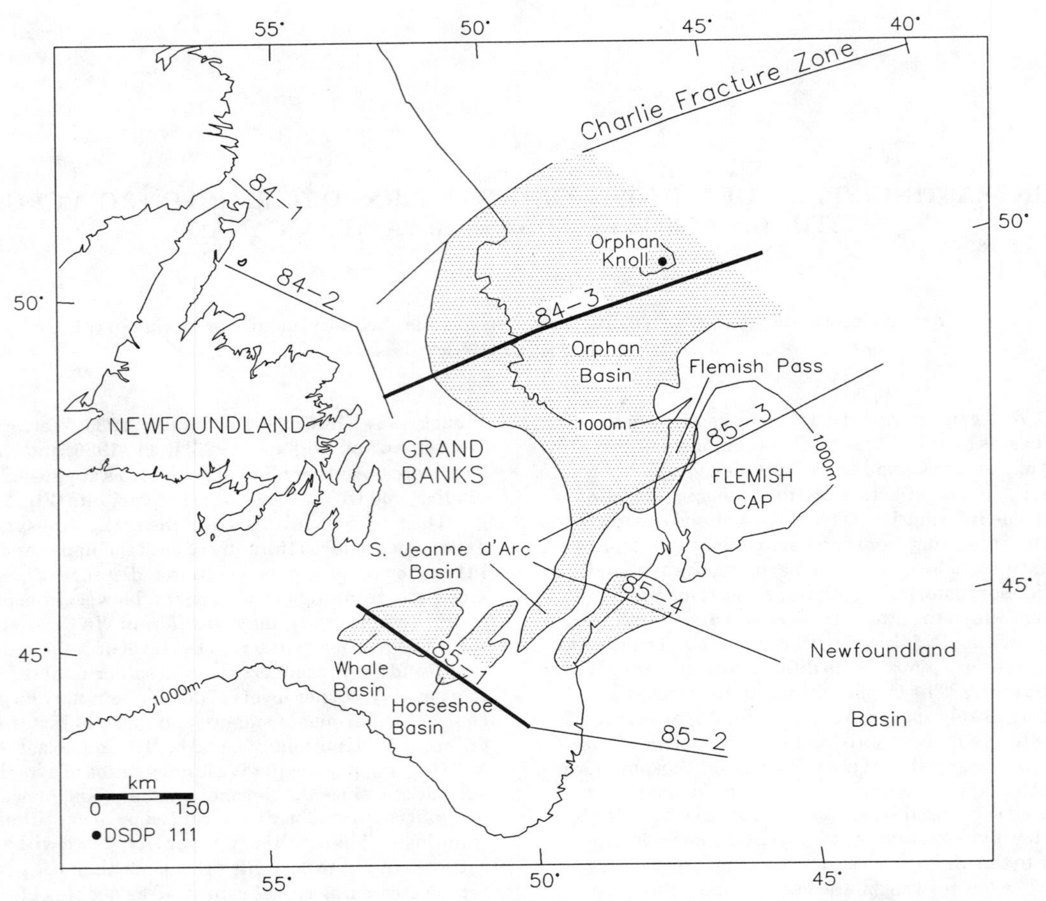

Fig. 1. Location of the region of study. The major sedimentary basins are stippled and the location of deep seismic reflection lines are shown. Note in particular the heavier lines representing line 85-1 crossing Whale and Horseshoe Basins on the Southern Grand Banks and line 84-3 crossing Orphan Basin in the north. The cross-sections shown in Fig. 3 lie along these lines.

asymmetry provides the major differences in basin evolution between these fault related models and the simple necking models [Wernicke, 1985].

In the remainder of this paper these various types of extensional geometries and their predicted subsidence histories are compared with observations. The observations include deep seismic reflection data and refraction data, and the amount of post- and syn-rift subsidence deduced from these data and associated boreholes. These observations are described in detail in Keen et al. [1987 a,b].

Crustal Extension from Seismic Observations

The seismic data crosses basins exhibiting two distinct styles (Fig. 1). In the north, the Orphan Basin is a wide basin, some 400 km across. In the south, narrow, elongate half graben basins occupy the Grand Banks, which shall be referred to here as "Grand Banks basins". All of these basins are filled with Mesozoic and younger sediment. The Grand Banks experienced a complex rifting history, consisting of at least three major episodes in Triassic, Early Jurassic, and mid Cretaceous time [Tankard and Welsink, 1987; Enachescu, 1987], while Orphan Basin may not have been affected by the Triassic event [Keen et al, 1987b]. Syn-rift sediments, corresponding to all the rifting episodes and spanning 100 Ma, are separated from younger post-rift sediments by a mid-late Cretaceous breakup unconformity (the Avalon seismic event on the Grand Banks; Enachescu, 1987).

The structures beneath Orphan Basin and beneath the Grand Banks basins to the south are illustrated in Fig. 3 where schematic cross-sections, based primarily on the deep seismic reflection lines, are shown. The Grand Banks basins (Fig. 3A) are filled with syn-rift sediments. They are overlain by the breakup unconformity and by a thin layer of post-rift sediments which blanket the entire Grand Banks region and are not confined to the basins, as are the syn-rift sediments. Thus syn- and post-rift subsidence exhibit very different lateral distributions. This observation, which is supported by an extensive grid of industry multi-channel seismic data in the area, suggests different underlying causes for the evolution of these syn-rift basins and for the later development of the breakup unconformity and the overlying post-rift sediments.

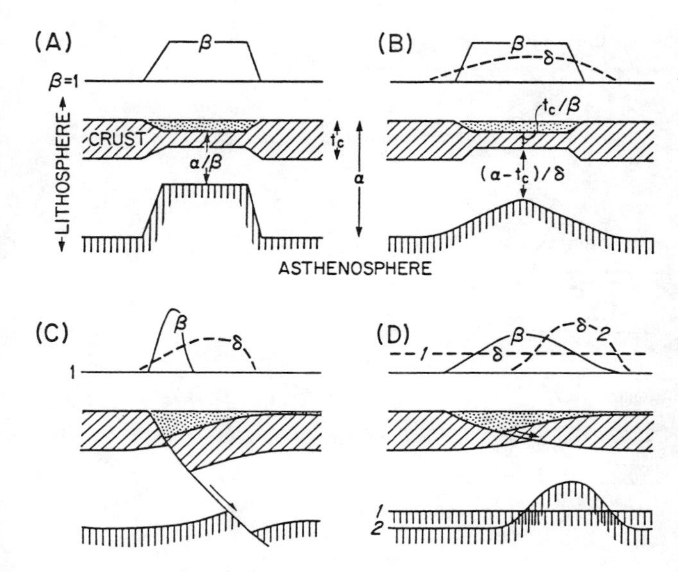

Fig. 2. Schematic illustrations of four different extensional geometries for the lithosphere. The drawings are not to scale.In Fig. 2a the uniform extension model of McKenzie [1978] is shown, in which both the upper and lower lithosphere stretch and thin by an amount β. β = 1 indicates no stretching. Fig. 2b illustrates a variant of the uniform extension model in which the crust and the lower lithosphere are allowed to thin by different amounts, given by β and δ, respectively. Fig. 2c illustrates extension which is accommodated along a low angle shear zone which offsets the lithosphere [Wernicke, 1985]. This also gives different amounts of thinning in the upper and lower parts of the lithosphere, which are skewed with respect to each other. Fig. 2d is a similar model, only with the shear zone confined to the crust. Below the mantle lithosphere extends by ductile necking whose geometry is uncertain. Therefore, two possible shapes for lower lithosphere thinning are shown, labelled "1" and "2" at the lithosphere-asthenosphere boundary. The lateral distributions of β and δ are shown above each illustration. The amount of crustal thinning is an important factor in determining syn-rift subsidence,as the light crustal material is replaced by heavier mantle material. Thinning of the lower lithosphere is also important as hot, light material rises from the asthenosphere and tends to cause uplift. During the post-rift phase, the lithosphere cools toward thermal equilibrium and thermal subsidence results. Thus the amount of post-rift subsidence is primarily controlled by δ, the amount of thinning of the lower lithosphere.

Large fault blocks bound these Grand Banks basins. The faults extend into the deep crust or to the Moho, where they may flatten along a decollement [Keen et al, 1987a], and do not appear to cut the Moho. The latter is an undulatory boundary, particularly below the basins. In this region the crust thins locally to about 50 % of the thickness observed away from the basins , although there is no straightforward relationship between basin subsidence and crustal thinning.

In contrast, the cross-section of Orphan Basin (Fig. 3B) exhibits a very thick post-rift sedimentary section, underlain by relatively thin syn-rift sediments. Basement is dissected by numerous, closely-spaced normal faults which are interpreted to terminate in a decollement at mid-crustal depths. A few faults may terminate at deeper levels, as indicated by the dashed lines between 300 and 400 km. Moho was not observed on the reflection line, but seismic refraction measurements place the Moho at a depth of 22-24 km, equivalent to crustal thinning by a factor of 2 [Keen and Barrett, 1981]. Preliminary results of more detailed refraction studies across the basin indicate (I.Reid, personal communication, 1988) that neither the Moho depth nor the lower crustal layer thickness are as smooth as shown in Fig. 3B.

The Orphan Basin cross-section has been extended out beyond the continent-ocean boundary, although a detailed discussion of that region is beyond the scope of this paper [see Keen and de Voogd, 1987]. It is interesting, however, to note how the oceanic crust appears to dip down under the continental crust and merges with the lower (mafic?) crustal layer under Orphan Basin [Keen and Barrett, 1981]. This mafic layer may result from magmatic underplating of the thinned continental crust during rifting.

These differences in tectonic style and in the relative amounts of post- and syn-rift subsidence can be related to the different extensional geometries presented. A discussion of geometries which best fit these observations is presented below.

Modes of Extension

The thin layer of post-rift sediments which covers the entire Grand Banks, and is not confined to the basins can be explained by uplift of basement due to broad regional thinning of the lower lithosphere over the entire Grand Banks region. This would delay deposition of post-rift sediment until thermal subsidence returned basement to sea level. Futhermore, the amount of thinning of the lower lithosphere need not be large, so that thermal subsidence is small. Keen et al. [1987a] present a quantitative thermo-mechanical model which fits the observed subsidence and crustal structure in which $\delta < \beta$ and values of δ range from 1.08 (below platforms) to 2 (below basins). This model is conceptually similar to that reported by Kuznir et al. (1987). One argument favouring small δ values is the requirement for roughly equal amounts of stretching in the crust and in the lower lithosphere, given by β and δ, and averaged over the Grand Banks region. Otherwise mass is not conserved. Alternatively, mechanisms other than extension must be postulated to thin the lower lithosphere. An extensional geometry with crustal thinning associated with major normal faults in the crust below the basins and broad regional thinning of the underlying mantle lithosphere corresponds to the geometry shown in Fig. 2d (curve 1).

In order to obtain the larger ratio of post- to syn-rift subsidence in Orphan Basin, thinning of the lower lithosphere must be larger, and greater than that in the crust. Crustal thinning by factors of 1.5 to 2 is consistent with seismic reflection and refraction observations. To match the observed subsidence ratio, the extensional models predict that the lower lithosphere must thin by factors of 4 to 8, over a horizontal distance of 400 km [Keen et al., 1987b]. Extension in the Orphan Basin is closest to the geometry shown in Fig. 2b, but with δ much greater than β. There is no apparent asymmetry in the rifting geometry, as both β and δ increase monotonically toward the east. However, only one side of the rift is observed; its conjugate now occupies the eastern Atlantic margin.

The major difference in extensional geometry between the Orphan Basin and the Grand Banks basins is the amount of

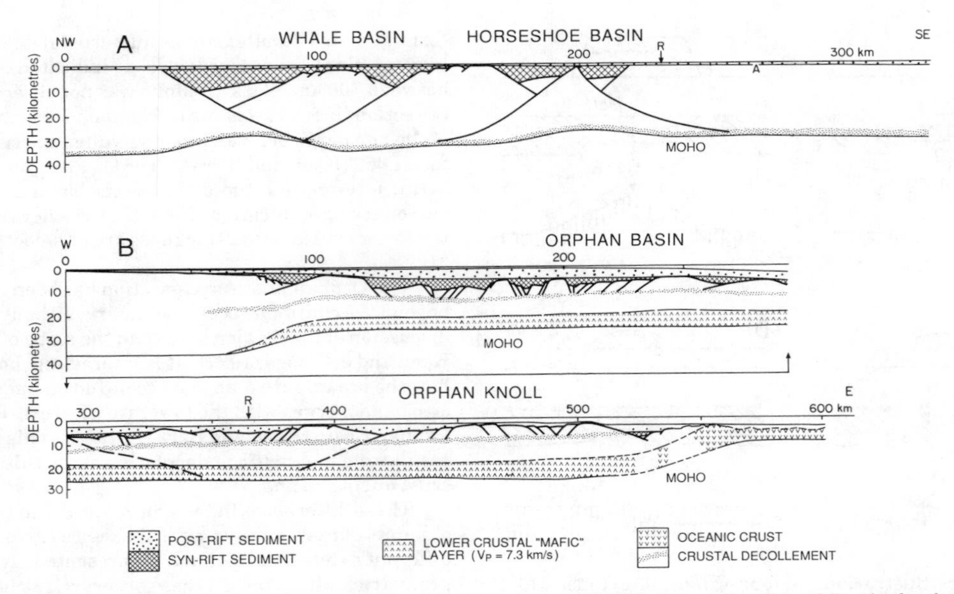

Fig. 3. Crustal cross-sections, based on seismic reflection data. Fig. 3A shows the Grand Banks basins along line 85-1, while Fig. 3B crosses Orphan Basin along line 84-3 (see Fig. 1). The sections were constructed by converting the reflection sections from two-way travel time to depth, using the following velocities: mean velocity of the continental crust = 6.4 km/s, velocity of syn-rift sediments = 4.0-4.5 km/s (depending on depth of burial); velocity of post-rift sediments = 2.5-3.0 (depending on depth of burial). An explanation of the patterns is given in the legend.

lower lithospheric thinning, which results in different subsidence histories. The other significant differences are the size and number of the fault blocks and the depth at which faults terminate in the crust.

Discussion

The observations strongly suggest that the lower lithosphere has responded differently to extension below the Orphan Basin and the Grand Banks basins. This result is based primarily on the observed subsidence history as determined from seismic and borehole data. Some factors which might modify this result are briefly examined below.

First, high ratios of syn- to post-rift subsidence (Fig. 3A) could arise from thermal cooling during the long rift phase on the Grand Banks, leaving little thermal subsidence for the post-rift phase. However, simple model calculations show that, even when the rift phase is 100 Ma long, as it was in the Grand Banks-Orphan Basin regions, significant post rift thermal subsidence would occur [Jarvis and McKenzie,1980; Keen et al., 1987b]. This predicted post-rift thermal subsidence directly related to the basins is not observed on the Grand Banks and therefore rift stage thermal subsidence is discounted as an alternative explanation.

Second, the issue of uplift during rifting must be addressed. Uplift of the lithosphere, such as that which occurred on the Grand Banks during continental breakup in mid- to late Cretaceous time, elevated the basement above sea level, and syn-rift sediments may have been removed by erosion while post-rift sediments might not be deposited until the thermal subsidence returned basement to sea level. This would change estimates of the ratio of syn- to post- rift sediments on the Grand Banks. However, the thickness of syn-rift sediment is so much greater (factor of 10) than the thickness of post-rift sediment that the ratio of syn-to post- rift sediment is not likely to be modified sufficiently to change the conclusions of this paper. Similarly, in Orphan Basin, significant erosion of the syn-rift sediments due to uplift of basement would provide a decrease its estimate. However, even if 5 km of erosion had occurred, the basic conclusion would not change, although estimates of thinning in the lower lithosphere would be less below Orphan Basin.

Third, volcanism and magmatic underplating of the lower continental crust such as that which may have occurred in the Orphan Basin will tend to reduce syn-rift subsidence. However, simple model calculations [Keen, 1987b] suggest that the reduction would not be significant in the present context.

Finally, these different amounts of subsidence could be due to differences in lithospheric properties, such as age , initial thermal gradient, and physical properties beneath the Orphan Basin and the Grand Banks basins. However, the regions are less than 300 km apart and both apparently lie within the Avalon tectonostratigraphic terrane of the Appalachian Orogen [Haworth and Lefort, 1979]. It would therefore be unlikely that lithospheric properties are significantly different in the two regions.

Given the above discussion, the difference in subsidence observed in these basins is probably directly related to the different amounts of thinning of the lower lithosphere, relative to those in the upper lithosphere. While variable amounts of lower lithospheric thinning have been previously interpreted in different basins and reported in the literature, there are very few studies where the deep extensional geometry is as well constrained as that reported here. Therefore these results warrant further comment. A cartoon which illustrates the lithospheric structure below both of these two basin types at the end of the rifting phase is shown in Fig. 4. The base of the lithosphere rises sharply beneath the Orphan Basin, compared to its position beneath the Grand Banks basins.

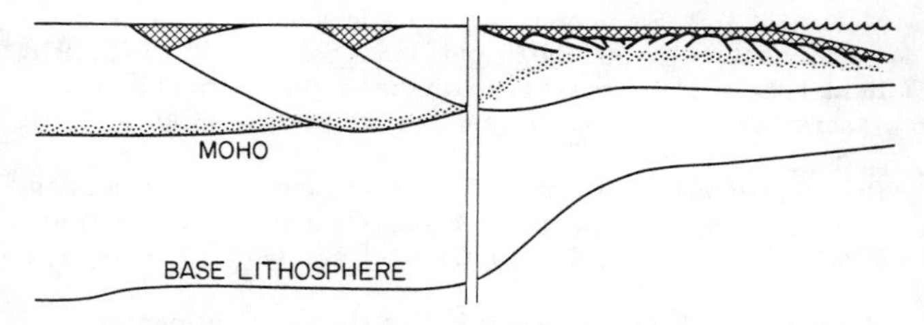

Fig. 4. Cartoon which sumarizes the seismic results and the lower lithospheric thinning geometry determined from the subsidence histories of Orphan Basin and the Grand Banks. It represents the end of the rift stage. The characteristics of the Grand Banks basins are represented by the left half of the cross-section, while those of Orphan Basin are shown on the right half. Syn-rift sediments are cross-hatched, the crustal decollement is shown as a gray band.

Differences in crustal geometry in the two regions include the number of faults , the depth at which the faults terminate, and the size of fault blocks. The occurrence of numerous, closely spaced faults above a shallow detachment as seen in Orphan Basin is consistent with the rheological models of Kuznir and Park [1987], when the thermal gradient is high.Higher thermal gradients might be expected in Orphan Basin due to the thermal anomaly created by thinning of the lower lithosphere. Furthermore, the likelihood that the lower crustal layer in this region results from magmatic underplating during rifting is increased when the lower lithosphere is thin and hot asthenosphere rises below it [Keen, 1987b].

Conversely, deeper detachment levels near the Moho would be compatible with the lower heat flow implied by the lithospheric model for the Grand Banks basins. There is no evidence for a lower crustal mafic layer beneath the Grand Banks [Reid, 1987].

Thus, the geologic style exhibited by Orphan Basin versus that exhibited by the Grand Banks basins can ,to first order, probably be explained by differences in the thermo-mechanical properties of the lithosphere which are in turn associated with the differences in lower lithospheric thinning. However, the reasons for the differences in thinning is not well understood.

It is important to note that these differences in the mode of lithospheric stretching are widespread. Other examples of the "Grand Banks" type are found in the Fundy graben, the Triassic basins of the eastern United States, the English Wessex Basin [Karner et al., 1987], and the Ridge Basin, Southern California [Karner and Dewey, 1986]. It has been suggested that many of these formed within the reactivated hanging walls of older thrust sheets. The Grand Banks basins appear to be on strike with older structures within the Avalon Terrane of the Appalachians [Haworth and LeFort, 1979]. Conversely, the Mesozoic faults in the Orphan Basin trend perpendicular to these older structures. Therefore, a contributing factor in the differences apparent in Orphan Basin and the Grand Banks basins may be the direction of Mesozoic extension relative to the trend of pre-existing structure.

Orphan Basin also has many analogues in terms of lower lithosphere extension, although in general the basins are not as wide. Most of the sedimentary basins on the continental shelf of eastern North America, such as the Scotian Basin and the Baltimore Canyon Trough are similar, in that the lower lithosphere was thinned significantly during rifting [Beaumont et al., 1982; Sawyer et al., 1982].

Thus the observations set out in this paper can be placed in a worldwide context. However, the key, unanswered question is: what controls the amount of thinning of the lower lithosphere in a given region? While the answer to this question has yet to be found, the preceeding description of these differing tectonic styles may help us to better formulate its solution.

Acknowledgements. K. Dickie, W. Kay, B. de Voogd and J. Verhoef critically read the manuscript and made many useful comments. Geological Survey of Canada Contribution Number 41587.

References

Beaumont, C., C.E. Keen, and R. Boutilier, Evolution of rifted continental margins; comparison of models and observations for the Nova Scotia margin, Geophys. J. Roy. astr. Soc., 70, 667-715, 1982.

Buck, W. R., Small-scale convection induced by passive rifting: the cause for uplift of rift shoulders, Earth and Planet. Sci. Lett., 77, 362-372, 1986.

Enachescu, M.E. The Tectonic and Structural Framework of the Northeast Newfoundland Continental Margin, in Sedimentary Basins and Basin-Forming Meechanisms edited by C. Beaumont and A.J. Tankard, Canadian Society of Petroleum Geologists, Memoir 12,in press, 1987.

Haworth, R.T., and J.P. Lefort, Geophysical Evidence for the extent of the Avalon Zone in Atlantic Canada, Can. J. Earth Sci., 16, 552-567, 1979.

Hellinger, S.J., and J.G. Sclater, Some comments on two layer extension models for the evolution of sedimentary basins, J. Geophys. Res., 88, 8251-8270, 1983.

Houseman,G., and P. England, A dynamical model of lithosphere extension and sedimentary basin formation, J. Geophys. Res.,91, 719-729,1986.

Jarvis, G.T., and D.P. McKenzie, Sedimentary Basin formation with finite extension rates, Earth and Planet. Sci. Lett., 48,42-53, 1980.

Karner, G.D., and J.F. Dewey, Lithospheric versus crustal extension as applied to the Ridge Basin of Southern California, in Future Petroleum Provinces of the World edited by M.T. Halbouty ,pp. 317-337, Am. Assoc. Pet. Geol. Mem. 40, 1986.

Karner, G.D., S.D. Lake, and J.F. Dewey, The thermal and mechanical development of the Wessex Basin, southern England, in Continental Extensional Tectonics Edited by M.P. Coward, J.F. Dewey, and P.L. Hancock, pp. 517-536, Geological Society Special Publication no, 28, 1987.

Keen, C.E., Dynamical extension of the Lithosphere during rifting: Some numerical model results, in Composition, Structure and Dynamics of the Lithosphere-Asthenosphere System, pp. 189-203, Edited by K.Fuchs and C. Froidveaux, Am. Geophys. Union, Geodynamics Series 16, 1987a.

Keen, C.E., Some important consequences of lithospheric extension, in Continental Extensional Tectonics edited by

M.P. Coward, J.F. Dewey, and P.L. Hancock, pp. 67-73, Geological Society Special Publication no. 28, 1987.

Keen, C.E. and D.L. Barrett, Thinned and subsided continental crust on the rifted margin of eastern Canada: crustal structure, thermal evolution and subsidence history, Geophys. J. R. astr. Soc., 65, 443-465, 1981.

Keen, C.E., R. Boutilier, B. de Voogd, B. Mudford, and M.E. Enachescu, Crustal geometry and models of the evolution of the rift basins on the Grand Banks off Eastern Canada: Constraints from deep seismic data, in Sedimentary Basins and Basin-Forming Mechanisms edited by C. Beaumont and A.J. Tankard, Canadian Society of Petroleum Geologists Memoir 12, in press, 1987a.

Keen, C.E. and B. de Voogd, The Continent-Ocean boundary at the rifted margin off eastern Canada: New results from deep seismic reflection studies, Tectonics, in press, 1987.

Keen, C.E., G.S. Stockmal, H. Welsink, G. Quinlan, and B. Mudford, Deep crustal structure and evolution of the rifted margin northeast of Newfoundland: Results from Lithoprobe East, Can. J. Earth Sci., 24, 1537-1549, 1987b.

Kuznir, N.J., and R.G. Park, The extensional strength of the continental lithosphere: its dependence on geothermal gradient, and crustal composition and thickness, Continental extensional tectonics, pp. 35-52, edited by M.P. Coward, J.F. Dewey, and P.L. Hancock, Geological Society Special Publication No. 28, 1987.

Kuznir, N.J., G.D. Karner, and S. Egan, Geometric, Thermal and Isostatic Consequences of Detachments in Continental Lithosphere Extension and Basin Formation, in Sedimentary Basins and Basin-Forming Mechanisms, pp. 185-203, edited by C. Beaumont and A.J. Tankard, Can. Soc. Petrol. Geologists, Memoir 12, 1987.

Le Pichon, X., and J.C. Sibuet, Passive margins: a model of formation, J. Geophys. Res., 86, 3708-3720, 1981.

McKenzie, D.P., Some remarks on the development of sedimentary basins. Earth and Planet.Sci.Lett., 40, 25-32, 1978.

Reid, I., Crustal Structure beneath the southern Grand Banks: seismic refraction results and their implications, Can. J. Earth Sci., 25, (in press).

Royden, L. and C.E. Keen, Rifting process and thermal evolution of the continental margin of eastern Canada determined from subsidence curves, Earth and Planet. Sci. Lett., 51, 343-361, 1980.

Royden, L., J.G. Sclater, and R.P. Van Herzen, Continental margin subsidence and heat flow, Important parameters in formation of petroleum hydrocarbons, Am. Ass. Petrol. Geol. Bull., 64, 173- 187, 1980.

Sawyer, D.S., M.N. Toksoz, J.G. Sclater, and B.A. Swift, Thermal evolution of the Baltimore Canyon Trough and Georges Bank Basin, in Studies of Continental Margin Geology, pp. 743-762, edited by J.S. Watkins and C.L. Drake, Am. Assoc. Petrol. Geol. Memoir 34, 1982.

Tankard, A.J. and H.J. Welsink, Extensional Tectonics and stratigraphy of Hibernia oil field, Grand Banks, Newfoundland, Am. Assoc. Petrol. Geol. Bull., 71, 1210-1232, 1987.

Wernicke, B., Uniform sense of simple shear of the continental lithosphere , Can. J. Earth Sci., 22, 108-125, 1985.

BASEMENT FEATURES UNDER FOUR INTRA-CONTINENTAL BASINS IN CENTRAL AND EASTERN AUSTRALIA

D. M. Finlayson, C. Wright, J. H. Leven, C. D. N. Collins, K. D. Wake-Dyster, and D. W. Johnstone

Bureau of Mineral Resources, Geology and Geophysics, P.O. Box 378, Canberra City, ACT 2601, Australia

Abstract. Differences identified in deep seismic data from within basement under four basins in central and eastern Australia illustrate the diversity of structures and processes that have to be reconciled in any modelling of basin development. No single unifying process is identified from the deep seismic data. The four basins are the Ngalia and Amadeus Basins in Precambrian central Australia and the Adavale Basin and Taroom Trough from Phanerozoic eastern Australia.

All four basins are now structurally asymmetric with the thickest sequences being adjacent to a steeply faulted margin. This asymmetry, however, is a product of events late in each basin's history. Older events have led to quite different characteristics of the present-day crust under the four basins.

In particular, we emphasize that 1) the pre-depositional history of the various regions and in particular the earlier major crustal faulting plays an important part in determining the structure of the basins; 2) much of the post-depositional thrust faulting is confined to the upper crust; 3) some high-angle faulting is interpreted as penetrating the entire crust; 4) under the Adavale Basin in eastern Australia, data from the lower crust and mantle suggest that, in places, underplating/intrusion has played an important role in basin history; 5) in Phanerozoic eastern Australia the Moho is clearly identified, but any topography on it associated with basin development has largely relaxed to give a Moho with little relief; 6) the Moho is not apparent at all in the seismic fabric of the older, thicker crust of the Precambrian central Australian Arunta Block and the adjacent Amadeus Basin.

Introduction

There are intra-cratonic basins in the central and eastern parts of continental Australia which, taken together, provide about 80 percent of onshore oil and gas production. They lie in central Australia, northern South Australia - western Queensland, and eastern Queensland. In this paper we discuss four basins (Fig.1), the Amadeus and Ngalia Basins in central Australia and, in eastern Australia, the Adavale Basin (a Devonian basin under the Eromanga Basin) and the Taroom Trough (a Permo-Triassic basin under the Surat Basin).

Published in 1989 by
International Union of Geodesy and Geophysics
and American Geophysical Union.

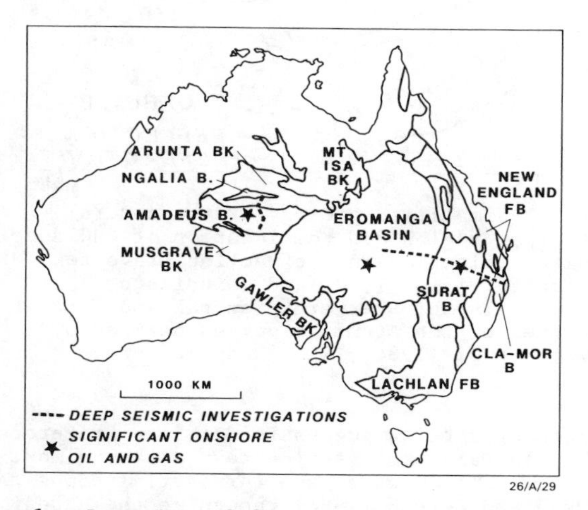

Fig. 1. Location of four Australian basins discussed in this paper. The Adavale Basin underlies the central Eromanga Basin and the Taroom Trough underlies the Surat Basin.

The origin and evolution of sedimentary basins is controlled by the tectonics of the basement rocks and underlying lithosphere. We examine the structures seen in seismic data from within the basement rocks under the basins listed above. Common and contrasting features are highlighted which must be borne in mind as present-day end products when the evolutionary history of these basins is compiled and modelled. The paper contains current research results from a number of projects by the Bureau of Mineral Resources, Geology and Geophysics (BMR).

Central Australia, Geological Background

The Amadeus Basin in central Australia (Fig.2) is described by Kennard et al. [1986] as the remnant of an intra-cratonic depression

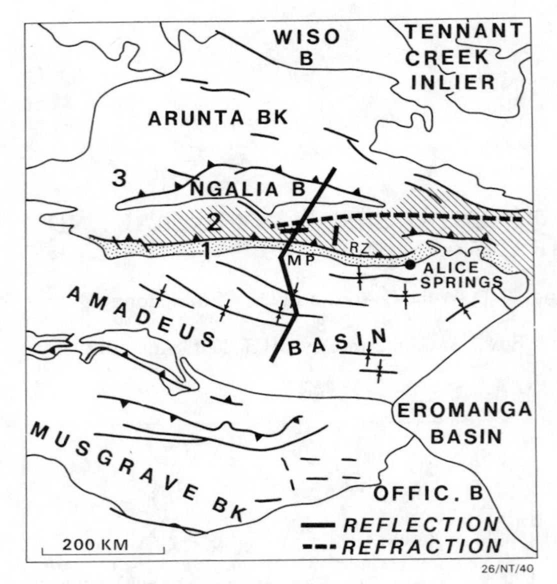

Fig. 2. Simplified geology of the central Australian region and the location of BMR deep seismic reflection and refraction traverses. MP = Missionary Plain; 1, 2, and 3 indicate respectively the Southern, Central and Northern Provinces of the Arunta Block as defined by Stewart et al.[1984]; RZ = Redbank Zone.

containing a thick succession of Late Proterozoic to Middle Palaeozoic sedimentary rocks, almost entirely shallow water and terrestrial deposits. Lindsay and Korsch [1988] recognize the oldest units to be basalts and clastics of a rift sequence with an unconformity between these and overlying Dean and Heavitree Quartzites. The basin sequences are thought to rest unconformably on much older, highly deformed Proterozoic rocks of the Arunta and Musgrave Complexes but since basement core is almost non-existent this supposition must remain speculative.

Shaw et al [1986] indicate that there are possibly nine depositional episodes which can be recognised in the Amadeus Basin sequences, all linked to tectonic events affecting the larger continental craton. Lindsay and Korsch [1988] summarised the basin evolution in three major episodes, the first two involving crustal extension and the third involving convergence with southward-directed thrust sheets shedding sediment into a downward flexure.

Shaw et al. [1986] indicate that there are depositional links between the Amadeus and Ngalia Basins, and Wells and Moss [1983] describe in detail the correlations between events recognised in all the basins within the Proterozoic provinces of central Australia. In the Ngalia Basin up to 5 km of sedimentary rocks are preserved towards the northern faulted margin of the basin and lie unconformably on Proterozoic basement. In the Amadeus Basin up to 14 km of basin sequences are preserved in a series of sub-basins towards its northern margin with the exposed Arunta Block [Lindsay & Korsch, 1988]. These sub-basins are separated by a central ridge from the platform area to the south where the maximum thickness of basin sequences is 5 km but is usually much less.

The history of the Arunta basement is described by Stewart et al. [1984] and Shaw et al. [1984]. The Arunta Block is interpreted as a Proterozoic ensialic mobile belt floored by continental crust which evolved in the period 2015 Ma to 300 Ma. Following a volcanic and sedimentary episode at 2015 - 1980 Ma, five major cycles of extension and convergence have been identified in the geological record. Figure 3 shows four of the evolutionary stages envisaged by Shaw et al [1984]. An important aspect of the Proterozoic development in central Australia interpreted from surface geology is that there are northward dipping deformation zones which are considered to extend deep within the crust and that these have persisted since the earliest known history of the region, e.g. the Redbank and Delny-Mount Sainthill Deformed Zones in Figure 3.

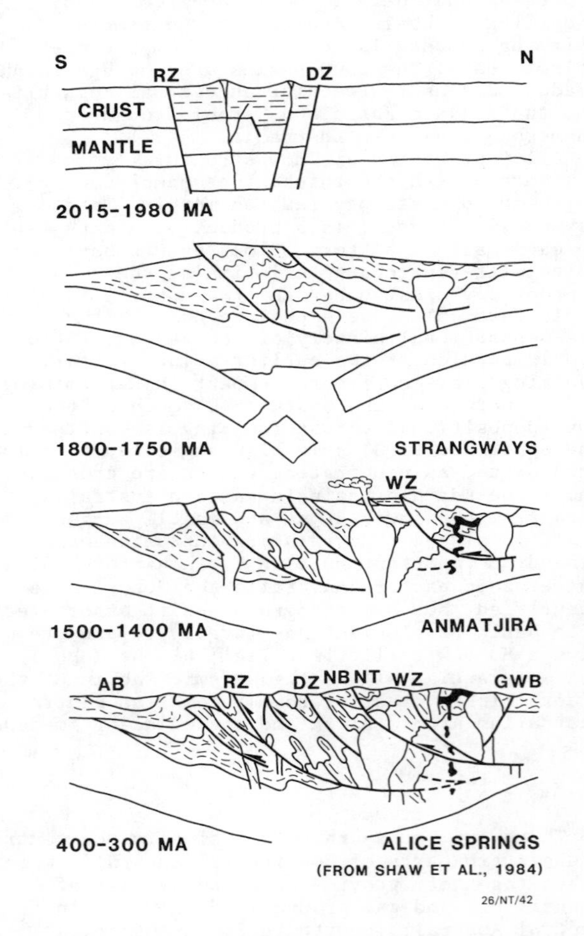

Fig. 3. Four of the Arunta Block evolutionary stages as described by Shaw et al.[1984]. AB = Amadeus Basin; RZ = Redbank Deformed Zone; DZ = Delny-Mount Sainthill Deformed Zone; NB = Ngalia Basin; NT = Napperby Thrust; WZ = Weldon Tectonic Zone; GWB = Georgina/Wiso Basins.

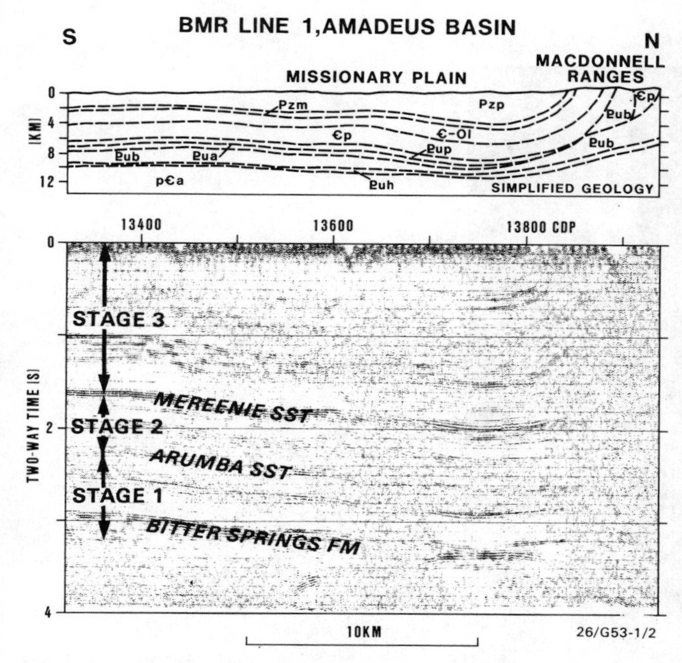

Fig. 4. BMR seismic reflection section across the Missionary Plain area of the Amadeus Basin and the southern margin of the Arunta Block, showing some of the main depositional sequences [Goleby et al., 1988b]. Geological legend: Pzp = Dev.-Carb. Pertnjara Group; Pzm = Siluro-Dev. Mereenie Sandstone; C-Ol = Cambro-Ord. Larapinta Group; Cp = Cambr. Pertatataka Fm; Pua = Prot. Areyonga Fm.; Pub = Prot. Bitter Springs Fm.; Puh = Heavitree Quartzite.

Central Australia, Seismic Data

The locations of the major reflection and refraction traverses in central Australia are shown in Figure 2. A seismic section from the BMR reflection traverse where it crosses from the Missionary Plain area onto the Arunta Block is shown in Figure 4 [Wright et al., 1987]. The deformation of the Amadeus Basin sequences at the Arunta Block boundary is manifest in the seismic section as a sudden break in reflectivity, with upturned units being identified from surface mapping. But if we are to gain a better understanding of the processes of basin formation, we should examine data from deep within basement structures of the region.

Ngalia Basin

Figure 5 shows deep seismic reflection data to 12 s TWT from the Northern Province of the Arunta Block under part of the Ngalia Basin [Goleby et al., 1987]. The data illustrate the predominance of northerly dipping reflectors within the deeper parts of the Arunta Block in that area. Apparent dips in this section are up to 18 degrees which will become about 19 degrees on migration. Many of these deep features can be correlated with structures which can be clearly mapped at the surface e.g. the Napperby Thrust (Fig.3). The northerly-dipping events continue throughout the Northern and Central Provinces of the Arunta Block to the Redbank Zone which forms the boundary between the Central and Southern Provinces (Fig. 2) [Wright et al., 1987; Goleby et al., 1988b]. Shaw et al. [1984] interpret this boundary, the Redbank Zone, as an overthrust in which rocks of the Central Province have been pushed over those of the Southern Province. The expression of the Redbank Zone (shown schematically in Figure 3) is apparent on seismic sections (not shown here) to depths of at least 30 km; the feature is remarkably linear with a migrated dip of about 50 degrees and is interpreted as a major crustal boundary. The overall character of the seismic reflections within the Northern and Central Provinces support a model of the tectonic evolution in the region in which older, deeply penetrating, steeply dipping, fault systems interpreted from geological mapping were reactivated during the most recent (Alice Springs) orogeny [Goleby et al., 1988a,b]. Because of the major, steeply-dipping structures associated with the Redbank Zone the seismic data do not support a structural model which includes a single low-angle thrust extending from the Amadeus Basin under the Arunta Block such as that proposed by Teyssier [1985].

It appears that the features in the top 10-12 s TWT of the seismic reflection sections involve processes affecting well-established zones in the crust, many of which are considered, from surface geological considerations, to have been in existence since the earliest known history of the region [Shaw et al., 1984; Stewart et al., 1984]. There is quite good correspondence between the deep seismic features and the type of structures postulated by Shaw et al., [1984] from their study of surface geology (Fig. 3). The regions of reflectivity are likely to represent zones of deformation, involving contacts between different rock types, characterised by the micro- and macro- scale cracking and chemical alteration effects that Mooney and Ginzburg [1986] have proposed to explain fault zone reflectivity.

However, this deformation/faulting process is not the only one applying in the Arunta Block crust. Across the central part of the Arunta Block we also have good reversed seismic refraction traverses (Fig. 2) which indicate that, in the top 9 s TWT of the reflection profiling records, the average velocities are in the range 6.2 to 6.5 km/s. These are usually regarded as upper crustal velocities. At two-way times greater than 12 s there are very few reflectors and the velocity from the refraction data increases to about 7.2 km/s at depths greater than 31 km [Wright et al.,1987]. The character of the seismic reflection profiling records within the Arunta Block is therefore providing information only on processes that have operated in the top 30 km of the crust. Later arrivals at shot offsets of between 200 and 240 km on the seismic refraction records are tentatively interpreted as wide-angle reflections from the top of the mantle; if this is so, the

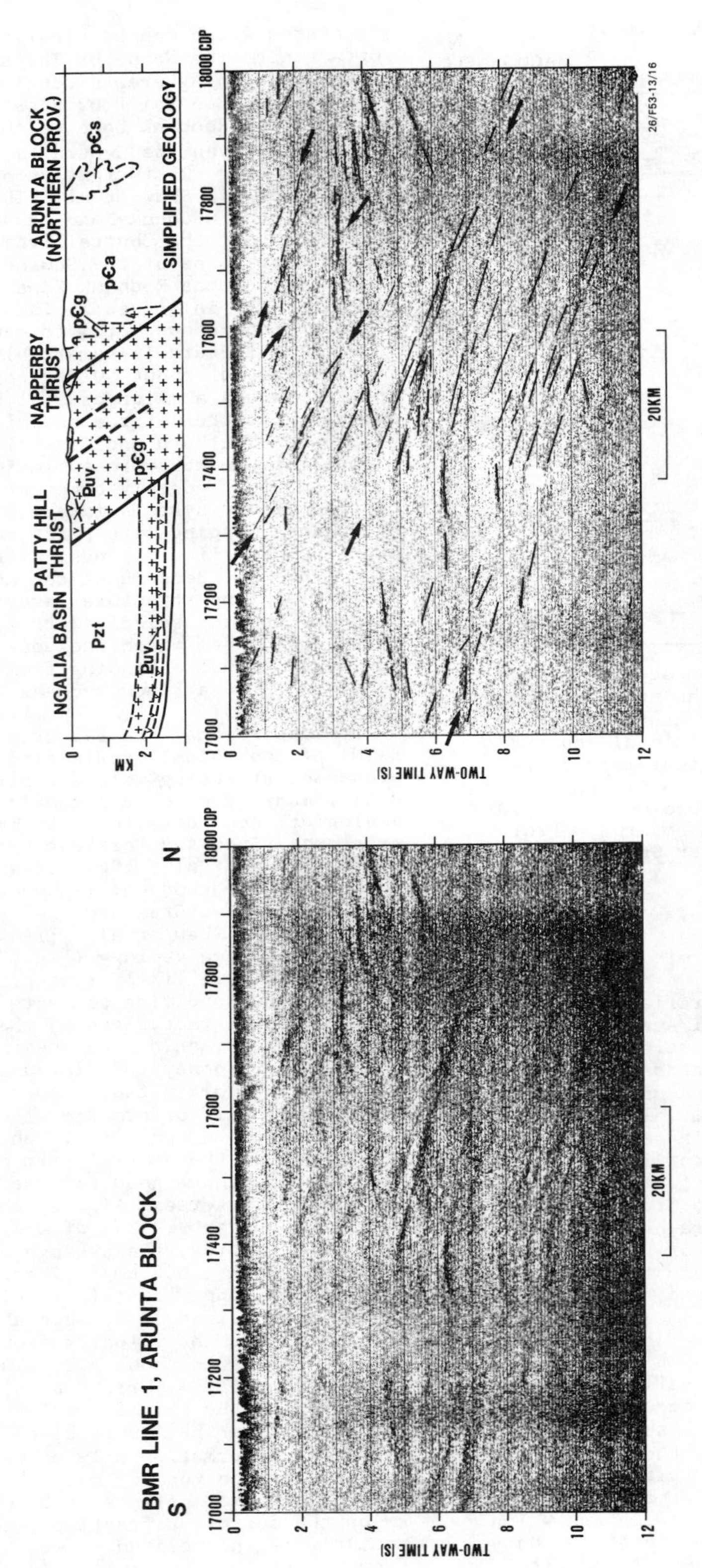

Fig. 5. BMR deep seismic reflection section across the Arunta Block underlying the northern part of the Ngalia Basin. V:H exaggeration approximately 1.4. The geology is a sketch diagram only [Goleby et al., 1988a]. Geological legend: Pzt = Camb. Mt. Eclipse Sandstone; Puv = Precamb. Vaughan Springs Quartzite; p€g = Precamb. granite; p€a = Precamb. orthogneiss; p€s = Precamb. schist.

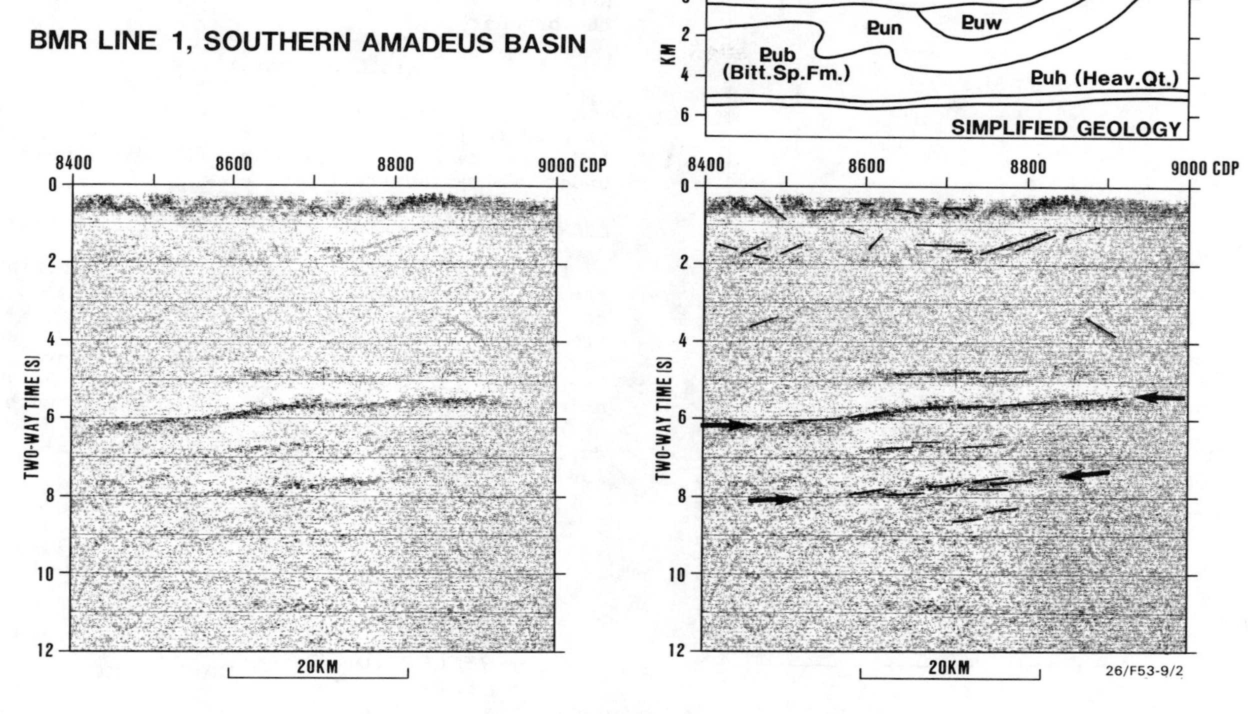

Fig. 6. BMR deep seismic reflection section across the southern Amadeus Basin near the southern end of the BMR traverse (Fig.2) [Goleby et al., 1988b]. V:H exaggeration approximately 1.6. Geological legend: Puw = Prot. Winnall Beds; Pun = Prot. Inindia Beds; others as in Figure 4.

crust-mantle boundary would be at a depth of at least 55 km. This agrees well with the older refraction work of Finlayson et al. [1974] which suggested a crustal thickness of 55-60 km within the Arunta Block and an upper mantle velocity of 8.18 km/s. Although the region from 30 to 55 km depth is largely non-reflective, it appears to comprise mainly crustal rather than upper mantle rocks.

Amadeus Basin

Within the Amadeus Basin to the south of the exposed Arunta Complex the seismic image of features within basement is quite different from that under the Ngalia Basin. Figure 6 is a deep seismic section from the platform area to the south of the central ridge in the basin, near the southern end of the BMR reflection traverse(Fig.2) [Goleby et al., 1988a]. The uppermost basement to depths of 17 km is largely non-reflective. Bands of reflections do, however, extend from about 17 to 28 km depth, but with significant dip to the south, suggesting (speculatively) northward thrusting of the Musgrave Block over the southern part of the Amadeus Basin on a mid-crustal ramp dipping southward.

Further north, under the thickest sequences of the Amadeus Basin in the Missionary Plain area, there are very few prominent deep events. Reflections at times greater than 4 s TWT appear to be 'peg leg' multiples generated by the highly reflective, 10 km thick, sedimentary sequence (Fig. 4). Beneath much of the Amadeus Basin events interpreted as possible basement reflections are very weak making it impossible to interpret structure within basement under the area of thickest sedimentary rocks.

However, just to the north of the area of the deepest Amadeus Basin sequences is the Southern Arunta Province. Under this region deep reflections are seen with a small apparent dip to the north at 6.5 to 10 s TWT (Fig. 7) [Wright et al, 1987; Goleby et al, 1988b]. The continuity of these bands of reflections suggests that they might originate from rocks of sedimentary origin or from mafic sills. The interesting possibility arises that the crust of the Arunta Block has been thrust over the crust that originally formed the northern margin of the Amadeus Basin. We may be seeing some form of crustal doubling beneath the southern Arunta Block where large gravity anomalies occur. This model provides a solution to the problem of reconciling the gravity and teleseismic travel-time data with the seismic reflection and refraction data [Goleby et al., 1988a; Lambeck et al., 1988].

All the indications suggest that the processes at 6 to 10 s TWT (18 to 30 km depth

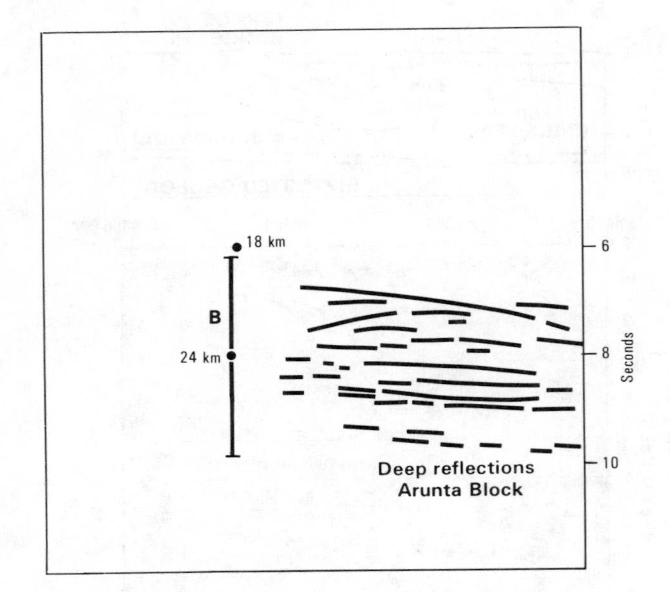

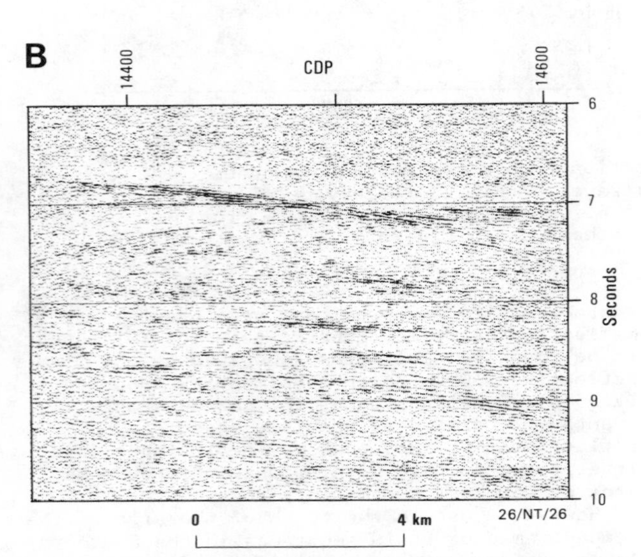

Fig. 7. BMR deep seismic reflections under the southern margin of the Arunta Block showing prominent events at 6.5-10 s two-way time. Depth estimates are based on interval velocities derived from move-out velocities and some associated wide-angle recordings [Goleby et al., 1988b]. V:H exaggeration approximately 1.5.

approximately) under the Amadeus Basin and Southern Arunta Province are different from those at similar depths farther north under the adjacent exposed Central and Northern Arunta Provinces and under the Ngalia Basin. At the boundary between the Central and Southern Provinces (the Redbank Zone) there is a basic change in the seismic fabric suggesting that, althought the areas to the north and south are linked, there are fundamental differences in the processes affecting the crust on either side of the boundary.

Eastern Australia.

In Phanerozoic eastern Australia the character of deep reflections under the central Eromanga and Surat Basins contrasts markedly with those under central Australia.

Adavale Basin, Geology and Shallow Seismic Data

Figure 8 shows the location of BMR reflection traverses in the central Eromanga Basin region mentioned in this paper. The oldest known rocks from the region comprise early Palaeozoic basement rocks of the Thomson Fold Belt [Murray & Kirkegaard,1978]. It is believed that the early structures within the fold belt controlled much of the later structuring in the basin sequences during reactivation periods [Finlayson et al., 1988].

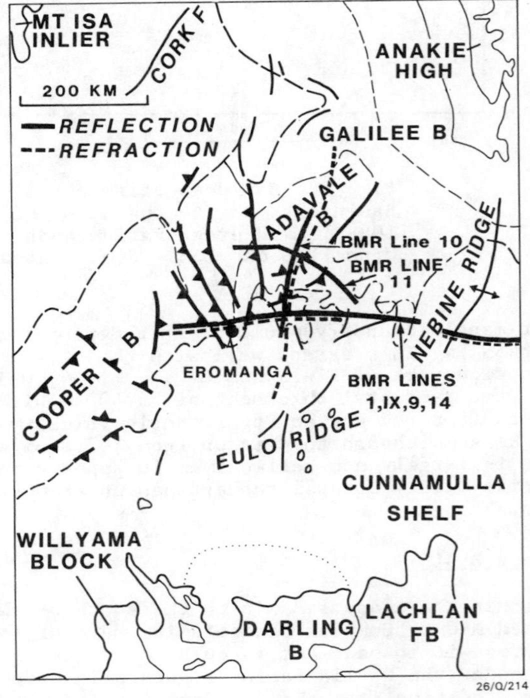

Fig. 8. Simplified geology of the central Eromanga Basin and the location of major BMR seismic reflection and refraction traverses.

During the Early Devonian the evolution of the Adavale Basin and its associated troughs (the Warrabin, Barcoo, Quilpie, Cooladdi, and Westgate Troughs) began. Passmore and Sexton [1984] describe the Adavale Basin as a foreland basin initiated, probably, in response to tectonic events on the edge of the Devonian craton to the east. Two major compressional events in the Late

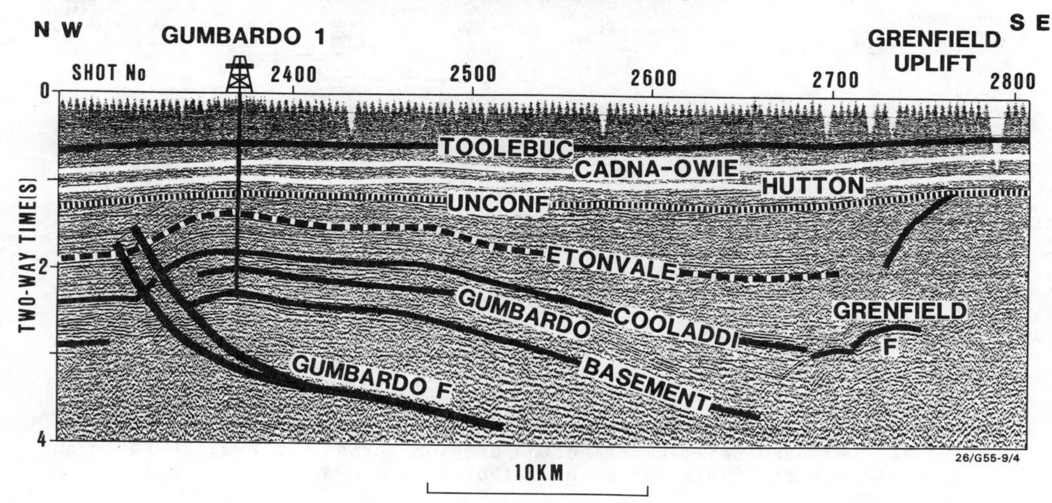

Fig. 9. BMR seismic reflection section across the Adavale Basin and Grenfield Uplift. The Eromanga Basin sequences overly the Adavale Basin, separated by a regional unconformity [Leven and Finlayson, 1987]. V:H exaggeration approximately 2.5.

Devonian and Early Carboniferous were identified by Leven and Finlayson [1987]. The first event was a north-south crustal shortening which resulted in faulting of the Devonian sequences. This was then followed by an east-west crustal shortening event which resulted in folding of the Devonian sequences. There was then an extensive glacial erosion in the region which partitioned the Devonian Adavale Basin into a main depocentre and a number of associated troughs separated by basement highs.

In Permo-Triassic times there were basin-forming episodes both west and east of the central Eromanga region, resulting in the Cooper and Galilee Basin sequences respectively. The sequences are comparatively thin in the central Eromanga region. During Jurassic - Cretaceous times there was a major regional sag which resulted in a blanket cover of Eromanga Basin sedimentary rocks 1-2 km thick.

Some of the thickest Devonian sequences are preserved in the Adavale Basin which will be discussed here. Figure 9 shows a seismic reflection profile from BMR Line 11 across the main depocentre of the Adavale Basin (Fig.8). The asymmetry of the basin and the major faulting across the Gumbardo and Grenfield Faults are evident. Analysis of the deformation geometry across a number of thrust faults in the region, including the Gumbardo Fault, indicates that the fault planes must be listric and that the sole depth must be less than 20 km, i.e. within the upper part of the crust [Leven and Finlayson,1987]. The upper crust is largely "transparent" to seismic energy. From drillhole information the basement rocks include Early Devonian Gumbardo Volcanics overlying Ordovician-Silurian basalts, andesites and granites [Finlayson et al., 1987a]. It is speculated that the rocks within upper crustal basement must be either highly folded with consistently steep dips, or be uniformly plutonic to frustrate most attempts to image their features using the seismic reflection method.

Adavale Basin, Deep Seismic Data

It is not until deep seismic sections are examined that other major features influencing the basin structure are evident. Figure 10 is a 16 second seismic reflection section from BMR Line 11 across the Adavale Basin. The outstanding feature of the deep data is that there is a clear difference in character between the reflections from the upper and lower parts of the crust. The upper crust is largely "transparent" but below 8 seconds TWT there are strong reflection events that can be correlated on crossline data e.g. BMR Lines 10 and 1, 1X, 9, 14 (Fig.8). There is a significant increase in the amplitude and continuity of crustal reflections between about 8 and 14 seconds TWT.

From wide-angle reflection and refraction data across the same area we know that there is a significant increase in velocity at mid-crustal depths. Finlayson and Collins [1986] and Drummond and Collins [1986] have argued that the increase in average velocity in the lower crust is caused by an increase in the mafic content of lower crustal rocks associated with underplating and intrusion. A more recent analogue of this process may be that described by Allmendinger et al. [1987] for the lower crustal layered fabric in the Basin and Range Province of western USA.

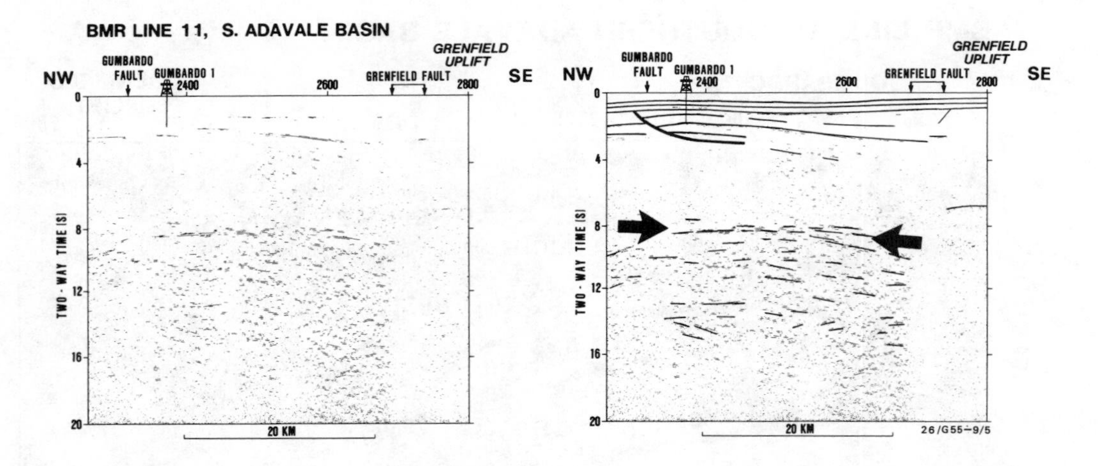

Fig. 10. BMR deep seismic reflection section across the Adavale Basin and Grenfield Uplift [Finlayson et al., 1987a]. V:H exaggeration approximately 0.6.

Finlayson et al.[1984] indicate that the Moho depth determined from wide-angle reflection and refraction data under the central Eromanga region is the same as that determined from coincident reflection profiling, within the uncertainties of the seismic interpretation methods. The reflection Moho is taken to be where there is a general extinction of sub-horizontal reflections as two-way time is increased, usually in the 12-14 s TWT range.

Also along BMR Line 11 there is another major feature that must be emphasized in any model of the evolution of the Adavale Basin. Under the Grenfield Uplift (Fig. 10) there is a marked decrease in the number of lower crustal reflections. The near-surface conditions in the Eromanga Basin sequences do not change significantly across the uplift, so the change is attributed to changes in geological basement which may extend from the surface through the entire crust.

The upper crustal Thomson Fold Belt is considered to have been formed during Late Ordovician - Silurian times. The strong semi-continuous reflectors at mid-crustal depths under the Devonian basin sequences and the associated velocity increase at such depths suggest that there is a mid-crustal discontinuity with the character of the lower crustal reflections being established after the formation of the Thomson Fold Belt as the upper crustal basement.

In the lower crust in other parts of the central Eromanga Basin region lower crustal reflections dip strongly towards the interpreted Moho but are not observed to penetrate it. The location of these dipping reflectors near basement highs suggests that they are a feature of lower crustal structuring associated with crustal blocks. Some of these blocks are shown to have different seismic reflection and velocity characteristics from the intervening basins [Finlayson et al.,1987b]. The Moho boundary is interpreted as being where the lower crustal reflections terminate, in good agreement with the Moho determined from wide-angle reflection and refraction data. It varies only slightly in depth over long distances. This suggests that either there never was any major Moho deformation associated with the basin formation or that it is mobile and has established its present level after the lower crustal reflections were established. We may therefore be seeing a younging of events as we penetrate the upper and lower parts of the crust and into the upper mantle.

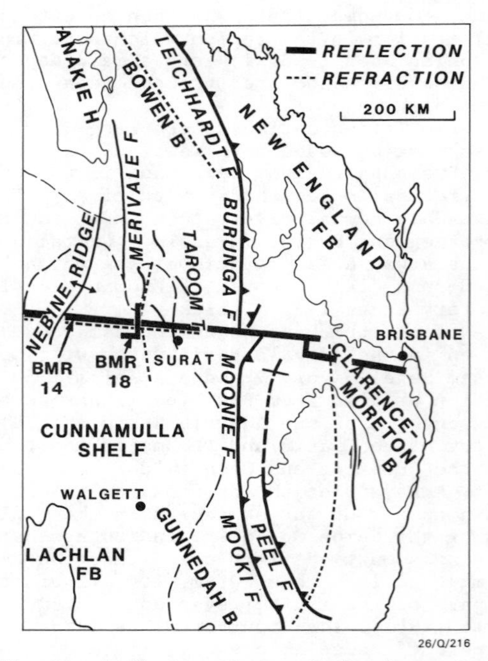

Fig. 11. Simplified geology of the Surat Basin and the location of BMR deep seismic traverses.

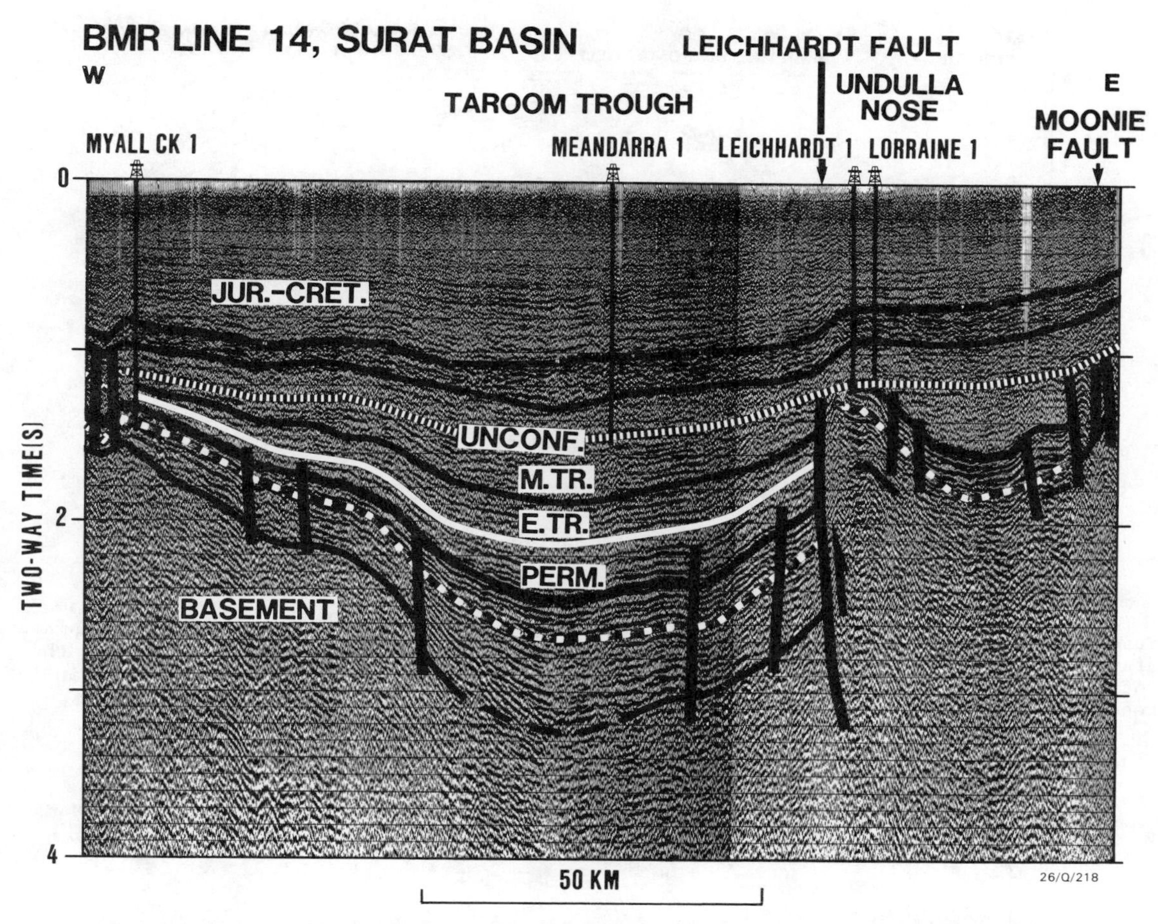

Fig. 12. BMR seismic reflection section (compressed) across the Surat Basin/Taroom Trough. Surat Basin sequences overly the trough above a regional unconformity (vertical exaggeration approx. x 9).

Taroom Trough, Geology and Shallow Seismic Data

The last basin considered in this paper is the Taroom Trough, a sub-basin of the Permo-Triassic Bowen Basin underlying Jurassic-Cretaceous Surat Basin sequences. Figure 11 shows where BMR seismic lines cross the area. Basin development began in the early Permian within the region between the Nebine Ridge in the west and the New England Fold Belt in the east. Cosgrove and Mogg [1985] describe the Bowen Basin as a product of back-arc spreading and subsidence on continental crust to the west of the New England Orogenic Zone.

Basin development is intimately tied up with the Late Palaeozoic tectonics of eastern Australia. The formation of an island arc, the Calliope Arc, during the Siluro-Devonian and its subsequent collision with the Australian craton during the Late Devonian - Early Carboniferous were major early events [Veevers,1984]. On the eastern margin of the basin another significant event was the formation of a major oroclinal structure and dextral shear within the New England Fold Belt during the Late Carboniferous - Early Permian, just prior to the first major Early Permian deposition in the region [Murray et al.,1987].

Figure 12 is a seismic section across the Taroom Trough from BMR Line 14 located in Figure 11 [Wake-Dyster et al., 1987]. The section extends from the Nebine Ridge in the west to the New England Fold Belt in the east. The present-day asymmetry of the basin is clearly illustrated, the thickest preserved sedimentary rocks being towards the east. The Jurassic - Cretaceous Surat Basin sequences at 0 - 1.5 second TWT are separated by an unconformity from the Permo-Triassic sequences of the Taroom Trough at greater two-way times. Within the Taroom Trough basement consists of Permo-Carboniferous volcanics and indurated sediments of the Kuttung Formation [Thomas et al.,1982].

On the western margin of the trough basement comprises steeply-dipping Devonian metasediments of the Timbury Hills Formation intruded by Late Carboniferous plutons (Roma Granite) and volcanics (Combarngo Volcanics). Episodes of

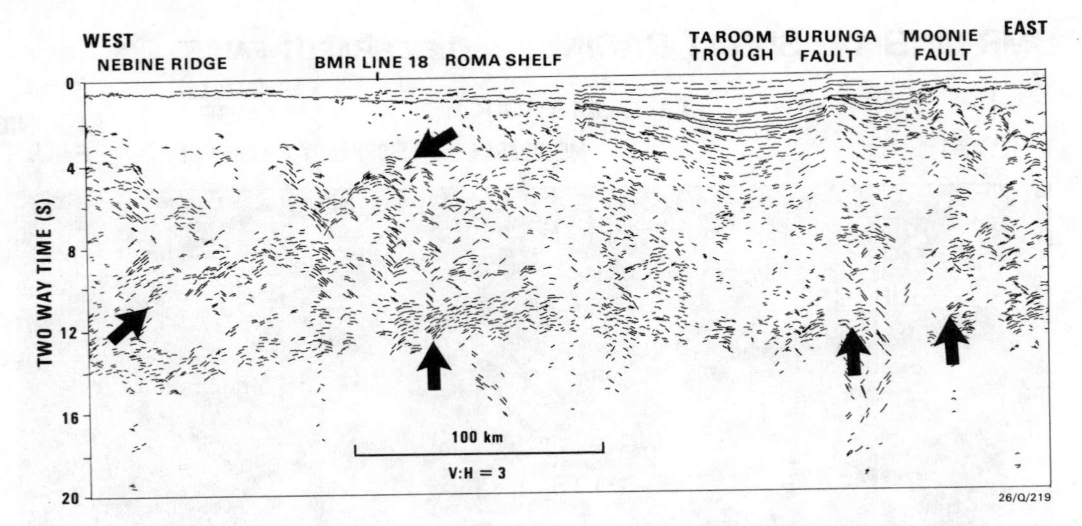

Fig. 13. Digitized deep seismic reflection events derived from BMR Line 14 across the Taroom Trough. Arrows indicate features mentioned in the text.

Early Permian half-graben formation and later wrench faulting are identified on this western margin of the trough. On the eastern margin of the Taroom Trough there was Late Permian to Middle Triassic uplift and granite emplacement accompanied by thrust movements along the Leichhardt-Burunga and Moonie-Goondiwindi Faults (Fig.11).

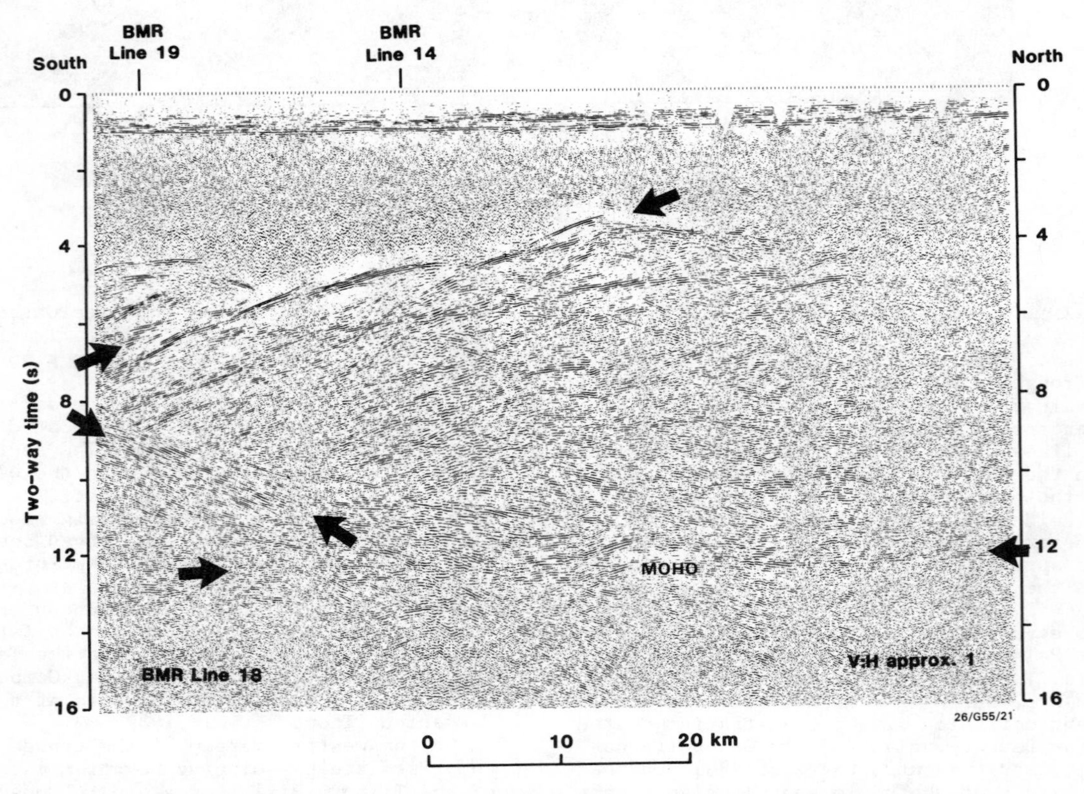

Fig. 14. BMR deep seismic reflection section across the western Surat Shelf area on BMR Line 18. Arrows indicate features mentioned in the text.

Taroom Trough, Deep Seismic Data

The deep features under the Taroom Trough can be seen in Figure 13, a diagram of digitized reflection events on the 20 second reflection records from BMR Line 14. There are major differences between events under the trough and those under other basins discussed in this paper. Relatively large numbers of reflections are evident throughout the whole crust to the west and east of the Taroom Trough but there are relatively few reflections from deep within basement under the trough itself. There is no prominent increase in the strength and continuity of reflections at mid-crustal levels nor a strongly reflecting lower crustal zone under the trough as there is under the central Eromanga Basin. At Moho depths, however, there is a well developed series of reflections at 12-13 seconds TWT which can be traced right across the basin and its margins.

The depth to the Moho within the Taroom Trough itself is not well constrained by coincident wide-angle reflection and refraction data. However, Collins [1988] derived a model for the northern Bowen Basin with a velocity transition to upper mantle velocities over a depth range of 34-44 km, corresponding to 11-13.5 s two-way time.

Some prominent features in Figure 13 include the inflection point in the Moho reflectors under the crossover with BMR Line 18 (arrowed) located in Figure 11. Also there is an apparent series of crustal reflectors dipping westward from about 3 seconds TWT under BMR Line 18 to near the Moho at 12 s TWT (arrowed). On this western margin of the Taroom Trough there is a "transparent" upper crustal zone which is not evident further east than the Roma Shelf. This is interpreted as being the eastern limit of the Thomson Fold Belt. East of this point is the hinge/ramp zone for the Taroom Trough.

Also from this western margin of the Taroom Trough Figure 14 shows a seismic section from north-south BMR Line 18 (located in Figure 11). Under the Jurassic - Cretaceous Surat Basin sequences (0-1 s TWT) lie the Devonian - Carboniferous Timbury Hills Formation and Roma Granite which are largely "transparent". However, at 3-5 seconds TWT there are a series of prominent reflectors which have an associated increase in velocity from about 6.0 km/s above to about 6.3 km/s below [Wake-Dyster et al.,1988]. These reflections seem to form an unconformity with the overlying sequences and, assuming that these latter sequences form part of the Thomson Fold Belt, we may be seeing a dramatic shallowing of the "mid-crustal horizon" evident at about 7-8 seconds TWT further west under the central Eromanga Basin.

On the eastern margin of the Taroom Trough is the New England Fold Belt (Fig.11) and it is evident that the character of deep basement reflections (Fig.13) under the fold belt to the east of the Moonie Fault is different from that under the trough, there being prominent refections throughout the entire crust under the fold belt. The exact nature of the structures at the Burunga and Moonie Faults (arrowed in Fig. 16) is not clear yet but it is known that Early

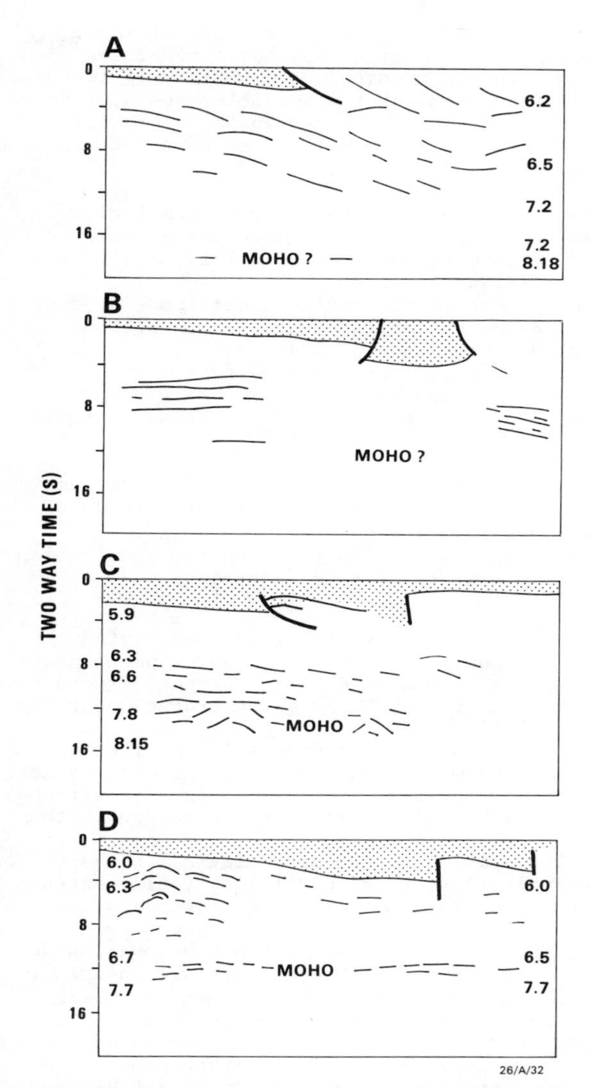

Fig. 15. Sketches of the seismic fabric within the crust under four Australian basins; a) Ngalia Basin, b) Amadeus Basin, c) Adavale Basin, and d) Taroom Trough.

Permian sequences exist east of the Moonie Fault. On shallow seismic sections both faults are interpreted as thrust faults reactivated late in the basin's development. However, from the deep seismic profiling the current indications are that both faults are associated with deep seismic features which have considerable depth extent and may have a shear component associated with the Carboniferous developments within the New England Fold Belt.

Discussion

This paper has presented seismic features from deep within the basement underlying four Australian basins, all of which are significantly different from each other (Fig.15) indicating that a variety of pre- and syn-depositional processes were acting. Any basin modelling must be able to accommodate the diversity of features identified. Some of the possible causes of deep crustal reflections listed in the scientific literature are given below [Matthews & Cheadle, 1986];

a) Fault zones, often involving mylonites.
b) Intrusions/sills of more basic rocks or alternating layers of felsic, basic or ultrabasic rocks.
c) Anastomosing ductile shear zones at depths below the brittle/ductile transition.
d) Metasediments.
e) Hydrated mineral assemblages.
f) Magma chambers.
g) Fluids trapped in sutures or major fault zones.

Nearly all of these may be sources of reflections at different levels within the basement under the four basins discussed here.

From the seismic data in the four Australian basins discussed here we emphasize:

1. The tectonic history of the underlying basement rocks prior to basin development plays an important role in determining the structure of the basins, perhaps best demonstrated under the Ngalia Basin.
2. The sedimentary sequences in the four basins are all asymmetric but this present-day assymetry is largely the result of major post-depositional crustal shortening/uplift on one margin of the basin.
3. There is no strong seismological evidence for low-angle (less than 10 degrees) faults extending from the surface down into the lower crust and displacing the Moho. However, higher-angle faulting is interpreted at block boundaries and may be associated with crustal-scale shearing.
4. Much of the structuring within the basin sequences is associated with faulting in the upper crustal basement. This region is cooler than the deeper parts of the crust and deformation takes place along well established zones of weakness.
5. At mid-crustal levels under the Devonian Adavale Basin and its associated troughs there is a prominent discontinuity, with an increase in the strength and continuity of reflections and a higher P-wave velocity in the lower crust. The nature of the mid-crustal seismic boundary suggests that it may be younger than the folding and plutonism in the overlying upper crustal rocks, and may be associated with the basin-forming mechanism.
6. Strong sub-horizontal reflections are a prominent feature in the lower crust under the Adavale Basin and its associated troughs. These reflections can change in strength and continuity across major block boundaries. Their association with increased velocity suggests an intrusive underplating process in an inhomogeneous lower crust.
7. Under the Adavale Basin and its associated troughs there is no distinct Moho band of reflectors. Rather there is a sometimes broken-up extinction of lower crustal reflections at 13-14 s TWT which is interpreted as the Moho. Lower crustal reflections may shallow to depths of about 7 s TWT in places.
8. The Moho under the Permo-Triassic Taroom Trough has an associated band of gently undulating reflectors about 3 km (1 s TWT) thick below a relatively "transparent" lower crust. The Moho extends across the basin and its margins at 12-13 seconds TWT, suggesting that this feature may have been re-established since the basin-forming events, assuming that such events involved large Moho displacements.
9. Under the Late Proterozoic-Early Palaeozoic basins in central Australia the Moho is not apparent at all on reflection profiling records but, from other seismic data, must be at a depth in excess of 50 km under much of the Arunta Block.

Acknowledgements. This paper uses deep seismic reflection profiling data acquired and processed by many BMR seismic staff since 1980. In addition to two of the authors (K.D.W-D and D.W.J), Bruce Goleby, Tim Barton, Surendra Mathur and Mike Sexton have played major roles in data acquisition and processing, and their contribution is here acknowledged. Bruce Goleby developed computer software used in the preparation of the digitized sections and provided considerable information on central Australian data. This paper is published with the permission of the Director, Bureau of Mineral Resources, Geology and Geophysics, Canberra.

References

Allmendinger, R. W., K. D. Nelson, C. J. Potter, M. Barazangi, L. D. Brown, J. E. Oliver, Deep seismic reflection characteristics of the continental crust, Geology, 15, 304-310, 1987.

Collins, C. D. N., The nature of the crust-mantle boundary under Australia from seismic evidence, In; The Deep Lithosphere of Australia (editor, B. J. Drummond), Geological Society of Australia Special Publication, in press, 1988.

Cosgrove J. L., and W. G. Mogg, Recent exploration of the Roma Shelf, Queensland, Aust. Petrol. Expl. Ass. J., 25, 216-234, 1985.

Drummond, B. J., and C. D. N. Collins, Seismic evidence for underplating of the lower continental crust of Australia, Earth Planet. Sc. Lett., 79, 361-372, 1986.

Finlayson, D. M., J. P. Cull, and B. J. Drummond, Upper mantle structure from the trans-Australia seismic refraction data, J. Geol. Soc. Aust., 21, 447-458, 1974.

Finlayson, D. M., C. D. N. Collins, and J. Lock, P-wave velocity features of the lithosphere under the Eromanga Basin, eastern Australia, including a prominent mid-crustal (Conrad?) discontinuity, Tectonophysics, 101, 267-291, 1984.

Finlayson, D. M., and C. D. N. Collins, Lithospheric velocity beneath the Adavale Basin, Queensland, and the character of deep crustal reflections, BMR J. Aust. Geol. Geophys., 10, 23-37, 1986.

Finlayson, D. M., J. H. Leven, S. P. Mathur, and C. D. N. Collins, Geophysical abstracts and seismic profiles from the central Eromanga Basin region, eastern Australia, Bur. Miner. Resour. Aust., Report, 278, 1987a.

Finlayson, D. M., J. H. Leven, M. J. Sexton, and K. D. Wake-Dyster, 3-Dimensional image of lower crustal lenticles under an intra-continental basin in eastern Australia, Abstracts, IUGG XIX General Assembly, Vol.1, 59, 1987b.

Finlayson, D. M., J. H. Leven, and M. A. Etheridge, Structural styles and basin evolution in the Eromanga region, eastern Australia, Am. Ass. Petrol. Geol. Bull., 72, 33-48, 1988.

Goleby, B. R., C. Wright, and B. L. N. Kennett, Preliminary deep reflection studies in the Arunta Block, central Australia, Geophys. J. Roy. Astr. Soc., 89, 437-442, 1987.

Goleby, B. R., R. D. Shaw, C. Wright and B. L. N. Kennett, Deep seismic reflection profiling in central Australia: support for a 'thick-skin' model of tectonic evolution. Submitted to Nature, 1988a.

Goleby, B.R., C. Wright, C. D. N. Collins and B. L. N. Kennett, Seismic reflection and refraction profiling across the Arunta Block and the Ngalia and Amadeus Basins. Aust. J. Earth Sc., 35, 275-294, 1988b.

Kennard, J. M., R. S. Nicoll, and M. Owen, Late Proterozoic and Early Palaeozoic depositional facies of the northern Amadeus Basin, central Australia, 12th Int. Sed. Cong. Field Guide 25B, 1986.

Lambeck, K., G. Burgess, and R. D. Shaw, Teleseismic travel-time anomalies and deep crustal structure in central Australia, Geophys. J. Roy. Astro. Soc., 94, 105-124, 1988.

Leven, J. H., and D. M. Finlayson, Lower crustal involvement in upper crustal thrusting, Geophys. J. Roy. Astr. Soc., 89, 415-422, 1987.

Lindsay, J. F., and R. J. Korsch, Interplay of dynamics and sea-level changes in basin evolution: an example from the intracratonic Amadeus Basin, central Australia, Submitted to Basin Research, 1988.

Matthews, D. H., and M. J. Cheadle, Deep reflections from the Caledonides and Variscides west of Britain and comparison with the Himalayas, In, Reflection Seismology: a Global Perspective, (editors, M. Barazangi and L. Brown), Am. Geophys. Un. Geodynamics Series, 13, 5-19, 1986.

Mooney, W. D., and A. Ginzburg, Seismic measurement of the internal properties of fault zones, Pageoph., 124, 141-157, 1986.

Murray, C. G., and A. G. Kirkegaard, The Thomson Orogen of the Tasman Orogenic Zone, Tectonophysics, 48, 299-326, 1978.

Murray, C. G., C. L. Fergusson, P. G. Flood, W. G. Whitaker, and R. J. Korsch, Plate tectonic model for the Carboniferous evolution of the New England Fold Belt, Aust. J. Earth Sc., 34, 213-236, 1987.

Passmore, V. L., and M. J. Sexton, The structural development and hydrocarbon potential of Palaeozoic source rocks in the Adavale Basin region, Aust. Petrol. Expl. Ass. J., 24, 393-411, 1984.

Shaw, R. D., A. J. Stewart, and L. P. Black, The Arunta Block: a complex ensialic mobile belt in central Australia. Part 2: tectonic history, Aust. J. Earth Sc., 31, 457-484, 1984.

Shaw, R. D., K. Lambeck, and M. A. Etheridge, Intracratonic basin formation in central Australia, Research Sch. Earth Sc., Aust. Nat. Univ., Annual Rep., 24-26, 1986.

Stewart, A. J., R. D. Shaw, and L. P. Black, The Arunta Block: a complex ensialic mobile belt in central Australia. Part 1: stratigraphy, correlations and origin, Aust. J. Earth Sc., 31, 445-455, 1984.

Teyssier, C., A crustal thrust system in an intra-cratonic tectonic environment. J. Struct. Geol., 7, 689-700, 1985.

Thomas, B. M., D. G. Osborne, and A. J. Wright, Hydrocarbon habitat of the Surat/Bowen Basin, Aust. Petrol. Expl. Ass. J., 22, 213-226, 1982.

Veevers, J. J. (editor), Phanerozoic Earth History of Australia, Clarendon Press, Oxford, 1984.

Wake-Dyster, K. D., M. J. Sexton, D. W. Johnstone, C. Wright, and D. M. Finlayson, A deep seismic profile of 800 km length recorded in southern Queensland, Australia. Geophys. J. Roy. Astron. Soc., 89, 423-430, 1987.

Wake-Dyster, K. D., M. J. Sexton, D. W. Johnstone, and D. M. Finlayson, Seismic features of the Thomson Fold Belt under the western Surat Basin, Geol. Soc. Aust., Abstract Series, 21, 406-407, 1988.

Wells, A. T., and F. J. Moss, The Ngalia Basin, Northern Territory: stratigraphy and structure, Bur. Miner. Resour. Aust. Bull, 212, 1983.

Wright, C., B. R. Goleby, C. D. N. Collins, B. L. N. Kennett, S. Sugiharto, and S. Greenhalgh, The central Australian seismic experiment, 1985: preliminary results, Geophys. J. Roy. Astr. Soc., 89, 431-436, 1987.

DEEP CRUSTAL STRUCTURAL CONTROLS ON SEDIMENTARY BASIN GEOMETRY

D. J. Blundell, T. J. Reston, and A. M. Stein

Department of Geology, Royal Holloway and Bedford New College, University of London, England

Abstract. Deep seismic reflection surveys around Britain have shown deep seated faults bounding sedimentary basins. The geometry of these faults has influenced the internal structure and stratigraphic evolution of the basins. The faults are commonly derived from older structures partially reactivated during basin development. At depth, seismic reflections from these bounding faults merge into a lower crustal reflective zone. Modeling the seismic reflection character of the mid to lower crust suggests that lower crustal extension relating to basin development develops along anastomosing shear zones through the lower crust. This geometry can be viewed in simple terms to give basin development resulting from simple shear along individual faults in the upper crust above a region of bulk pure shear in the lower crust which can extend beyond the confines of the basin, possibly leading to the development of secondary shears in mid-crust at their periphery. Variations in the relative amounts of extension in upper and lower crust, together with fault geometry, can give rise to a variety of basin geometries and styles.

Basin Development on Deep-seated Faults

BIRPS deep seismic reflection profiles around Britain [Matthews et al, 1987] have traversed a number of Mesozoic sedimentary basins. Most, though not all, of them are seen to be bounded by deep seated faults. These are recognised as dipping reflections through the upper crust and are identifiable with fault outcrops at surface. The best example is the DRUM profile offshore from the north coast of Scotland [McGeary and Warner, 1985]. An interpreted line drawing of the stack record section is shown in Figure 1. In this, it is seen that a series of wedge shaped basins containing westerly dipping strata are bounded at their western ends by easterly dipping faults that can be traced to mid-crustal levels.

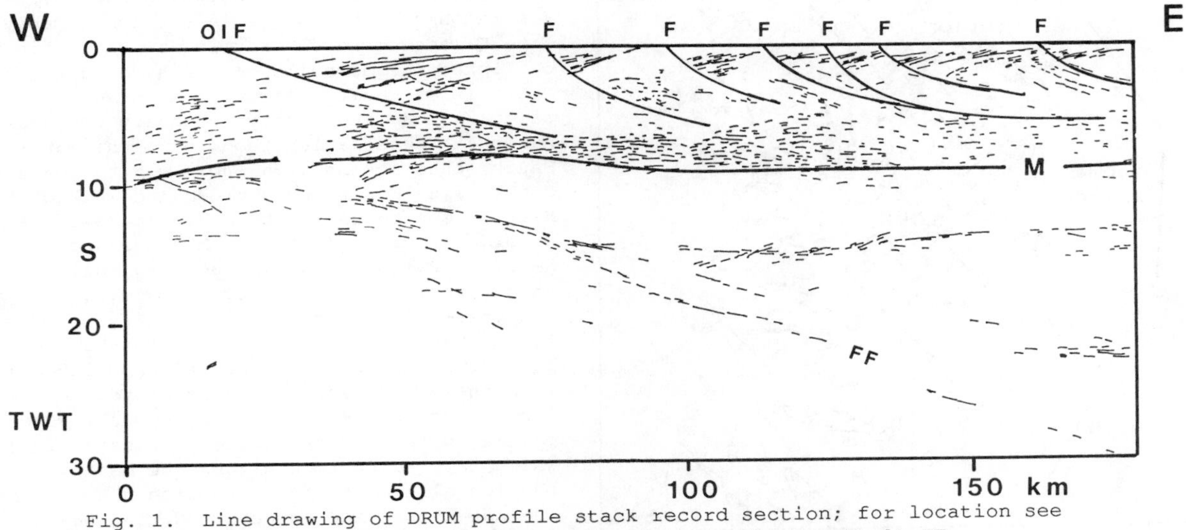

Fig. 1. Line drawing of DRUM profile stack record section; for location see Figure 2. Fault F, Moho M, Outer Isles Fault OIF, Flannan Fault FF

The most westerly of these faults, marked OIF, has been correlated at surface with the Outer Isles Thrust [Brewer and Smythe, 1984] which is well known on the Outer Hebrides (Figure 2). This extends as a linear feature nearly to the base of the crust and may reach the Moho, which is recognised as a strong reflector, marked M in Figure 1. The Moho is evident across most of the profile at around 9s two-way time (TWT), and matches the Moho determined from a cross-cutting refraction profile, LISPB [Bamford et al, 1978]. Within the lower crust there is a zone of numerous short, sub-horizontal reflection segments into which the fault reflections merge. The lower bound of this zone is at the Moho. Similar features have been found on many other BIRPS profiles around Britain although a variety of reflection characteristics have been recognised [McGeary, 1987]. Unique to the area north of Scotland is the presence of strong reflectors within the mantle, shown in Figure 1, one of which (FF) dips eastwards and can be followed to 30s TWT, from a depth of around 20 km. The reality of this reflection has been confirmed from a grid of reflection profiles in the area to map it in three dimensions. These mantle reflections are best explained by Warner and McGeary [1987] as faults or shear zones but there is no corroboratory evidence available as yet. Similar features in the mantle may be present elsewhere around Britain but their presence has not yet been firmly established.

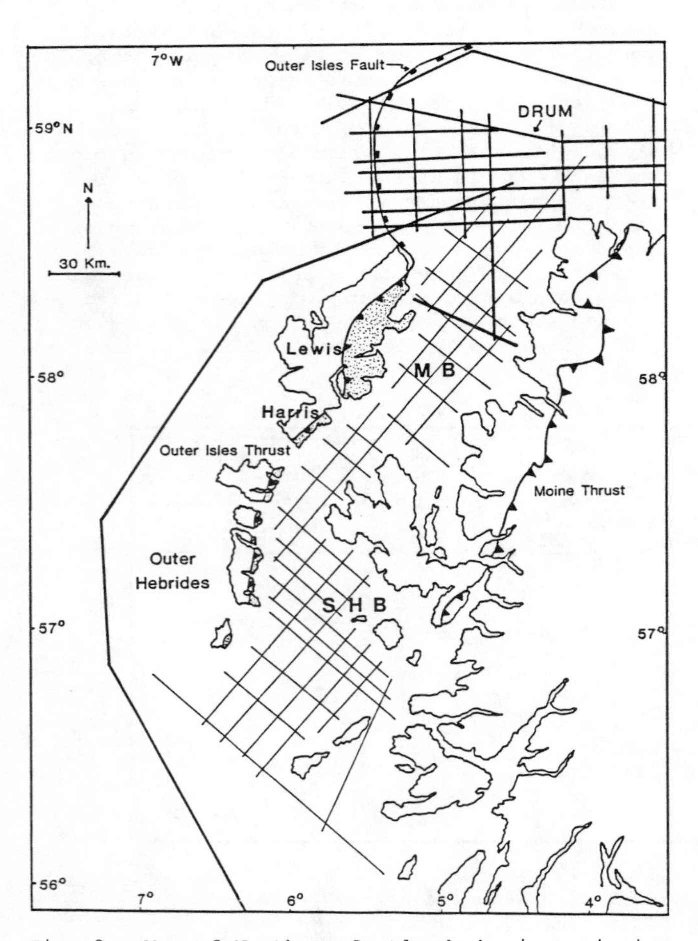

Fig. 2. Map of Northern Scotland showing seismic line locations offshore and major structural features : Outer Isles Thrust Zone stippled, Minches Basin MB, Sea of Hebrides Basin SHB

Basin development on inherited structures

Stein [1987] has examined the geometry of the Outer Isles Fault, Figure 1, in three dimensions, and its relationship to the stratigraphy and fault geometry within the Minches and Sea of Hebrides Basins, Figure 2. Field observations on Harris, part of the northernmost island of the Outer Hebrides (M. Lailey, A. Stein and T Reston, unpublished, 1986), have established that the Outer Isles Fault was originally active during the Early Proterozoic. It probably acted as a normal fault during the Late Proterozoic, and that although in part reactivated as a thrust during the Caledonian Orogeny it has remained locked up on the Hebrides subsequently. In the Early Proterozoic, the Outer Isles Fault probably acted as a linking element between a number of NW-SE trending ductile shear zones which were at that time the dominant structural features. Offshore, in the Minches, a network of seismic reflection profiles shot as a survey for commercial purposes by JEBCO Seismic Ltd has been interpreted by Stein. He has recognised a sedimentary succession including Late Proterozoic (Torridonian) strata overlain unconformably by Late Palaeozoic and Mesozoic strata within a set of linked basins built on the hanging wall of the Outer Isles Fault. A set of near-vertical normal faults act as the western boundary to these basins. These faults root on to the Outer Isles Fault, thus acting as a short cut to leave the Outer Isles Fault undisturbed and locked up on Harris as a thrust, Figure 3, but reactivated as a normal fault beneath the Minches and Sea of Hebrides Basins, thus influencing their overall shape and internal geometry. Mapping the base of the post-Caledonian succession, Figure 4, Stein found that the Outer Isles Fault is offset and its geometry altered at locations which coincide with the NW-SE trending Early Proterozoic shear zones. He concluded that these, too, were partially reactivated to form transfer faults which have compartmentalised the basins. The en echelon alignment of faults within the basins indicates that the extension during basin development was oblique to the bounding faults. Thus, as illustrated in Figure 3, the basin architecture has developed from the partial reactivation of inherited structures and from the creation of new faults. Associated block

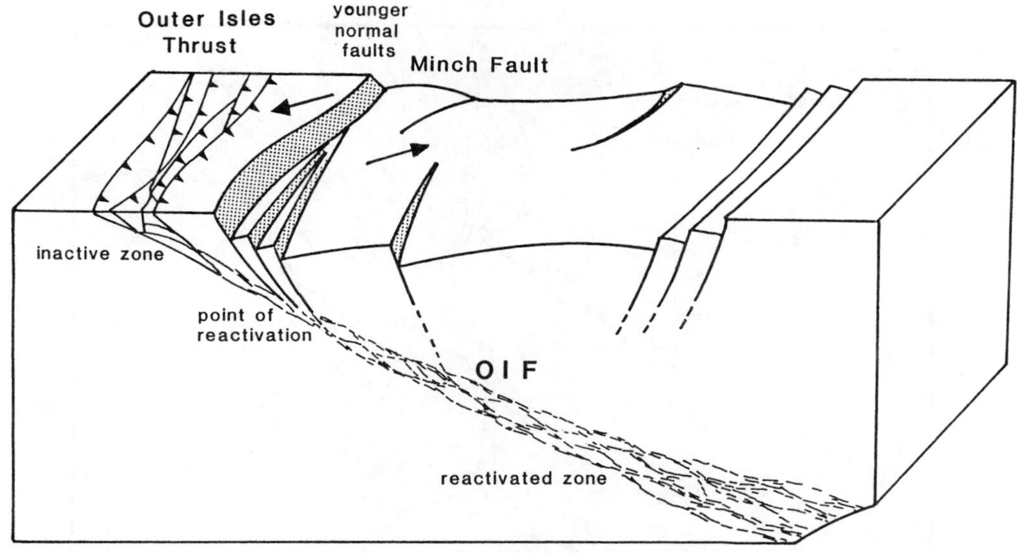

Fig. 3. Sketch of structural style of faults controlling the Minches Basin: Outer Isles Fault OIF

rotation has accommodated in three dimensions the extensional regime in which the basins formed. Basin development through the partial re-use of inherited structures, observed in detail by Stein in the Minches and Sea of Hebrides, is found generally elsewhere around Britain, for example in the Celtic Sea [BIRPS and ECORS, 1986] and receives confirmation from analogue modelling [McClay and Ellis, 1987] in which it can be demonstrated that, during inversion from a former compressional to an extensional regime, pre-existing structures are used initially but they become increasingly overprinted with new structures relating to the new stress regime.

Extension within the crust as a whole

Seismic profiles such as DRUM, Figure 1, indicate basin development occurs as a consequence of extension in the upper crust accommodated by displacement and rotation of blocks along a system of linked faults. The geometrical consequences for basin development have been well illustrated by Gibbs [1983, 1987]. As seen in Figure 1, and on other deep reflection profiles, the fault reflections merge into a lower crustal reflective zone. The possible origins of the lower crustal reflective zone are currently a matter for debate [Hobbs et al, 1987; Matthews and Cheadle, 1986]. The first consideration, however, is to understand how numerous reflection segments with the observed character can arise. Blundell and Raynaud [1986] pointed out that, just as in the case with reflections from sequences of sedimentary strata, reflections from deeper within the crust are also likely to arise from the combination of reflections from closely spaced interfaces through constructive interference. They also demonstrated the likelihood that reflections also come from interfaces out of the vertical plane of section, to add to the interference and complexity of the record section. Reston [1987] took this further, using the AIMS forward modeling computer software developed by GeoQuest International Inc. to match the characteristics of the lower crustal reflective zone. He realised that the multitude of short reflection segments observed were constrained by the limitations of spatial resolution of the seismic waves reaching and returning from the lower crust. The vertical resolution is limited to about one quarter of the wavelength of the seismic waves, whilst the horizontal resolution is limited to the diameter of the first Fresnel zone, which is the minimum size of reflecting surface required for a coherent reflection. Typical values for these at various crustal levels are presented in Table 1. He used as model a distribution of lenticular bodies with

TWT (s)	f (Hz)	V (km/s)	λ (m)	Z (km)	R (km)
1	50	2.5	50	1.25	.18
3	40	4	100	6	.55
6	30	6	200	18	1.34
9	20	7	350	31.5	2.34

TABLE 1. Values of wavelength (λ) and first Fresnel Zone radius (R) for seismic p-wave signals of typical velocity (V) and frequency (f) at various two-way times (TWT) and equivalent depths (Z)

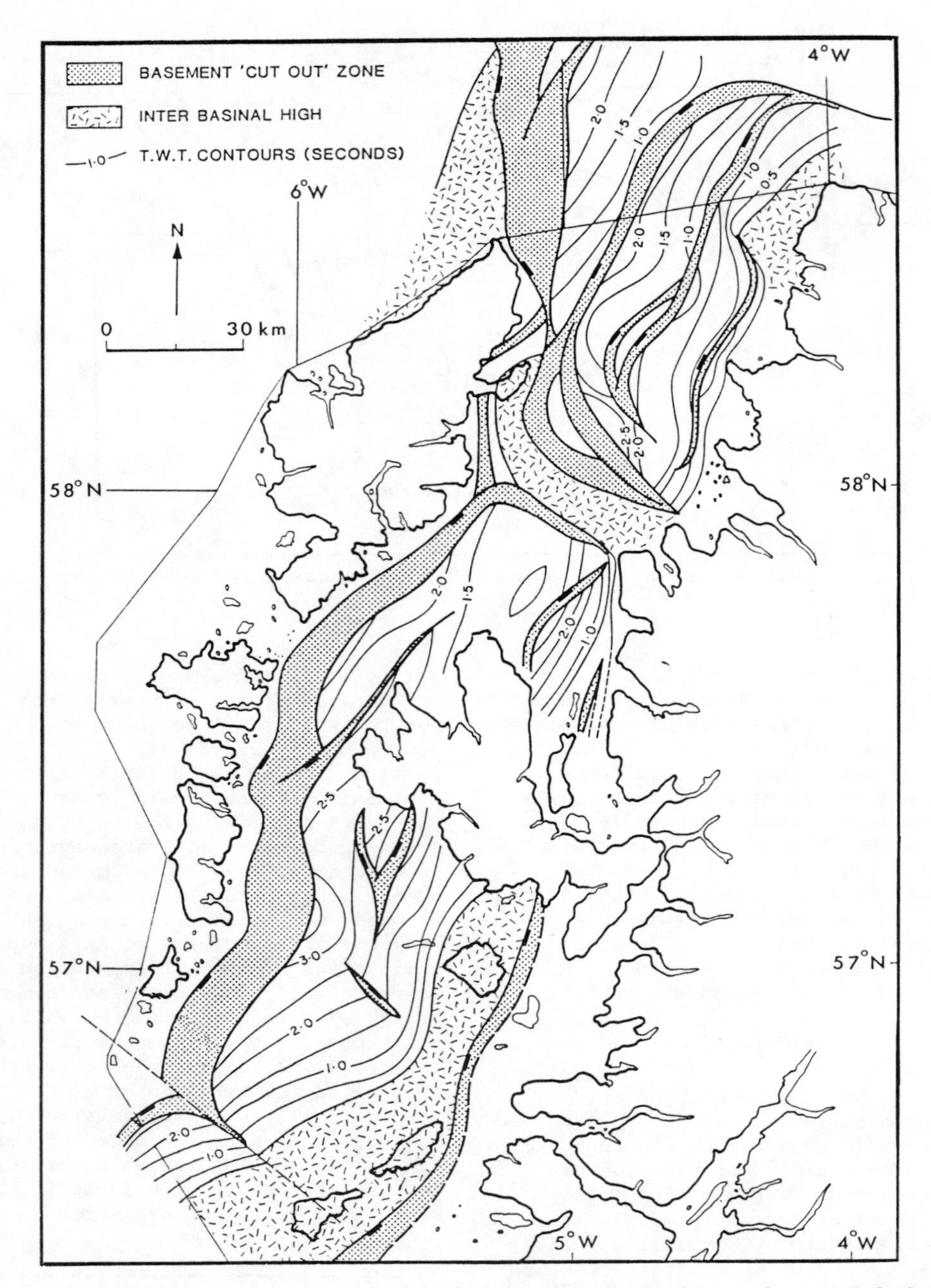

Fig. 4. Contour map of two-way-time in seconds to the base of post-Caledonian sediments in the Minches and Sea of Hebrides Basins

lengths smaller than the diameter of the first Fresnel zone. If these are closely packed he could produce short, sub-horizontal reflection segments through the mixture of constructive and destructive interference between their individual responses, Figure 5, known as spatial interference. He also showed how by increasing the aspect ratios of the lenses, the reflection pattern contains more intersecting, convex upward reflection segments which are observed on some profiles. His work demonstrated that the character of the lower crustal reflective zone must arise from some element of spatial interference from interfaces having sizes or topography with wavelengths smaller than the diameter of the first Fresnel zone, including

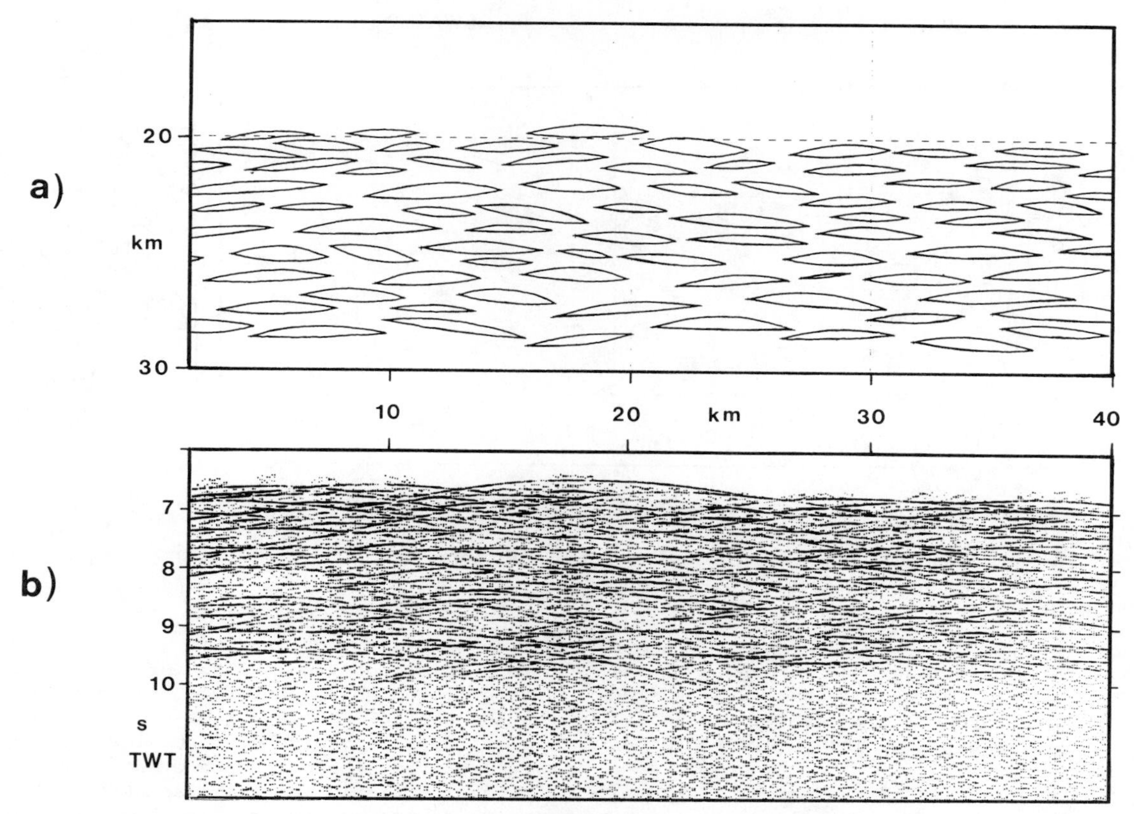

Fig. 5a. 2-D Lower crust model composed of closely packed lenses
b. Synthetic zero-offset seismic section derived from this model, with random noise added

effects from out of the vertical plane of section.

More debatable is the geological cause of such interface geometries within the lower crust. Reston [1987] saw evidence in some of BIRPS profiles of larger scale variations in the character of the lower crust, in which it is sometimes divided into highly reflective bands separated by non-reflective zones. Because BIRPS profiles, along with most other deep reflection profiles around the world, are individual lines across major features of surface geology, they are generally dip lines with little or no three-dimensional control. However, the BIRPS lines across the south Irish Sea, SWAT 1 and SWAT 2 [BIRPS and ECORS, 1986] act as strike and dip lines, respectively, relative to the Cardigan Bay Basin [Blundell et al, 1971]. From his line drawings of the stack record sections of these lines, Figure 6, Reston recognised a distinction in reflection character between them. Both show reflection segments within the lower crust occurring in bands, but reflection segments are longer, straighter and nearer to horizontal along the strike direction, SWAT 1, than along the dip direction, SWAT 2. Because this distinction relates to the direction of extension in the upper crust, he concluded that it indicates a structurally derived cause of the lower crust reflections. The bands of lower crustal reflections suggest that these are shear zones. Small lenticular bodies within them such as shear pods or boudins create the reflections. The larger lenticular shaped unreflecting zones represent undeformed regions between the shear zones. The boudins are elongated along the strike direction, giving rise to longer, sub-horizontal reflection segments, many from out of the plane of section. They are shorter in the extensional direction, giving rise to shorter, more curved reflection segments. This conclusion led directly to the model illustrated in Figure 7 of crustal response to extension during basin formation in which the upper crust deforms by simple shear along individual faults, together with block rotation. The lower crust deforms through a network of anastomosing shear zones, within which localised simple shear strain can create fabrics and lithological contrasts capable of creating seismic reflections, but the overall effect of which is a bulk pure shear. This is essentially the model initially proposed by

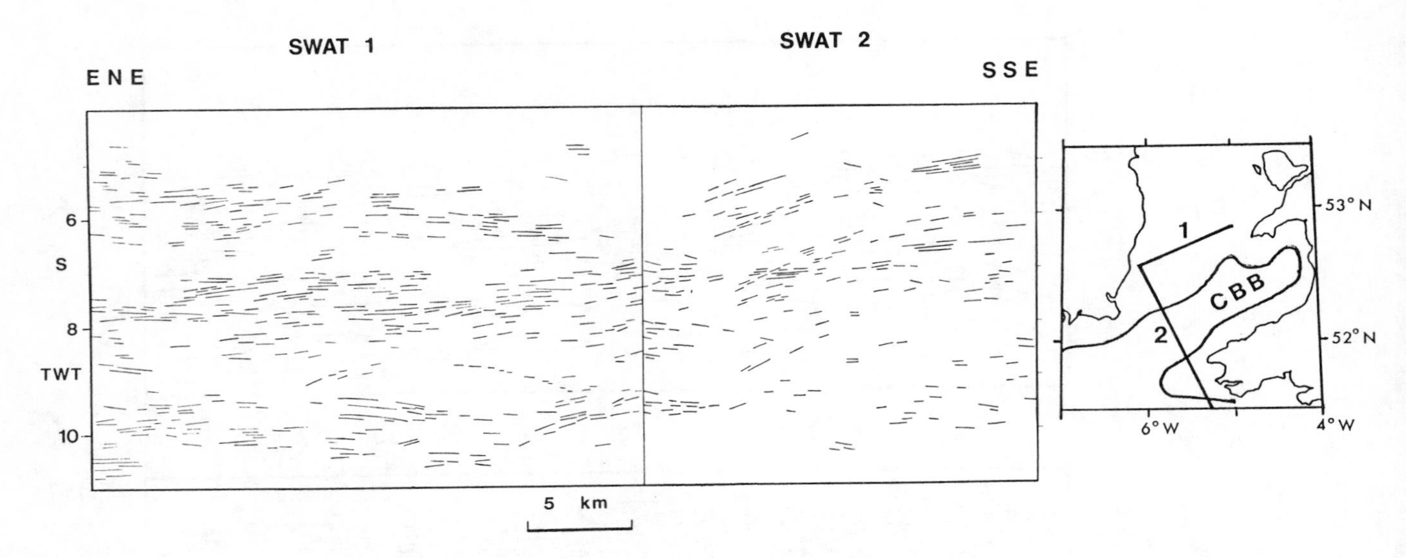

Fig. 6. Line drawings of stack record sections for SWAT 1 and SWAT 2; inset map of line locations

Hamilton [1982] to account for extension in the Basin and Range Province of western USA, where geological circumstances are not too dissimilar from those around Britain, though arrived at on different evidence.

The same distinction in character of lower crust reflections between extensional and strike directions has been observed on several profiles elsewhere [Wever et al, 1987] although the number of appropriate intersecting lines is still rather small. Further evidence of structural influence on lower crustal reflectivity comes from locations where regions of more recent crustal deformation abut undeformed crustal blocks. Along the COCORP 40°N transect [Allmendinger et al, 1987], strongly developed lower crustal reflectivity found beneath evident upper crustal extension within the Basin and Range Province stops at the eastern margin and is not observed beneath the undeformed block of the Colorado Plateau. Neither is there any Moho reflection. Similarly in Northern France, the ECORS deep reflection profile [Bois et al, 1986] shows a strongly developed lower crustal reflective zone

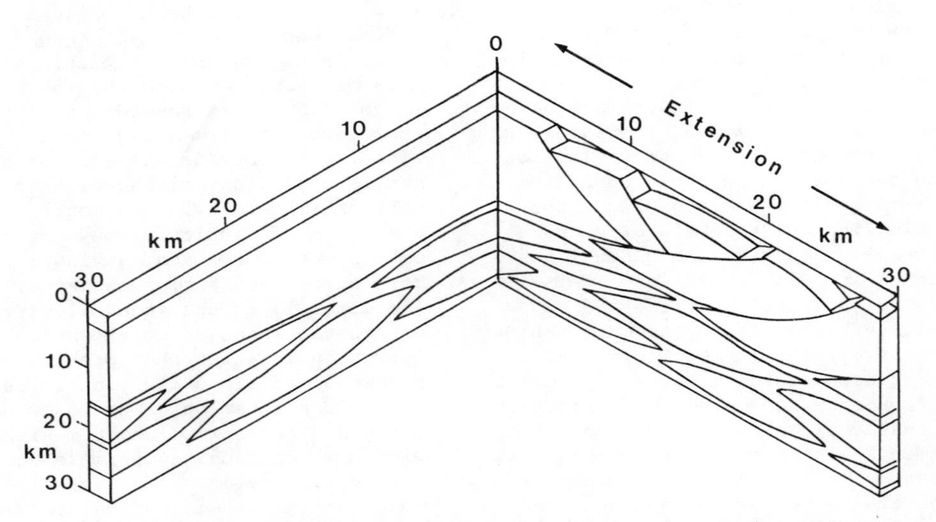

Fig. 7. Sketch of model of extensional deformation in continental crust, after Reston [1987]

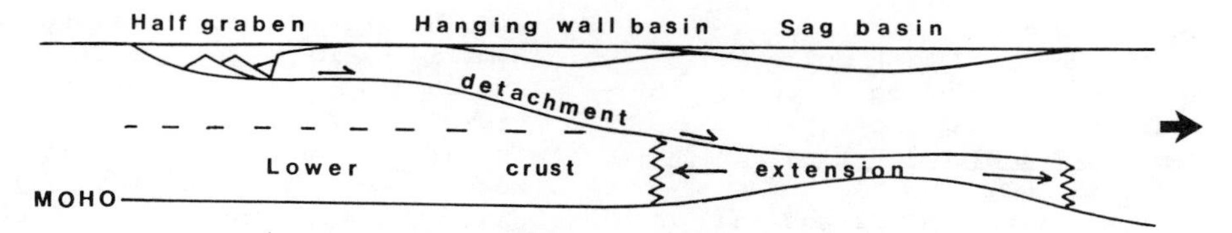

Fig. 8. Sketch to illustrate styles of sedimentary basin developed from extension at various crustal levels, after Gibbs [1987]

beneath the Paris Basin which terminates against the London-Brabant massif, where no Moho reflection is observed.

Crustal extension and basin geometry

If crustal extension occurs in the manner just described, then basin geometry should relate to the manner and amount of relative stretching within the crust. This is illustrated in Figure 8, following Gibbs [1987]. Wedge-shaped half graben basins develop through upper crustal extension by displacement and rotation of blocks sliding along individual faults. A basin can develop through subsidence resulting from displacement of an upper crustal block down the ramp of a detachment fault, to create a hanging wall basin. A sag basin can develop through subsidence over a region of stretching and thinning by bulk pure shear of the lower crust. In this diagram, the upper and lower crust deformations are displaced laterally, though this need not be the case. Deformation must also occur somewhere within the upper mantle to accommodate the extension, and this will have a further effect on basin geometry which is neglected in Figure 8. An implication of this model is that differential stretching occurs at various levels, requiring horizontal shear in some form. Dipping reflections observed in mid-crust near the NE margin of the London Platform [Reston and Blundell, 1987] may be due to secondary shears developed as a result of differential stretching of the upper and lower crust at the periphery of the southern North Sea Basin. Similarly, the presence of reflections close to or at the Moho in regions of crustal extension and their absence beneath stable, undeformed blocks suggests they may relate to shear between crust and upper mantle due to differential stretching. There is the possibility that deformation in the mantle occurs along narrow faults or shear zones which give rise to the reflections such as FF observed on DRUM, Figure 1. This suggests that the lower crust may decouple deformation in the upper mantle from that in the upper crust unless mantle shear zones are colinear with upper crustal faults, and so far this has not been observed.

Also affecting basin geometry are the isostatic and flexural response of the crust and the thermal consequences of lithosphere stretching [McKenzie, 1978], none of which are shown in Figure 8. With so many variables, modeling basin geometry is difficult, although Kusznir et al [1987] have done so with considerable success. Using models such as theirs for guidance, it is now possible to use the network of deep reflection profiles around Britain to give the regional scale crustal deformation in conjunction with more detailed seismic surveys over individual basins or sub-basins to work out their architecture and evolution. However to do so effectively it is vital to have three-dimensional control on both the local and the regional scales.

Acknowledgements. Modelling the lower crustal reflection zone was done by T Reston using the AIMS package donated by GeoQuest International Inc. and the DISCO processing software developed by Digicon whilst he was at the Univeristy of Wyoming at the kind invitation of Prof S Smithson.

Seismic survey data across the Minches Basin were kindly provided by JEBCO Seismic Ltd. BIRPS seismic data were provided by the British Institutions Reflection Profiling Syndicate under the authority of the Deep Geology Committee of the Natural Environment Research Council (NERC).

T Reston and A Stein gratefully acknowledge receipt of NERC studentships during the course of this research.

References

Allmendinger, R.W., T.A. Hauge, E.C. Hauser, C.J. Potter, and J. Oliver, Tectonic heredity and the layered lower crust in the Basin and Range Province, western United States, in Continental Extensional Tectonics edited by M.P. Coward, J.F. Dewey, and P.L. Hancock, Geol. Soc. Spec. Publn. No.28, 223-246, 1987.

Bamford, D., K. Nunn, C. Prodehl, and B. Jacob, LISPB IV. Crustal structure of Northern Britain, Geophys. J. R. Astron. Soc., 54, 43-60, 1978.

BIRPS and ECORS, Deep seismic reflection profiling between England, France and Ireland, J. Geol. Soc. London, 143, 45-52, 1986.

Blundell, D.J., F.J. Davey, and L.J. Graves, Geophysical surveys over the South Irish Sea and Nymphe Bank, J. Geol. Soc. London, 127, 339-375.

Blundell, D.J., and B. Raynaud, Modeling lower crust reflections observed on BIRPS profiles, in Reflection Seismology: a Global Perspective, edited by M. Barazangi and L. Brown, A.G.U. Geodynamics Ser., 13, 287-295, 1986.

Bois, C., M. Cazes, B. Damotte, A. Galdeano, A. Hirn, A. Mascle, P. Matte, J.F. Raoult, and G. Toreilles, Deep seismic profiling of the crust in northern France: the ECORS Project, in Reflection Seismology: a Global Perspective, 21-29, 1986.

Brewer, J.A., and D.K. Smythe, MOIST and the continuity of crustal reflector geometry along the Caledonian-Appalachian orogen, J. Geol. Soc. London, 141, 105-120, 1984.

Gibbs, A.D., Balanced cross-section constructions from seismic sections in areas of extensional tectonics, J. Struct. Geol., 5, 152-160, 1983.

Gibbs, A.D., Development of extension and mixed-mode sedimentary basins, in Continental Extensional Tectonics edited by M.P. Coward, J.F. Dewey, and P.L. Hancock, Geol. Soc. Spec. Publn. No.28, 19-33, 1987.

Hamilton, W., Structural evolution of the Big Maria Mountains, Northeastern Riverside County, southeastern California, in Mesozoic- Cenozoic tectonic evolution of the Colorado River region, California, Arizona and Nevada (Anderson-Hamilton volume) edited by E.G. Frost and D.L. Martin, Cordilleran Publishers, San Diego, 1-27, 1982.

Hobbs, R.W., C. Peddy, and the BIRPS group, Is lower crustal layering related to extension? Geophys. J. R. Astr. Soc., 89, 239-242, 1987.

Kusznir, N.J., G.D. Karner, and S. Egan, Geometric, thermal and isostatic consequences of detachments in continental extension and basin formation, in Sedimentary basins and basin-forming mechanisms, edited by C. Beaumont, and A.J. Tankard, Can.9 Soc. Petrol. Geol. Mem. 12, 185-203, 1987.

Matthews, D.H., and M.J. Cheadle, Deep reflections from the Caledonides and Variscides west of Britain and comparison with the Himalayas, in Reflection Seismology: a Global Perspective, edited by M. Barazangi and L. Brown, A.G.U. Geodynamics Ser., 13, 5-19, 1986.

Matthews, D., and the BIRPS group, Some unresolved BIRPS problems, Geophys. J. R. Astr. Soc., 89, 209-215, 1987.

McClay, K.R., and P.G. Ellis, Analogue models of extensional fault geometries, in Continental Extensional Tectonics edited by M.P. Coward, J.F. Dewey, and P.L. Hancock, Geol. Soc. Spec. Publn. No.28, 109-125, 1987.

McGeary, S., Nontypical BIRPS on the margin of the northern North Sea: The SHET Survey, Geophys. J. R. Astr. Soc., 89, 231-238, 1987.

McGeary, S., and M.R. Warner, Seismic profiling the continental lithosphere, Nature, 317, 795-797, 1985.

McKenzie, D.P., Some remarks on the development of sedimentary basins, Earth Planet. Sci. Lett., 40, 25-32, 1978.

Reston, T.J., Spatial interference, reflection character and the structure of the lower crust under extension: Results from 2-D seismic modelling, Ann. Geophys., 5B, 339-348, 1987.

Reston, T.J., and D.J. Blundell, Possible mid-crustal shears at the edge of the London Platform, Geophys. J.R. Astr. Soc., 89, 251-257, 1987.

Stein, A., Basement controls upon basin development in the Caledonian foreland, NW Scotland, Terra Cognita, 7, 203, 1987.

Warner, M., and S. McGeary, Seismic reflection coefficients from mantle fault zones, Geophys. J. R. Astr. Soc., 89, 223-229, 1987.

Wever, T., H. Trappe, and R. Meissner, Possible relations between crustal reflectivity, crustal age, heat flow and viscosity of the continents, Ann. Geophys., 5B, 255-266, 1987.

MECHANICAL MODELS OF TILTED BLOCK BASINS

H. W. S. McQueen[1,2] and C. Beaumont[1]

[1]Department of Oceanography, Dalhousie University, Halifax, N.S., Canada B3H 4J1
[2]Research School of Earth Sciences, Australian National University, GPO Box 4, Canberra, ACT 2601, Australia

Abstract. Tilted block basins develop as a consequence of crustal or lithospheric rotations during imbrication under horizontal in-plane compression. We develop the basic mechanics of the block interactions, consider the consequences of elastic flexure and outline how more realistic elastic-plastic rheologies may explain decoupling and deformation at depth.

Introduction

The crustal features we refer to as tilted block structures occur in many parts of the world. Probably the best documented case is in the Wyoming province of the Rocky Mountain foreland (Figure 1) where Late Cretaceous and Early Tertiary thrusting produced a series of commonly asymmetrical basins bounded by linear ranges and structural highs which consist of exposed and often deeply eroded Precambrian basement. Early drilling and shallow seismic surveys pointed to thrust faults dipping at around 30° beneath one side of some of the basement uplifts [Berg, 1962; Gries, 1983], but the nature of the faults at depth remained the subject of lively debate [Matthews, 1978 ; Sales, 1968] until deep seismic profiling clarified the form of the Wind River fault in the late 1970's [Smithson et al., 1978; Lynn, 1983; Sharry et al., 1986]. While the dip of a few of the faults is now better constrained, these areas present interesting challenges to geodynamic modelling as they represent the response of stable continental crust to the most intense inplane compression it encounters.

The Wyoming structures in plan view (Figure 1) show two major trends, one NW–SE and another more E–W, which may represent two different stages of deformation or the effect of pre-existing tectonic 'grain' striking at an angle to the applied stress field. This illustrates a three dimensionality to the structures which has not yet been approached in modelling. Another three dimensional aspect is the existence of shear zones which offset basins and thrusts perpendicular to strike, playing a similar role to transfer faults in extensional terranes. The style of deformation of these broken foreland structures is quite distinct from the thin skinned thrust belt to the west. It is characterised by:

1) major basin bounding thrust faults dipping at 30–45°, of which at least one penetrates deep into the crust and perhaps through it. The similarity of surface style between this and other thrusts in the area suggests they also plunge to great depth;
2) breakup of the crust into blocks 50–200 km in length between major thrusts and somewhat longer along strike;
3) average tilting and imbrication of the fault bounded blocks by up to 5–7° to allow overall shortening and thickening of the lithosphere; tilts near the faults may be larger as a result of flexure or local deformation;
4) development of syn-deformational basins that subside and accumulate sediments as the blocks tilt;

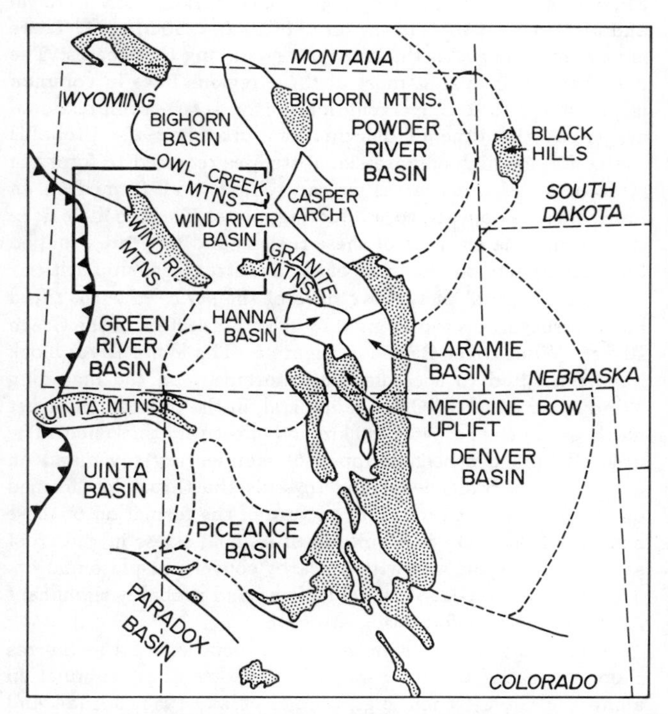

Fig. 1. Wyoming Province of the Rocky Mountain foreland showing approximate outlines of basins and arches. Exposed basement uplifts are stippled, and the sawtooth line marks the front of the thin skinned Sevier thrust belt to the west.

5) a gravity anomaly signal correlated with the basin structures indicating that they are out of local isostatic adjustment;
6) two distinct styles of basin margin; a gradational one with progressive exposure of stratigraphic levels over tens of km from an uplifted basement core towards the centre of the basin, and a narrow one with basement juxtaposed with the youngest strata, sometimes separated by a thin belt of steeply upturned or overturned older strata.

These structures are not limited to the Wyoming province. We see in several parts of the world examples which, we suspect, reflect the same mode of formation [Rodgers, 1987]. The deformation has occurred at very different times in a variety of crustal types in these cases. The Sierras Pampeanas of western Argentina [Jordan and Allmendinger, 1986] and the Tien Shan of central Asia [Tapponnier and Molnar, 1979] are currently deforming examples of this style, and provide evidence on two rate questions of importance to dynamic models. Firstly, the structures in the Sierras Pampeanas formed within the past 10 Myr, and in at least one area the major phase of deformation postdates a 3 Ma old ash bed [Jordan et al., 1983] so the basins may form in as little as 3–5 Myr. Secondly, it appears that syn-tectonic erosion and sedimentation is significant and must be included in dynamical models of the basins' formation.

The Wyoming deformation is Laramide, in the 50–90 Ma range, while examples from the Boothia Peninsula in northern Canada [Okulitch et al., 1986] and the Amadeus Basin of central Australia [Forman and Shaw, 1973; Lambeck, 1983] are 300–400 Ma old and the Kapuskasing region in central Canada [Percival and McGrath, 1986] is of the order of 1950–2250 Ma old. There is no common crustal thickness or age linking these cases. The one characteristic that most of these regions have in common is proximity to a convergent margin or active compressional orogen at the time of deformation, providing the probable source for the high compressional stresses required to form the structures. Also, as far as we can tell, they all formed on an overriding, as opposed to subducting, plate. There is little deep structural data on most of these regions and they are grouped together loosely on the basis of surface structural similarities.

The conceptual picture we have of the styles of such 'tilted block' structures is represented by the cross section of the Green River – Wind River Basins in Figure 5. The Wind River block has been tilted to accommodate movement on the bounding Wind River and Owl Creek faults and, in the process, the crust has been shortened by 20–30 km and effectively thickened. The uplifted limb has been stripped of sedimentary cover and an asymmetrical basin deepening towards the thrust has formed on the downthrust limb. Our picture of the formation of these basins is that there is a buildup of inplane stress in the crust associated in some way with a nearby convergent plate margin. The stress causes deep crustal faulting and probably significant internal ductile deformation at depth.

Several authors have noted the association of the Sierras Pampeanas basins with a region of shallow angle subduction along the Chile trench [e.g., Jordan et al., 1983; Jordan and Allmendinger, 1986] and of the Wyoming basins with a similar shallow angle subduction episode inferred from other evidence in the Laramide of the western U.S. [Lowell, 1974; Cross and

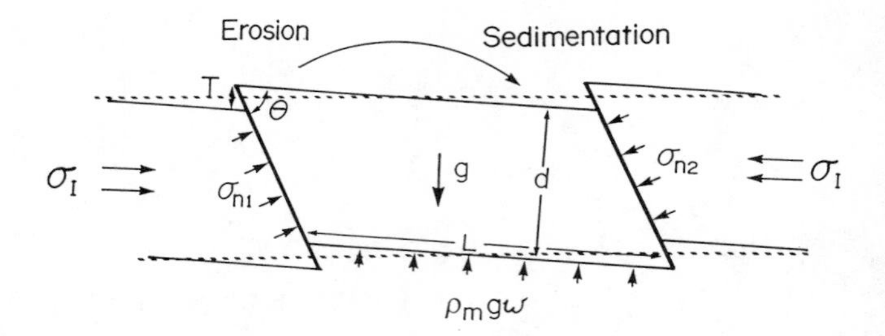

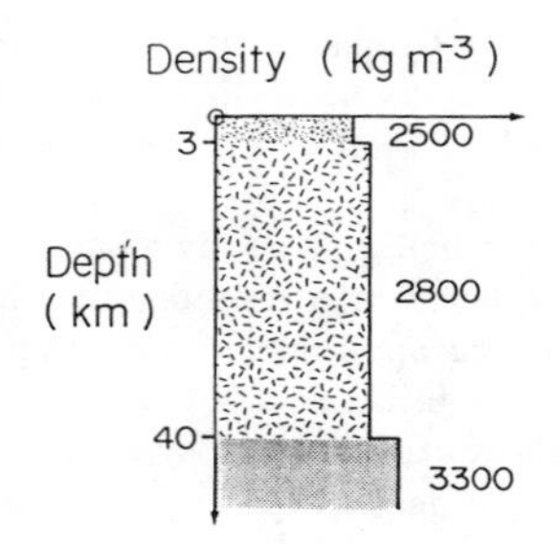

Fig. 2. Rigid block model of tilted block structures showing the density structure and parameters used in Figure 3. The upper 3 km is a pre-thrust sediment.

Pilger, 1978; Bird, 1984; Cross, 1986]. This probably highlights an important mechanism for transmission of high compressive stresses into the foreland, but the tectonic situation is quite different in the Himalayan orogen so the association of shallow angle subduction with tilted block structures is not general. For our models, we only require that there be some mechanism focussing particularly high compressional stresses on the region in question.

Rigid Block Models

The simplest conceptual model of tilted block basins is the rigid block picture sketched in Figure 2. The inplane stress, σ_I, required to produce a given vertical offset, T, on the bounding faults can be calculated by balancing the forces and moments on the blocks against the resistance of the pressure forces acting on the base of the block. A balance of moments between inplane compression and basal restoring forces leads to the approximate relationship

$$\sigma_I = \left(\frac{\Delta\rho\, g\, L \tan\theta}{12d}\right)T$$

where $\Delta\rho = (\rho_m - \rho_s)$ is the basal restoring force corrected for sediment loading, and L, d, and θ, the initial fault dip, are defined in Figure 2.

This formula is approximate because it does not include the effect of full self weight in the block and the changing geometry and orientation of basal pressure as the block tilts. A comparison of the more cumbersome complete calculation with the approximate result is shown in Figure 3 for the density structure shown. The load response is not particularly sensitive to the detailed vertical distribution of density, so the

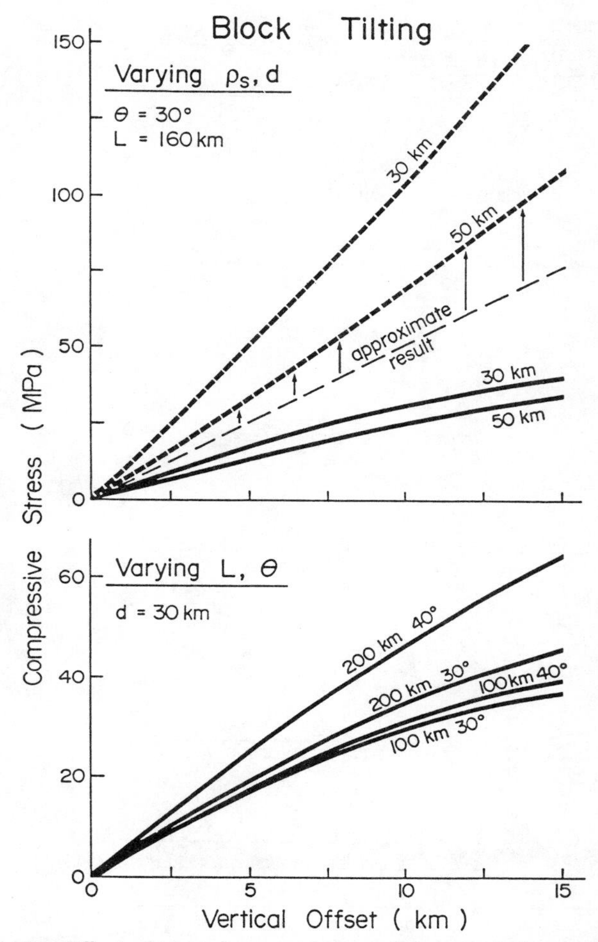

Fig. 3. Effect of several parameters on the compressive stress excess required to produce a given vertical offset across a dipping thrust separating two tilted crustal blocks. For comparison, the vertical offset across the Wind River fault is approximately 14 km [Smithson et al., 1979]. In all cases the two bounding faults have the same dip. a) The effects of varying plate thickness, and of syn-tectonic erosion and sedimentation. The upper pair of curves are calculated without erosion or sedimentation effects, and the d=50 km case is compared with the approximate formula given in the text. The lower pair of curves show the result when erosion and sedimentation keep the upper surface horizontal at the initial level and the sediment density (2300 kg/m^3) is used for the eroded material. b) The effects of varying block length and fault dip with erosion and sedimentation.

three layer structure is adequate. The principal extra physical consideration in the full calculation is the horizontal offset of the block's centre of mass from the centre of pressure on the base. In the full gravity model, σ_I is the stress in excess of the 'containment stress' required to prevent normal movement on the faults in the absence of friction. The containment stress is a depth-dependent compressional stress averaging several hundred MPa across the 30 or 50 km layers used here, which we take to be the background reference state of stress in the lithosphere. The major difference between the block thicknesses chosen is that the Moho is tilted when d=50 km, because the thickness exceeds that of the crust.

Figure 3 shows that the simple formula underestimates the stress required to tilt a crustal block compared to the full calculation (thicker dashed lines). However, the inclusion of syn-tectonic erosion and sedimentation (which removes 90% of the uplift) substantially reduces the required driving stress (solid lines). It appears from the modern examples, that the effect of partial syn-tectonic sedimentary infill must be included. This result is important: erosion and sedimentation aid the tilting process and reduce the required inplane stress for a given vertical offset, T.

Other systematic effects of varying block length L, thickness d, and fault dip θ are more minor, though increasing θ, L, and d all increase σ_I for a given T. The principal conclusion from this analysis is that the minimum stress required to produce offset of the scale of the Wind River structures is of the order of 50 MPa, assuming complete syn-tectonic erosion and sedimentation. We emphasize that this is the stress excess over the basic containment stress.

50 MPa is a moderate level of stress fluctuation for a proximal orogenic region and is significantly less than the stress excess needed to cause slip on such deep faults unless relatively low coefficients of friction and ductile shear strength apply [Sibson, 1986]. Consequently, the movement must be seen as being limited as much by fault resistance as by gravitational stresses set up by the deformation.

Elastic Flexural Models

In the rigid block picture, the condition that the fault surfaces remain in contact requires that the tilt of neighbouring blocks

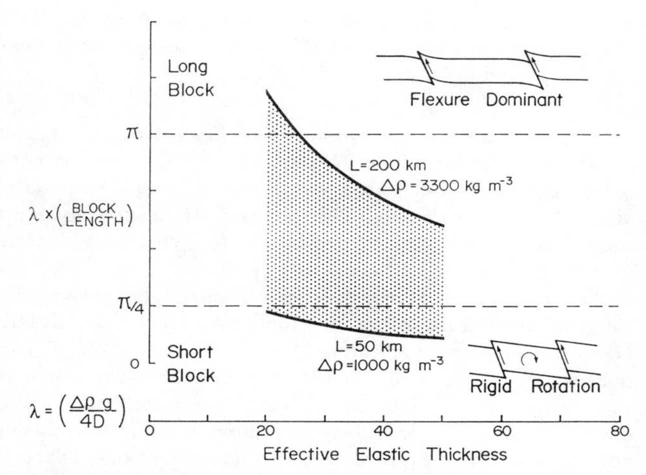

Fig. 4. Classification of finite beams on the basis of flexural parameter. For an assumed effective elastic thickness between 20 and 50 km, the range of lengths and sedimentation factors appropriate to the Wyoming basins puts them in the stippled region. Blocks are effectively short if $\lambda < \pi/4$ and long if $\lambda > \pi$.

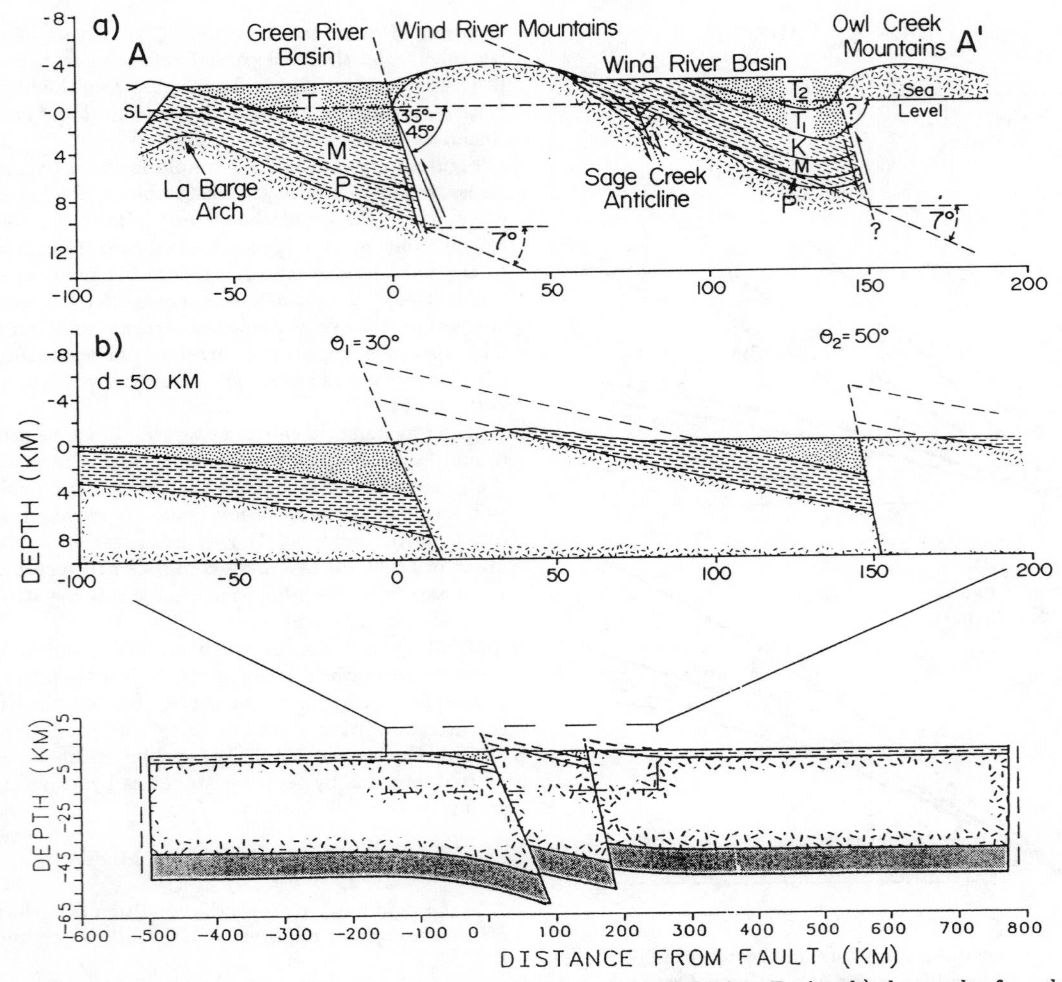

Fig. 5. a) A schematic cross section across the Green River and Wind River Basins, b) the result of an elastic finite element model of shortening across two faults with dips of 30° and 50° caused by horizontal compression (modified from Lloyd [1986]).

be the same. 'Heavy' blocks would therefore restrain movement of adjacent 'lighter' blocks. This limitation does not actually arise because elastic flexure will permit different amounts of tilting of adjacent blocks. This points to the next step in the modelling of these structures which is to incorporate internal elastic deformation, and consequently flexure.

We find that most tilted block structures fall into a size range where flexure is just becoming important. A standard classification from engineering theory [Hetenyi, 1946] divides beams into long, short and intermediate on the basis of their length relative to their flexural parameter λ. In this classification, short beams may be treated as rigid blocks; while with long beams, flexure at one end has a negligible effect at the other. In the size range we are concerned with here (50–200 km), and with a reasonable range of effective thickness and sedimentation factors, Figure 4 shows that the Wyoming basins are in the intermediate range with a few of the smaller basins behaving as nearly rigid blocks.

Analytical formulae for the deflection of tapering blocks are not available and Lloyd and Beaumont (unpublished manuscript, 1985) have used linear elastic finite element models to add a flexural component to the deformation. Figure 5 shows a model of this type, an elastic plate 50 km thick containing two faults dipping at 30° and 50°, subjected to inplane compression of the order of 100 MPa. Gravity forces are not explicitly included in this model, but buoyancy forces are applied at density interfaces in the crust to represent gravity forces as well as possible without requiring repeated iterations of the structural model. This approach corresponds to the 'density stripped' models normally used in simple elastic flexural models of the lithosphere [Turcotte and Schubert, 1982]. The right hand end is fixed horizontally and a 'no tilt' left hand end is free to move horizontally under the effect of the inplane load. The two faults are represented by low rigidity elastic elements, resulting in a near zero resistance to slip on the fault.

The inclusion of flexural effects reproduces more accurately the structure of the Wyoming basins, particularly the sediment thickness distribution and the existence of intrabasinal arches. By varying the relative fault orientations we can reproduce not only Wind River style basins but also uplifted block styles which may be appropriate to the Uinta Mts. and adjacent basins in NE Utah, and downdropped blocks of varying lengths which reflect the structure of the Hanna Basin in Wyoming, and possibly the Tarim Basin in Tibet and the Amadeus Basin in central Australia (Lloyd and Beaumont, unpublished manuscript, 1985).

Regarding the dynamics, the principal effect of flexure is to limit the effective length of uplifted blocks to the order of a flexural wavelength. Lithospheric shortening and fault offset between longer blocks is therefore easier to produce than the rigid block calculations indicate. The magnitude of inplane stress needed to create these structures in an elastic medium is also around 50–100 MPa when syn-tectonic erosion and sedimentation occur.

This is about as far as linear elastic models can be taken. They are limited by the lack of finite fault resistance, the absence of significant internal plastic deformation, and the incomplete representation of gravitational body forces. This last problem is due to the fact that, in the finite element models, the deformation of the low rigidity elements is so severe that the calculation cannot be iterated to properly establish equilibrium in the deformed configuration without a major regridding effort. This is not a critical problem in the classical analysis of the flexure of an unbroken plate, but is more significant here because fault motion causes a major redistribution of mass.

Elastic-Plastic Frictional Model

While there are many similarities among styles of deformation in the various tilted block regions, there are also some significant differences, one of which is the magnitude of the associated gravity signal. Because tilted blocks are out of local isostatic equilibrium, there is generally a gravity anomaly signal correlated with the basin structures, but in the Wind River case it has been found to be largely attributable to near surface and crustal structures [Smithson et al., 1979]. The contribution to the gravity anomaly implied by a tilted Moho and the development of 'crustal keels' (Figures 2 and 5) is just not seen in this case.

In contrast, the principal feature of the Australian gravity map [Wellman and Murray, 1979] is the Bouguer anomaly of up to 160 mgal over the Amadeus basin in central Australia; a basin that is superficially similar to the Green River basin [Malahoff and Moberly, 1968]. There is a correlation of the anomaly with the basin structure but the signal cannot be wholly reproduced by near surface features of reasonable density [Lambeck, 1983]. This has led to models invoking significant topography on deep density surfaces, and these are generally corroborated by the seismic travel time studies of Lambeck and Penney [1984] and K. Lambeck et al. (Teleseismic travel time anomalies and deep crustal structure in central Australia, 1987, unpublished manuscript).

An important factor in a geodynamic model is therefore the 'soleing out' of the fault and the effect of fault motion on the deeper structure. Current ideas on the depth-dependent rheology of the lithosphere may shed some light on this. A fundamental characteristic of the elastic and rigid block models is the creation of underthrust crustal 'keels' protruding down into the lower crust or into the mantle depending on the block thickness. It is not clear that these would form if ductile deformation were to occur in the crust, and a significant gravity anomaly does require the Moho to be offset. We therefore speculate that in some cases faults sole out in the mid crust, while in others they may offset the Moho (Figure 6). If faults do not offset the Moho there is a space problem in the lower crust which must be accommodated by distributed plastic strain, presumably within a mid to lower crustal weak layer. This layer would therefore have to be very weak to accommodate large strains over the few million years in which the modern examples seem to form.

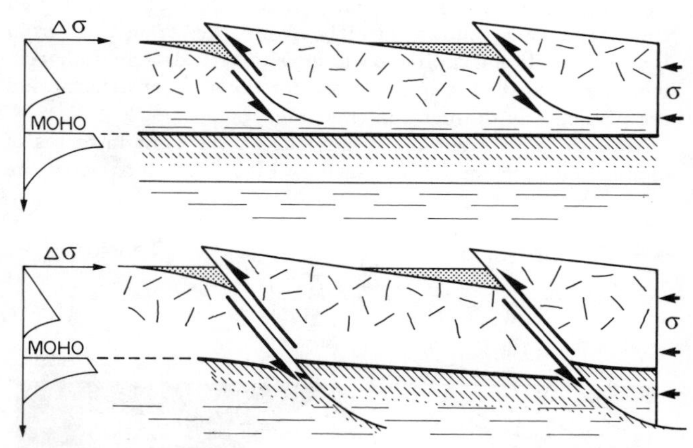

Fig. 6. Two hypothetical models of styles of subsurface deformation accompanying block tilting.

Our preliminary results from models which incorporate frictional sliding conditions and plastic yield into the tilted block picture suggest how this strain may occur. These more realistic models include the increasing strength of the faults with depth and plastic deformation as an alternative means of deformation where fault friction is large.

Calculations in this case are done with the mesh of quadratic elements shown in Figure 7, and full gravitational body forces are now incorporated with buoyancy forces acting only at the base. The two faults are friction limited contact surfaces with a coefficient of friction of 0.2, and the material is elastic-plastic with a uniform yield strength of 100 MPa. Below 50 km, the lithosphere is assumed to have no significant yield strength on the timescales appropriate to this style of deformation.

Applying several hundred MPa of inplane compressional stress excess to this model, we find that the greater strength of the fault at depth transfers much of the deformation to the distributed plastic yielding, so that fault slip decreases with depth. This produces a warping which is particularly pronounced at depth and contributes to broad arching at the surface. In other words, below a certain depth, the

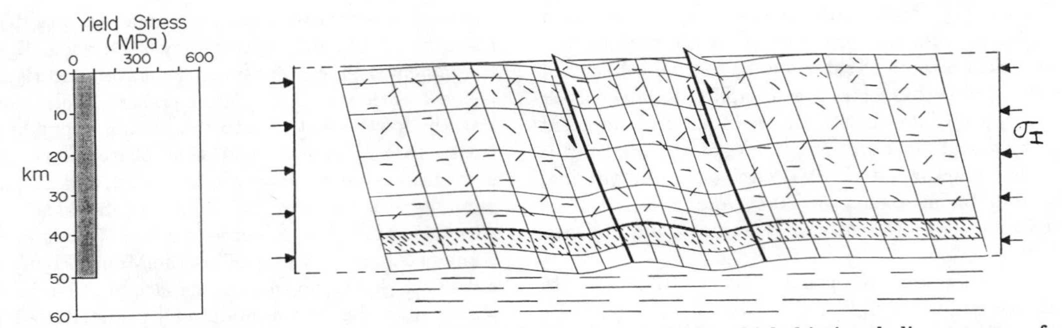

Fig. 7. Style of deformation in an elastic-plastic finite element model in which frictional slip occurs on faults near the surface but is prevented by increasing containment stress at depth.

faults, although embedded in the model, are 'stronger' than the surrounding plastically deforming crust. Although this behaviour follows directly from the transition from Byerlee's frictional sliding law to quasi-plastic deformation with increasing depth, the implications for the style of deformation are particularly dramatic in this case.

We can understand this behaviour on the fault in terms of incipient failure zone plots, such as Figure 8, which show the

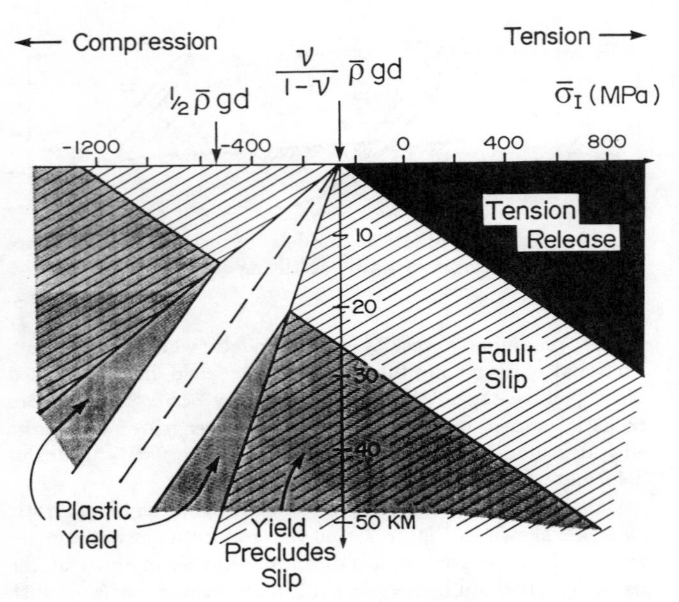

Fig. 8. Incipient failure zone plot for an elastic-plastic beam of uniform yield strength (100 MPa) containing a frictional fault (μ=0.2). The density structure used is that of Figure 2; ρ is the average density to the base of the block, and ν is the Poisson's ratio. Possible fault conditions are: tension release, slipping fault, surrounding material in yield (which may release enough stress on the fault to preclude slip), and locked fault with surrounding material in the elastic or plastic field. Deformation outside the locked core zone redistributes stress and reduces the size of the core zone until the whole fault can move, either by slip or by yield in the surrounding material.

interplay between fault slip and plastic yield. For a particular value of inplane stress (horizontal axis), the conditions at varying depths on a fault are reflected by a vertical section through the diagram. In some regions slip is possible, while in others plastic yield reduces the shear stress to below the slip threshold. From this initial state any slip or other deformation will redistribute stress and modify the boundaries in this figure, so it is only a guide to the kind of behaviour to be expected. For example, in the case shown in Figure 8, when the mean inplane stress acting in the lithosphere is equal to the containment stress, $\frac{1}{2}\rho gd$, the upper 15 km of the fault is able to slip in a reverse sense, the region from 15–35 km is locked, and below that deformation is dominated by plastic yield. Slip on the upper section of the fault will transfer shear stress to the core section causing slip at depths greater than 15 km, while plastic flow in the lower section will transfer part of the inplane compressional load to higher levels, moving the elastic-plastic boundary further up in the plate. In practice the plate comes to equilibrium in these conditions with about 100 m of throw on the fault at the surface, tapering to zero near 20 km depth.

The deformation in Figure 7 still displays protruding keels, but they are of a very different style, more like the regional isostatic roots classically envisaged beneath mountain ranges This observation may be of significance to the interpretation of gravity and travel time delays over these structures [e.g., Lambeck and Penney, 1984]. If the seismic evidence can distinguish between this and the underthrust keels we will have a better understanding of modes of deformation in the upper mantle, particularly the relative importance of strain localization and distributed viscoplastic flow.

Conclusions

A comparison of model results with observations leads to the conclusion that tilted block structures grow in response to high levels of inplane compressional stress within the lithosphere, sometimes but not always associated with shallow angle subduction. Thrusting and tilting on inherited or newly created faults allows brittle regions of the crust to shorten by imbrication. The creation of uplifts and subsidence, to form sedimentary basins, are part of the same process and they should be analyzed together. Syn-tectonic erosion and sedimentation reduce the inplane stress required to tilt blocks by a given amount. For the same reason, post-tectonic erosion and sedimentation will tend to amplify tilting if friction on the faults is low.

Blocks that are short by comparison with their flexural parameter may be modelled as rigid. Both rigid and elastic block models develop underthrust keels during deformation and the Moho is predicted to tilt unless the lower crust is capable of significant deformation on the time scale of the thrust movement. Gravity anomalies from the Wyoming province and central Australia indicate that cases exist where faults 'sole' or die out both within the crust and within the mantle.

A more realistic analysis that includes friction on the faults and plastic strain shows how fault offset may be transformed to distributed deformation at depth. Under these circumstances, the Moho will still warp, but distinct underthrust keels do not develop.

Acknowledgments. H.M. appreciates the assistance in this work provided by an Australian Bureau of Mineral Resources Postdoctoral Fellowship and by a Killam Postdoctoral Fellowship from Dalhousie University. Funding was provided through a Natural Sciences and Engineering Research Council grant to C.B. We acknowledge stimulating discussions with Bert Bally, Kurt Lambeck, Terry Jordan, Jim Steidtmann and Russell Shaw.

References

Berg, R. R., Mountain flank thrusting in the Rocky Mt. foreland, Wyoming and Colorado, *Am. Assoc. Pet. Geol. Bull., 48*, 2019–2032, 1962.

Bird, P., Laramide crustal thickening event in the Rocky Mountain foreland and great plains, *Tectonics, 3*, 741–758, 1984.

Cross, T. A., Tectonic controls of foreland subsidence and Laramide style deformation, western United States, *Spec. Publ. Int. Assoc. Sediment., 8*, 15–39, 1986.

Cross, T. A., and Pilger, R. H., Tectonic controls of Late Cretaceous sedimentation, western interior USA, *Nature, 274*, 653–657, 1978.

Forman, D. J., and R. D. Shaw, Deformation of the crust and mantle in central Australia, *Aust. Bur. Min. Res. Geol. Geophys. Bull., 87*, 1973.

Gries, R., North–south compression of Rocky Mountain foreland structures, in: *Rocky Mountain Foreland Basins and Uplifts*, ed. J. D. Lowell, Rocky Mt. Assoc. Geol., 9–32, 1983.

Hetenyi, M., *Beams on Elastic Foundation*, University of Michigan Press, Ann Arbor, 1946.

Jordan, T. E., B. L. Isacks, R. W. Allmendinger, J. A. Brewer, V. A. Ramos, and C. J. Ando, Andean tectonics related to geometry of subducted Nazca plate, *Geol. Soc. Am. Bull., 94*, 341–361, 1983.

Jordan, T. E., B. L. Isacks, V. A. Ramos, and R. W. Allmendinger, Mountain building in the central Andes, *Episodes, 1983*, No. 3, 20–26, 1983.

Jordan, T. E., and R. W. Allmendinger, The Sierras Pampeanas of Argentina: a modern analogue of Rocky Mountain foreland deformation, *Am. J. Sci., 286*, 737–764, 1986.

Lambeck, K., Structure and evolution of the intracratonic basins of central Australia, *Geophys. J. R. Astr. Soc., 74*, 843–886, 1983.

Lambeck, K., and C. Penney, Teleseismic travel time anomalies and crustal structure in central Australia. *Phys. Earth Planet. Inter., 34*, 46–56, 1984.

Lloyd, P., Mathematical modeling of tilted-block basins and uplifts, Unpublished M.Sc. thesis, Geology Department, Dalhousie University, Halifax, N.S., Canada, 256 p., 1986.

Lowell, J. D., Plate tectonics and foreland basin deformation, *Geology, 2*, 275–278, 1974.

Lynn, H. B., S. Quam, and G. A. Thompson, Depth migration and interpretation of the COCORP Wind River, Wyoming, seismic reflection data, *Geology, 11*, 462–469, 1983.

Malahoff, A., and R. Moberly, Effects of structure of the gravity field of Wyoming, *Geophysics, 33*, 781–804, 1968.

Matthews, V., (ed), Laramide folding associated with basement block faulting in the western United States, *Geol. Soc. Am. Memoir 151*, 1978.

Okulitch, A. V., J. J. Packard, and A. I. Zolnai, Evolution of the Boothia Uplift, arctic Canada, *Can. J. Earth Sci., 23*, 350–358, 1986.

Percival, J. A., and P. H. McGrath, Deep crustal structure and tectonic history of the northern Kapuskasing uplift of Ontario: an integrated petrological–geophysical study, *Tectonics, 5*, 553–572, 1986.

Rodgers, J., Chains of basement uplifts within cratons marginal to orogenic belts, *Am. J. Sci., 287*, 661–692, 1987.

Sharry, J., R. T. Langan, D. B. Jovanovich, G. M. Jones, N. R. Hill, and T. M. Guidish, Enhanced imaging of the COCORP seismic line, Wind River Mountains, in: *Reflection Seismology: a Global Perspective*, ed. M. Barazangi and L. Brown, AGU geodynamics series *13*, 223–236, 1986.

Sibson, R. H., Earthquakes and rock deformation in crustal fault zones, *Ann. Rev. Earth Planet. Sci., 14*, 149–175, 1986.

Smithson, S. B., J. A. Brewer, S. Kaufman, J. E. Oliver, and C. A. Hurich, Nature of the Wind River thrust, Wyoming, from COCORP deep-reflection data and from gravity data, *Geology, 6*, 648–652, 1978.

Smithson, S. B., J. A. Brewer, S. Kaufman, J. E. Oliver, and C. A. Hurich, Structure of the Laramide Wind River uplift, Wyoming, from COCORP deep-reflection data and from gravity data, *J. Geophys. Res., 84*, 5955–5972, 1979.

Tapponnier, P., and P. Molnar, Active faulting and Cenozoic tectonics of the Tien Shan, Mongolia, and Baykal regions, *J. Geophys. Res., 84*, 3425–3457, 1979.

Turcotte, D. L., and G. Schubert, *Geodynamics*, John Wiley, 1982.

Wellman, P., and A. S. Murray, Bouguer Gravity Anomaly Map, in: *BMR Earth Science Atlas of Australia*, 1979.

CRUSTAL STRUCTURE AND ORIGIN OF BASINS FORMED BEHIND THE HIKURANGI SUBDUCTION ZONE, NEW ZEALAND

T. A. Stern and F. J. Davey

Geophysics Division, DSIR, P.O. Box 1320, Wellington, New Zealand

Abstract. Two principal basin types have formed behind the Hikurangi Margin of New Zealand. Both basins formed in the past 4-5 Ma. An extensional back-arc basin has developed behind the northern portion of the subduction zone where the subducted Pacific plate extends to a depth of 300 km or more, and the subduction process is well established. Behind the southern portion of the Hikurangi Margin, on the other hand, subduction is a younger phenomenon, and the subducted plate has reached a depth of only about 200 km. Here a broad sedimentary basin - the Wanganui Basin - has developed in the back-arc region which is characterized by steeply dipping reverse faults and intense crustal seismicity. Multichannel seismic reflection data are interpreted to show that the reflection Moho has been flexed downward beneath the Wanganui Basin, with the half-wavelength of flexure being about 200 km. It is thus proposed that within the North Island of New Zealand we are observing a time progression in the development of back-arc basins. In the initial stages of subduction (0-8 Ma) a flexurally controlled, compressional basin is developed, followed by a fully fledged extensional back-arc basin with high heat flow after about 10-20 Ma of subduction history.

Introduction

Back- arc or marginal basins were once regarded as one of the great paradoxes of plate tectonics [Uyeda, 1977]. In the main, these basins display all the manifestations of extension and lithospheric spreading, yet they form adjacent to the main convergent boundaries of the tectonic plates. Extension, high heat flow, normal faulting and active volcanism are commonplace within most active back-arc basins, but mechanical models proposed by McKenzie [1969] and Davies [1981] predict the development of a broad flexural downwarp ("wedge flexure" in the terminology of Davies) behind a subduction zone. Both McKenzie and Davies note, however, that examples of such wedge-flexure type back-arc basins are rare, either behind present day active margins, or in the geological record.

In this study we document the occurrence of two geophysically contrasting, yet coeval, back-arc basins that have formed in continental lithosphere behind the Hikurangi Margin [Hatherton, 1970] of New Zealand. Whereas one of the basins shows all the features of an active back-arc basin, the other is more akin to the wedge-flexure type of back-arc basin discussed above. The emphasis in this paper is on crustal structure, particularly for the Wanganui Basin where the results and interpretation of a multichannel seismic reflection profile are presented.

The term " back-arc basin " is used here to describe continental analogues to oceanic back-arc basins. It is used in the sense outlined by Taylor and Karner [1983] : " those marginal basins located behind active or inactive trench systems and whose origin is inferred to be subduction related".

Structural Background and Subduction History of the North Island

Figure 1 shows the location and outline of the two basins to be discussed. North of about 39° South is an active back-arc spreading basin that has developed within continental lithosphere. This is the Central Volcanic Region (CVR) [Thompson, 1964; Stern, 1987]. South of 39° South, and as an apparent southwestward continuation of the CVR, is a broad sedimentary basin characterized by steeply dipping reverse faults and intense crustal seismicity; this is the Wanganui Basin [Lensen, 1959; Hunt, 1980; Anderton, 1981; Garrick and Gibowicz, 1983]. Both basins have developed in the past 4-5 Ma behind the Hikurangi Margin.

Subduction of the Pacific plate is occurring at the Hikurangi Margin, with the subducted plate extending at least as far south as 41° South [Adams and Ware, 1977]. Shown on Figure 1 is the boundary of the Pacific plate if it were to be unfolded to the surface today [Ansell and Adams, 1986]. A common element of most plate tectonic reconstructions for New Zealand during the Cenozoic is that there has been an active subduction regime present at the Hikurangi Margin during the past 20 Ma [Walcott, 1987; Kamp, 1986; Cole and Lewis, 1981]. K-Ar dating of low-potash andesites (Figure 2) shows that the "source zone" [Hatherton, 1969] for these low-potash andesites has migrated southward and eastward with time; here the "source zone" is identified with the 70-100 km isobath of the subjacent Benioff Zone.

Figure 2 shows postulated positions of past and the present low-potash andesite axes [after Calhaem, 1973; Stern, 1987]

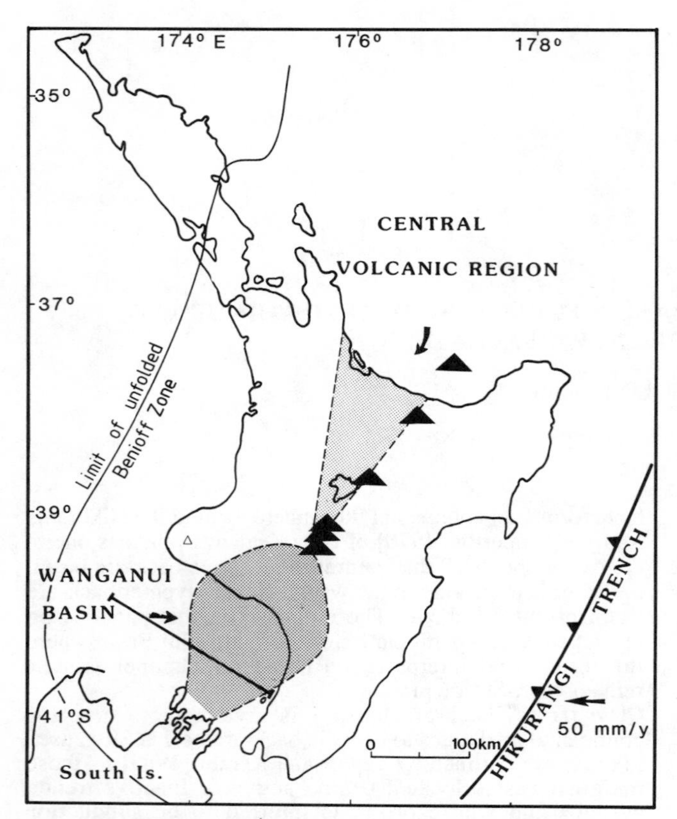

Fig. 1. Setting of the North Island of New Zealand showing the location of the two basins discussed in this paper. Limit of the Benioff Zone when unfolded to the earth's surface from Ansell and Adams [1986]. Filled triangles represent active to recently active low-potash andesites-dacites. The lone empty triangle is the recently active high-potash volcano of Mt. Egmont. Position of multichannel seismic line through the southern Wanganui Basin is shown by the heavy line with bars.

based on K-Ar ages. Over the past 4 Ma the axis appears to have undergone a 20 mm/y southeastward translation and a 30° rotation, and the southern apex of the axis, now at Mt Ruapehu, has migrated southeastwards at an indeterminate rate. Overall, the motion at the Bay of Plenty coast can be represented by a single southeast directed vector of magnitude 20-30 mm/y as shown in Figure 2. This rapid migration of the low-potash andesite axis in the past 4 Ma is thought to be linked to the development of the CVR by back-arc spreading [Calhaem, 1973; Stern, 1987; Walcott, 1987].

Also shown in Figure 2 are depocentres for the Wanganui Basin as determined by Anderton [1981] from seismic stratigraphy. The overall direction, rate and time span for the migration of the depocentres are almost identical to that for the low-potash andesite axis. This similarity in migration suggests that both the CVR and the Wanganui Basin are linked to the same driving mechanism - namely, the excess mass of the subjacent, subducted Pacific plate.

The Central Volcanic Region

Recent published work relating to the CVR includes a detailed analysis of ignimbrite stratigraphy and caldera structures [Wilson et al., 1984] as well as work on crustal structure and heat flow [Stern, 1987; Stern and Davey, 1987]. The main crustal structure features of the CVR relevant to this study are summarised in Figure 3.

Seismic and bore-hole data show at least 2-3 km of low-density, young volcanics within the CVR. These volcanics appear to be locally isostatically compensated by what is interpreted to be a low seismic-velocity upper mantle at a depth of 15 km beneath the CVR. Natural heat discharge from the CVR amounts to at least 4×10^9 W, which if averaged over the approximate 5000 km^2 area of the geothermal downflow area [Stern, 1987] represents an average heat flow of about 800 mW/m^2. This vast heat resource has been exploited over the past 40 years for geothermal power production.

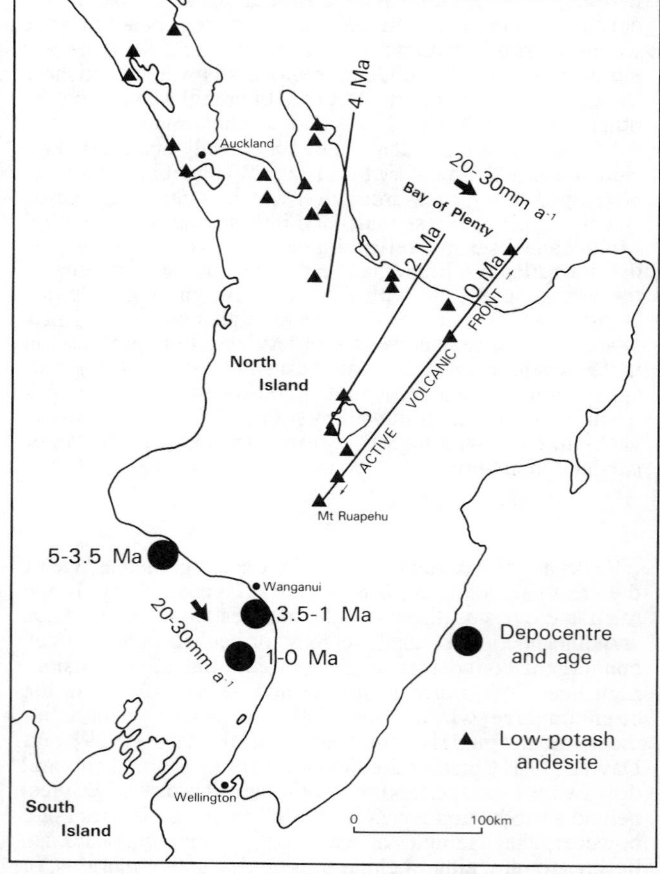

Fig. 2 Axes of low-potash andesites for epochs 4 Ma B.P., 2 Ma B.P. and the present day (after Calhaem, 1973 and Stern et al., 1987). The present day Active Volcanic Front is the 0 Ma axis. Also shown is the similar, in rate and direction, migration of the depocentres for the Wanganui Basin (Depocentre positions and ages from Anderton, 1981).

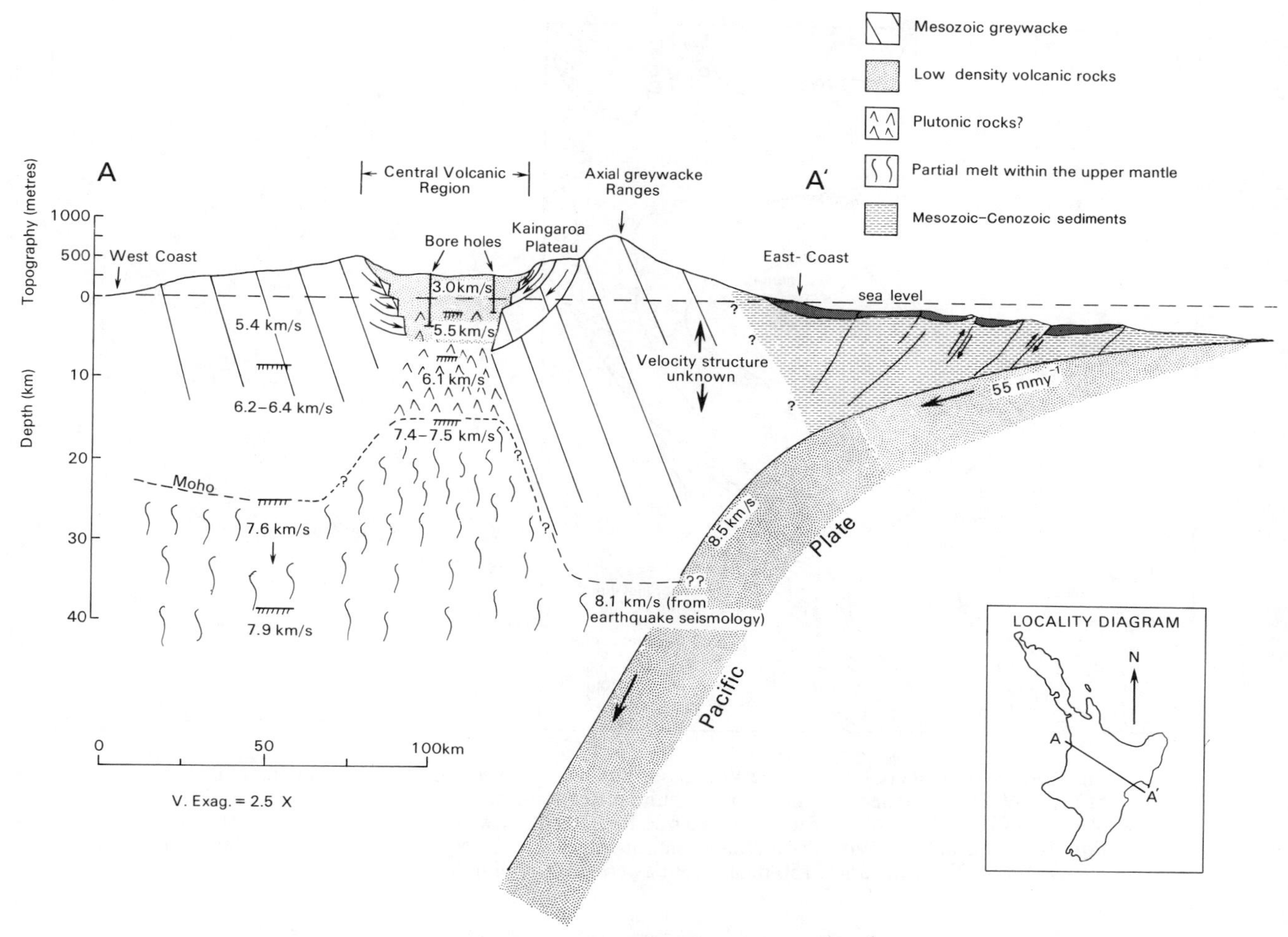

Fig. 3. Cross-section of the central North Island showing crustal and upper mantle structure as determined from explosion seismology [Stern and Davey, 1987; Stern et al., 1987] and earthquake seismology [Haines, 1979]. Vertical exaggeration = 2.5

Geodetic data [Walcott, 1987] are interpreted to show the CVR extending in a NW-SE azimuth at a rate of about 12 mm/y. Paleomagnetic data from Tertiary sediments east of the CVR are interpreted by Wright and Walcott [1987] to show an increase of rotation with age. In particular, Wright and Walcott propose a 7°/Ma rate for the past 5 Ma, which is consistent with the approximately 30° angle subtended by the 4 and 0 Ma andesite axes about an origin near Mt.Ruapehu (Figure 2). Hence, both short-term (geodetic) and long-term (paleomagnetic and andesite migration) indicators appear to be consistent with a complex fan-like opening for the CVR, which includes both rotation and translation within the past 4 Ma. A fan or wedge-shaped opening for slivers of continental lithosphere at active margins has been noted elsewhere, and may be a manifestation of the complex strength profile for continental lithosphere compared to that of oceanic lithosphere [Steckler and ten Brink, 1986; Otofuji and Matsuda, 1987].

We identify the "driving load" [Watts and Ryan, 1976; Sleep et al., 1980] for subsidence within the CVR to be isostatic adjustment associated with spreading and the rising of upper mantle material to anomalously shallow depths beneath the CVR.

The Wanganui Basin

The elliptical-shaped Wanganui Basin, southwest of the CVR (Figures 1 and 4a), has remained a geological and geophysical enigma to New Zealand earth scientists for some time [Robertson and Reilly, 1958; Eiby, 1964; Hunt, 1980]. Drill-hole data and seismic reflection work show that the Wanganui Basin to contain 4-5 km of Pliocene-Pleistocene, shallow - marine sediments [Cope and Reed, 1967; Anderton, 1981; Hunt, 1980]. Two geophysical properties of the basin have attracted particular interest. Firstly, dominating the initial 1:4 000 000 gravity map of New

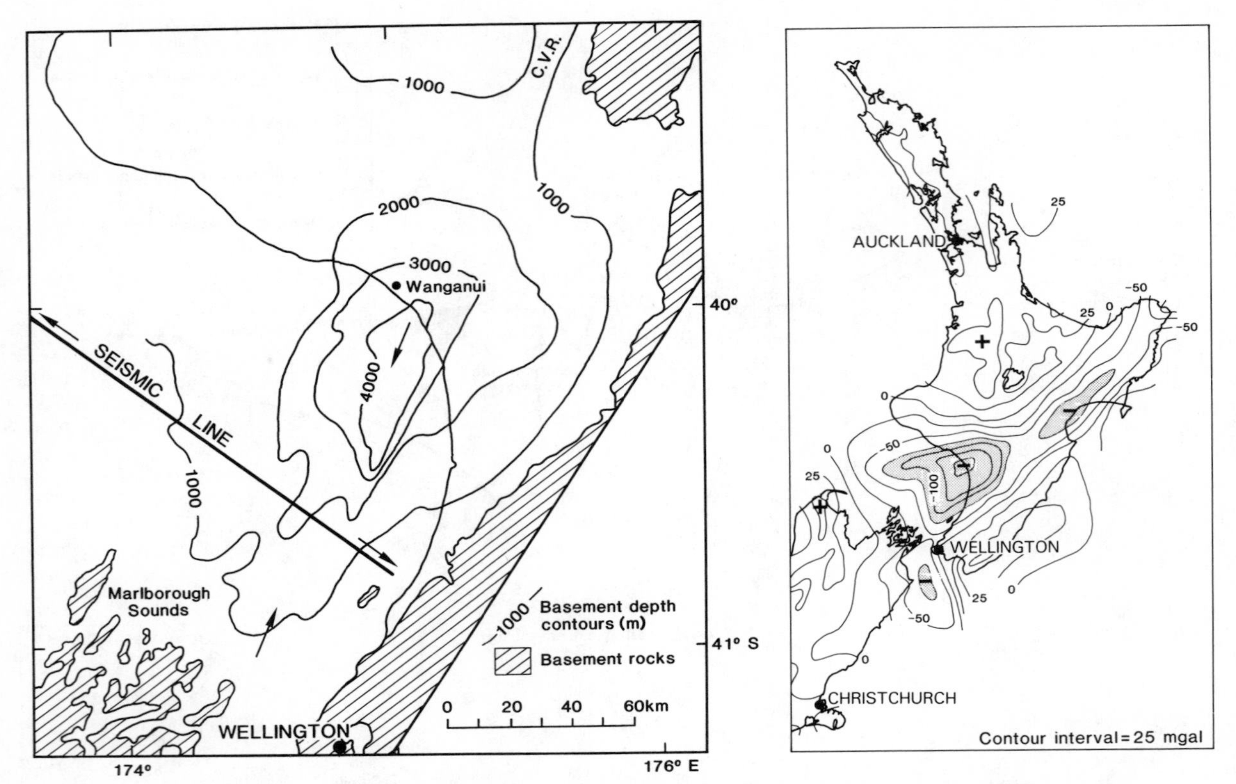

Fig. 4(a). Basement contours for the Wanganui Basin (after Anderton, 1981). Shown are positions of the NW-SE orientated multichannel seismic reflection line discussed in this paper and Davey (1987), and the SSW-NNE orientated seismic line 141 (shown by arrows) of Anderton (1981). Fig. 4(b). Isostatic gravity anomalies (contour interval = 25 mgal) for the North Island (after Reilly, 1965). Note the large -150 mgal anomaly coincident with the Wanganui Basin.

Zealand [Reilly, 1965] is a remarkably circular to elliptical-shaped negative gravity low, 300 km in wavelength, and centred near the city of Wanganui (see Figure 4b). Values of Bouguer and isostatic gravity are as low as -150 mgal yet the negative gravity effect of the 4-5 km of sediments within the basin is estimated to be only about -50 mgal [Hunt, 1980].

Crustal downwarping is likely to be partly responsible for this -150 mgal anomaly, because the base of the crust is the next largest density contrast boundary in the lithosphere after the solid surface and the basement-sediment interface. Moreover, unlike some large circular continental sedimentary basins, such as the Michigan Basin [Haxby et al., 1976], or basins formed at a continental edge [Walcott, 1972], there is no indication of positive gravity anomalies within the Wanganui Basin. Basins whose initiating mechanism is crustal thinning should show some mixture of positive and negative anomalies, the exact pattern being a function of the elastic properties of the lithosphere at the time the basin was formed [Watts et al., 1982]. Thus, given the entirely negative gravity anomaly field associated with the Wanganui Basin, and the manner in which the gravity contours mimic the basement contours (Figures 4a and b), the driving load for the basin, which will have a positive gravity effect, must be either widely distributed or it is deep. i.e. deeper than than the base of the crust.

The second noteworthy geophysical feature of the Wanganui Basin is its unusual seismicity. Seismicity occurs as persistent swarms of earthquakes distributed throughout the crust with few events reaching a magnitude higher than 5. In addition, two large earthquakes have occurred since European settlement - in 1843 and in 1897 - with magnitudes of 7.5 and 7.0 respectively [Garrick and Gibowicz, 1983]. A composite focal mechanism for crustal earthquakes occurring beneath the southeastern Wanganui Basin shows nearly pure strike-slip motion, with the compression axis orientated east-west [Robinson, 1986]. There is a conspicuous lack of interplate thrust events. Robinson interprets this lack of thrust events as evidence either that there is strong coupling between the plates, and the region is now accumulating strain, or that we may be observing a late stage in the seismic cycle.

A Flexural Model for the Wanganui Basin

Multichannel seismic reflection data have been collected in the Wanganui Basin by the oil industry since the early 1960's. These data have been collated and interpreted by

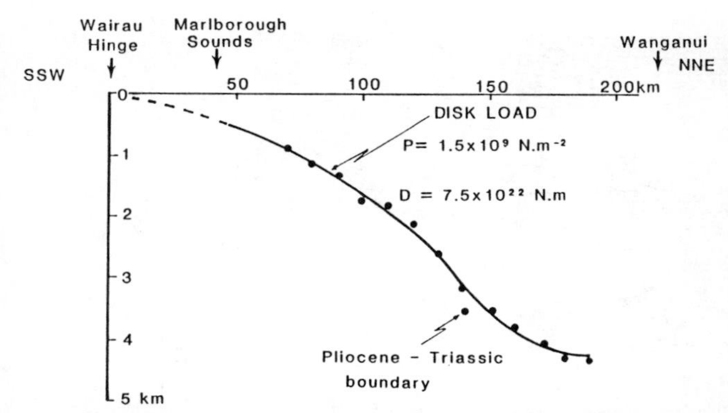

Fig. 5. Flexural model for a disk load compared to the digitised Pliocene- Triassic boundary beneath the Wanganui Basin. See Figure 4a for the position of line 141 [Anderton, 1981] from which it was digitised. Formulae for the disk load model are given by Brotchie and Silvester [1969]. A surface load of 1.5 x 10^9 N.m^{-2} acting on a 20 km radius disk was used. A density contrast of 1000 kg/m^3 was used between the displaced athenosphere and the infilling sediments. A value for the flexural rigidity (D) of the lithosphere of 7.5 x 10^{22} N.m was derived.

Anderton [1981]. Most of the data are 5 s two-way travel-time (TWTT) sections designed to image sedimentary stratigraphy and faulting within the basin. One of the lines, line 141, ran in a SSW-NNE azimuth from just north of Marlborough Sounds almost to Wanganui city (Figure 4). This line runs from the edge of the elliptical-shaped Wanganui Basin to its centre, along the largest axis of the ellipse, and is thus suitable for demonstrating the broad flexure of Triassic basement beneath the basin.

In Figure 5 we show a flexural model for the Wanganui Basin, using the Triassic -Pliocene interface as a reference boundary. The model used to simulate the driving force for the basin involves a disk load acting on an elastic sheet (the lithosphere) which in turn overlies a weak fluid (the athenosphere). Mathematically tractable expressions for the flexural displacement due to a disk load are available in terms of Kelvin functions [Brotchie and Silvester, 1969; Haxby et al., 1976]. A disk load of radius 20 km and amplitude 1.5 x 10^9 N.m^{-2} acting on an elastic plate of flexural rigidity 7.5 x 10^{22} N.m was found to give a satisfactory fit.

A disk-load model is used rather than a line load because the Wanganui Basin displays some axial symmetry. Although the precise geological meaning of the disk load is not important for the analysis at this stage, we will return to this question after examining some of the deep seismic data from the Wanganui Basin. The principal reason for conducting the flexural modelling is to demonstrate that the wavelength of the basin is largely controlled by lithospheric flexure, hence lithospheric strength, and to get an estimate for the magnitude of loading required to produce the basin. For example, we note that the above load amplitude of approximately 10^3 MPa exceeds by nearly an order of magnitude what is generally regarded as the maximum stress difference that the lithosphere can withstand over geological time periods [McNutt, 1980]. Such a large implied load suggests that the lithosphere beneath the Wanganui Basin will not be behaving perfectly elastically, but there will be a certain amount of brittle failure and ductile flow taking place. Evidence that at least brittle failure is occurring within the crust can be seen in the intense crustal seismicity of the Wanganui area as discussed previously. We also note that the derived value for flexural rigidity, 7.5 x 10^{22} N.m, appears to be low for continental lithosphere. The New Zealand lithosphere is relatively young (≈200 Ma), however, and when the value of 7.5 x 10^{22} N.m is superimposed on the plot of Karner et al., [1983] of lithospheric age at time of loading versus rigidity, it is broadly consistent with other global data.

As an aside, the flexural model provides a simple mechanical explanation for the recently drowned topography of the Marlborough Sounds [Cotton, 1913; Beck, 1964]. At the Wairau Valley, about 40 km south of the Marlborough Sounds (Figure 5), Cotton [1913] notes that the pattern of subsidence associated with the Sounds changes to uplift. The position of what is therefore termed here the "Wairau

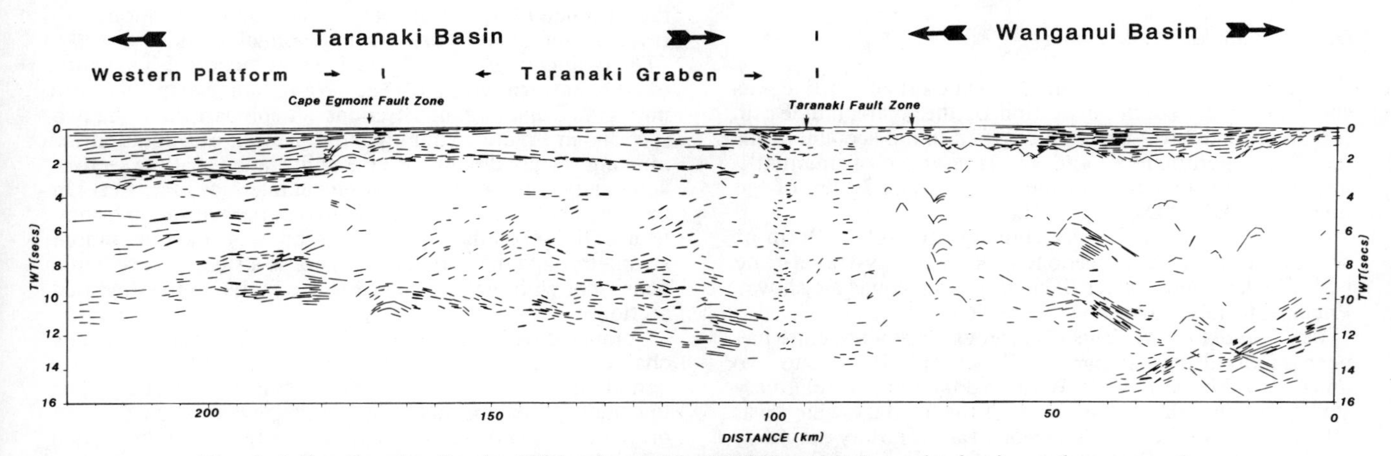

Fig. 6. A line drawing for the 220 km long, 16 s TWTT multi-channel seismic section across the Taranaki and Wanganui Basins. Shooting and recording parameters are given by Davey [1987]. Position of the line is shown on Figures 1 and 4a.

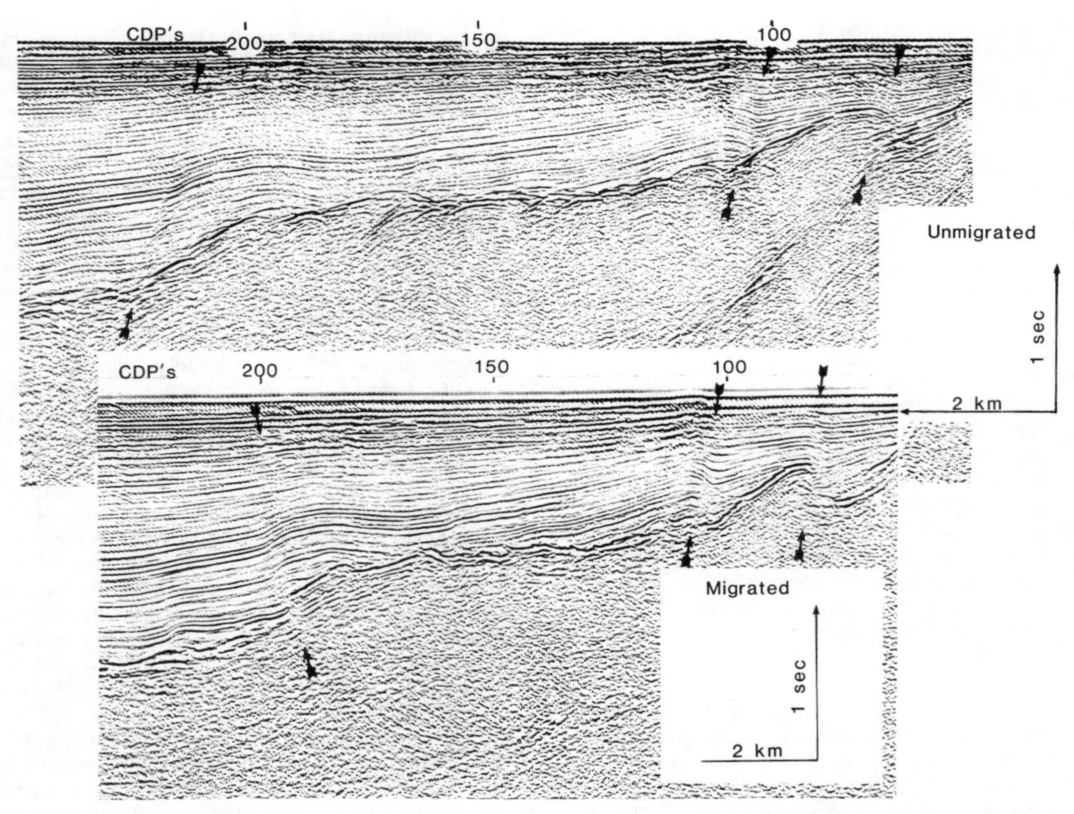

Fig. 7. Section of the top 2-3 s of TWTT for the eastern end of the Wanganui Basin. Migrated and unmigrated sections are shown. Note the enhanced reverse faults bought out in the migrated version.

Hinge" fits the zero node of the flexural model well (Figure 5). About 200 m of maximum subsidence at the northeastern end of the Marlborough Sounds is predicted by the model of Figure 5. This could be checked by searching for submerged beaches in the Sounds with seismic reflection methods.

Deep Seismic Reflection Data

A 220-km-long multichannel seismic reflection line was shot across the southern portion of the Wanganui Basin (Figures 1 and 4 for line location). An initial account of the recording parameters, and interpretation of manually migrated line segments for the eastern-most 70 km of the line, are given by Davey [1987].

Figure 6 shows a line drawing for the full 220 km of seismic line. Twenty seconds of two-way-travel-time (TWTT) data were collected, but only 16 seconds are shown. Principal features to note on this section are:

The Taranaki Basin. This Cretaceous-Recent structure has been subdivided by Pilaar and Wakefield [1978] into two units. To the west is the Western Platform, a relatively stable structure throughout most of the Tertiary which was affected only in the late Cretaceous-early Tertiary period by normal faulting. East of the Platform lies the Taranaki Graben which consists of a half-graben formed mainly during the Miocene. Seismic data indicate that up to 7000 m of Tertiary-Recent sediments accumulated adjacent to the Taranaki Fault. The Taranaki Graben had a similar Late Cretaceous-Early Tertiary history to the Western Platform of normal faulting on northeast-southwest trends [Knox, 1982]. By late Miocene, and extending into Plio-Pleistocene times the Taranaki Graben underwent structural inversion with reverse faulting along pre-existing, normal faults.

The Wanganui Basin. Figure 7 shows the top 2-3 s section of the eastern edge of the Wanganui basin on both unmigrated and migrated sections. A comparison of the two sections in Figure 7 highlights the benefit of migration for imaging steeply dipping reverse faults in the shallow section, and imaging the sediment - basement interface in general. In particular note that what appears to be a normal slump feature near CDP 200 on the unmigrated section is, upon migration, swung through the vertical to appear as a steep reverse fault, consistent with the two other reverse faults on the migrated section.

Compared with the Taranaki Basin, the Wanganui Basin is characterized by a highly reflective sedimentary sequence underlain by a relatively seismically transparent crust. The line drawing may be misleading in this regard because on the original section there are many diffractions and diffraction multiples that are not easy to reproduce on the line drawing. As noted by Davey [1987], the diffractions, the curved discontinuous events and the general lack of coherent energy in the middle crust beneath the Wanganui Basin may be due to the irregular basement-sediment interface (Figure 7), coupled with the high reflectivity of this interface.

The Cape Egmont and Taranaki Fault zones. These fault zones consist of en-echelon groupings of steep normal-to-reverse faults near the surface [Pilaar and Wakefield, 1978]. In their deeper portions, the faults are characterised seismically by many short-reflection segments, short diffraction tails and, in places, by seismically transparent columns (Figure 6). Similar patterns of seismic disruption to Moho depths have been noted as a feature corresponding to major strike-slip faults, such as the San Andreas Fault of California [Lemiszki and Brown, 1988; Cheadle et al., 1986]. The Taranaki and Cape Egmont fault zones have been discussed mainly in terms of their steeply dipping normal and reverse character [Knox, 1982; Anderton, 1981]. Both Pilaar and Wakefield [1978] and Anderton [1981] recognize, however, that these faults are also major strike-slip faults fitting into a regional pattern of dextral wrench deformation. We suggest, therefore, that the disrupted seismic character of the crust associated with the Cape Egmont and Taranaki fault zones (Figure 6) adds further, albeit circumstantial, support for the strike-slip character of these faults.

The character of the "reflection Moho". (here the definition of the reflection Moho follows that of Klemperer et al., [1986] as "the deepest, high-amplitude, laterally extensive reflection or group of reflections present at traveltimes approximately commensurate with other estimates of crustal thickness"). An intriguing feature of the complete seismic section shown in Figure 6 is the vertical relief shown on the reflection Moho. The reflection Moho reaches a local minimum at a depth of about 10 s TWTT immediately west of the Cape Egmont Fault Zone. For an average crustal velocity of 6.0 km/s this would correspond to a depth of about 30 km. The reflection Moho then displays a distinctive ramp-like character to reach a local maximum of nearly 13 s TWTT, or a depth of about 40 km, beneath the Taranaki Fault Zone. This apparent rapid thickening of the crust beneath the Taranaki Graben is unexpected. We might expect a graben-like feature to be extensional and therefore driven by stretching and thinning of the crust in a manner similar to that proposed by McKenzie [1978].

In the vicinity of the Taranaki Fault Zone the reflection Moho signal is degraded as mentioned previously. East of the fault zone the signal from the lower crust is still weak, although the reflection Moho is shallower here than beneath the Taranaki Graben. Directly beneath the deepest section of the Wanganui Basin the reflection Moho displays cross-dipping events that form a "bow-tie" in the TWTT interval of 11-15 s. Bow-tie structures that occur in shallow seismic data are usually associated with synclinal structures [Badley, 1985] and require a migration technique to resolve [Claerbout, 1985].

Interpretation of the Seismic Data

Two principal aspects of the data in Figure 6 pertinent to this study are the bow-tie structure beneath the Wanganui Basin and the anomalously deep reflection Moho beneath the Taranaki Graben.

Moho beneath the Wanganui Basin. Davey [1987] initially performed a manual migration of some of the line segments forming the bow-tie beneath the Wanganui Basin and "untied" the bow. He showed that the westward dipping reflections correlated in position with the top of the Benioff Zone as independently estimated by Robinson [1986] on the basis of earthquake hypocentres. A similar analysis has been performed here using a computer-aided constant velocity migration routine [Gallow, 1986]. Shown in Figure 8 is a line-migrated and depth-converted plot for the easternmost 70 km of the seismic line.

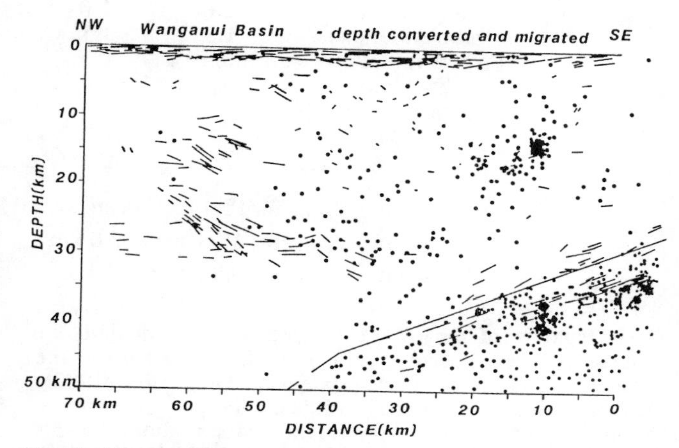

Fig. 8. Migrated line section drawing with superimposed hypocentres as given by Robinson [1986]. Heavy diagonal line is Robinson's pick for the top of the subducted plate. The section was migrated with a constant velocity migration (6 km/s) routine. The section was also converted to depth using a velocity model similar to that of Garrick [1968] for the southern North Island, but also included a 2 km deep, 2 km/s section to represent the southeastern Wanganui Basin.

Also shown in Figure 8 are the superimposed hypocentres for Robinson's northeastern profile as derived from the Wellington seismograph network. The top of the Benioff Zone, based on earthquake hypocentres [Robinson, 1986], is shown by the heavy diagonal line across the lower right of the figure. The constant velocity migration has resolved the northwestward dipping reflections as the top of the subducted Pacific plate, and the southeasterly dipping reflections as the Moho associated with the overriding Australian Plate.

An intriguing feature of the superposition of seismicity and reflections, shown in Figure 8, is what appears to be an almost inverse correlation between the density of earthquake hypocentres and that of seismic reflection segments. This is particularly so throughout the crust beneath the Wanganui Basin. An explanation advanced here is that where the crust is seismically active there occurs intense faulting and deformation, and thus there are few laterally continuous reflectors. In a similar vein, the apparent lack of reflection continuity between the reflection Moho of the overriding Australian Plate, and the reflections from the top of the subducted Pacific Plate, may be due to intense deformation occurring in the shear zone between the plates.

Over-thickened crust beneath the Taranaki Graben?. Our

TABLE 1. Velocity pull down analysis. The pull down effect of the Taranaki Graben relative to other sedimentary basins along the seismic line is estimated to be not greater than 0.4s.

	Plio-Pleistocene TWTT(s)	km*	Miocene TWTT (s)	km**	TWTT (s) t o 5 km***
Wanganui Basin	2.0	2.0	--	--	3.0
Taranaki Graben	1.5	1.5	1.5	2.25	3.4
Western Platform (adjacent to Cape Egmont Fault Zone)	0.5	0.5	2.5	3.75	3.25

* assumes V_{rms} for Plio-Pleistocene of 2.0 km/s [Anderton, 1981]
** assumes V_{rms} for Miocene of 3.0 km/s [" "]
*** assumes basement velocity of 6.0 km/s

understanding of the lower continental crust, particularly the source of layered seismic reflections from the base of the crust, is still rudimentary [Hale and Thompson, 1982 ; Klemperer et al., 1987]. Thus it is difficult to be definitive about the excessively deep reflection Moho (about 2.5 -3 s TWTT deeper) associated with the Taranaki Graben. The following possibilties are nevertheless considered :

1. "velocity pull down" [Badley, 1985] due to the overlying wedge of Tertiary-Recent sediments (Figure 6). An analysis of velocity pull down is given in table 1 for the relative effects of sediments in the Wanganui Basin, the Taranaki Graben and the eastern end of the Western Platform. This analysis shows that there is a velocity pull down effect for the Taranaki Graben, but no more than 0.4 s TWTT relative to the other two basins.

2. Juxtaposed crustal sections. If the Taranaki Fault Zone does have a significant dextral strike-slip component then it is feasible that the Taranaki Graben block has been bought in from the south where the crust may be thicker, and juxtaposed against the Wanganui crustal block. This proposal is difficult to test without knowledge of strike-slip rates along the Taranaki Fault Zone.

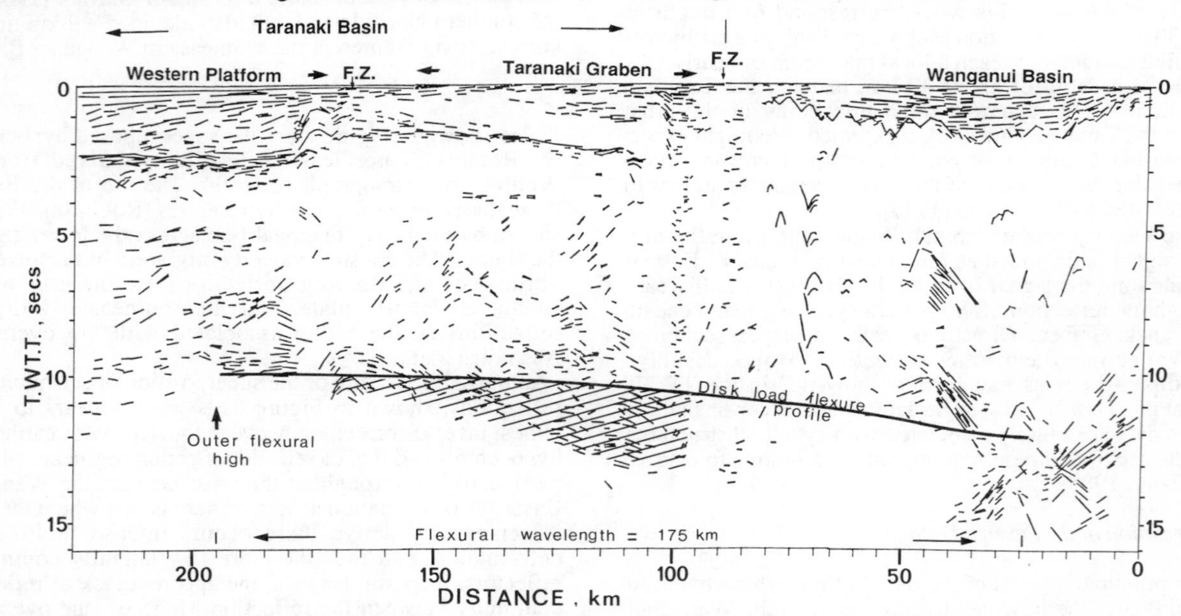

Fig. 9. The line drawing of Figure 6 shown on a 2X vertical exaggeration. Also shown is the superimposed flexure profile corresponding to the disk load, but placed at Moho depth in an attempt to model flexure of the reflection Moho. The shaded area beneath the Taranaki Graben highlights the 2-3 s TWTT -thick packet of reflectors below the flexure profile. F.Z. = fault zone.

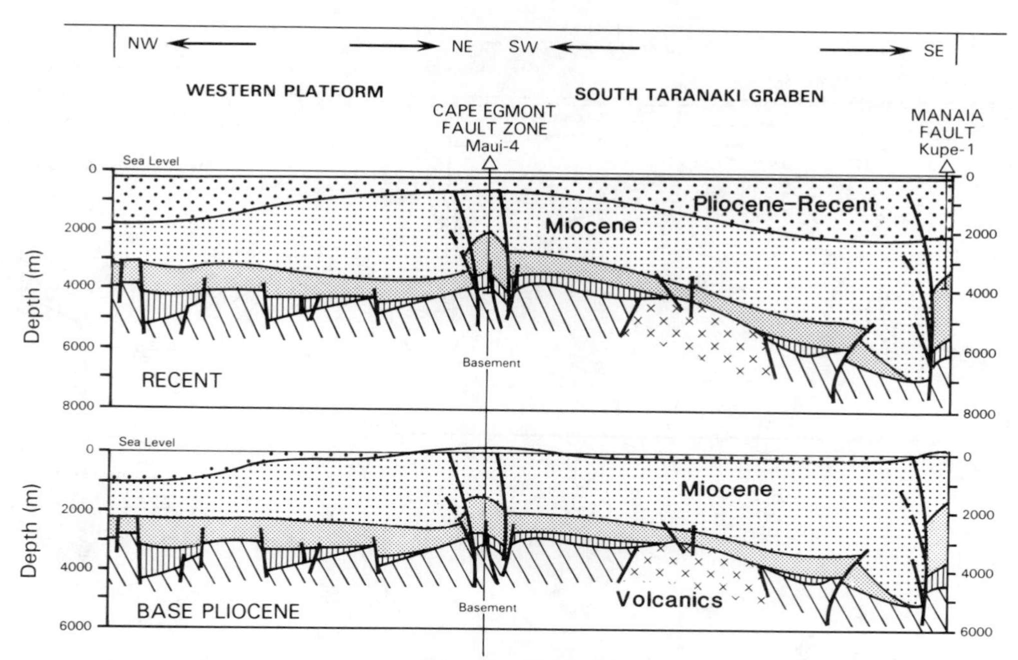

Fig. 10. Seismic stratigraphy interpretation of the Taranaki Graben and the western Platform as given by Pilaar and Wakefield [1978]. Note that the pre-Pliocene southeast dip on basement beneath the Taranaki Graben is about 0.8°. From Pliocene to the present day this dip has doubled to about 1.8°.

3. Upper-mantle underplating. It could be argued that the anomalously deep, coherent reflections defining the reflection Moho at the eastern end of the Taranaki Graben represents some form of "rift-pillow" or crust-mantle mixture that is thought to underlie many continental rifts [Mooney et al., 1983]. A difficulty with this explanation, however, is that the limited seismic reflection data from areas that have undergone significant extension are characterised by dominantly horizontal reflections marking the reflection Moho, quite unlike the dipping events seen on our data. For example, in the Basin and Range Province of the Western United States [Klemperer et al., 1986; Allmendinger et al., 1987] the reflection Moho displays such a dominant horizontal character.

4. Distributed flexure. We note that at least part of the apparently over-thickened crust beneath the Taranaki Graben is due to distributed flexure associated with the present-day formation of the Wanganui Basin. Figure 9 shows the seismic section of Figure 6, with a 2X vertical exageration, and the superimposed flexural profile of Figure 5 shifted down to Moho depths. The justification for fitting the flexural profile of the Pliocene-Triassic interface to the reflection Moho is that flexure should affect the whole lithosphere, and thus displacement of the sediment-basement interface should be approximately matched by displacement of the Moho. In Figure 9 there is a reasonable match between the flexural model and the position of the reflection Moho beneath the Wanganui Basin and the Cape Egmont Fault Zone. The half-wavelength of flexure on the reflection Moho is about 175 km, whereas on the south-north, sediment-basement profile (Figure 5) the half-wavelength is about 200 km. This difference in flexural wavelength highlights an anisotropy that probably arises from the approximately north-south striking faults in this region reducing effective flexural rigidity in the east-west azimuth.

Underlying the flexural curve shown in Figure 9 there is a packet of lower crustal reflectors that is triangular-shaped and about 2s TWTT thick. This region is beneath the eastern edge of the Taranaki Graben and suggests that the crust here was thicker prior to formation of the Wanganui Basin. For example there may have been crustal downwarping and shortening, similar to that occurring today beneath the Wanganui Basin, during the Miocene and earlier beneath the Taranaki Graben. Alternatively, a westward moving thrust sheet in the Miocene may have loaded the crust just west of the Taranaki Fault Zone, thus producing a flexural foreland basin in the style similar to that outlined by Beaumont [1981].

Support for both a pre and post-Pliocene flexure of basement beneath the Taranaki Graben comes from the stratigraphic sections supplied by Pilaar and Wakefield [1978] (shown in Figure 10). Pre-Pliocene there was a SE dip on basement of about 0.8°. This SE dip doubled to about 1.8° at the present day. Therefore the Taranaki Graben appears to have undergone two stages of SE-directed downwarping and, although we can document the post-Pliocene cause, the early phase of downwarping is due to causes unknown.

Discussion

Mass balance

We interpret the Central Volcanic Region to be a back-arc spreading basin where the basin, or graben structure, is a direct isostatic response of the thin crust beneath. Yet

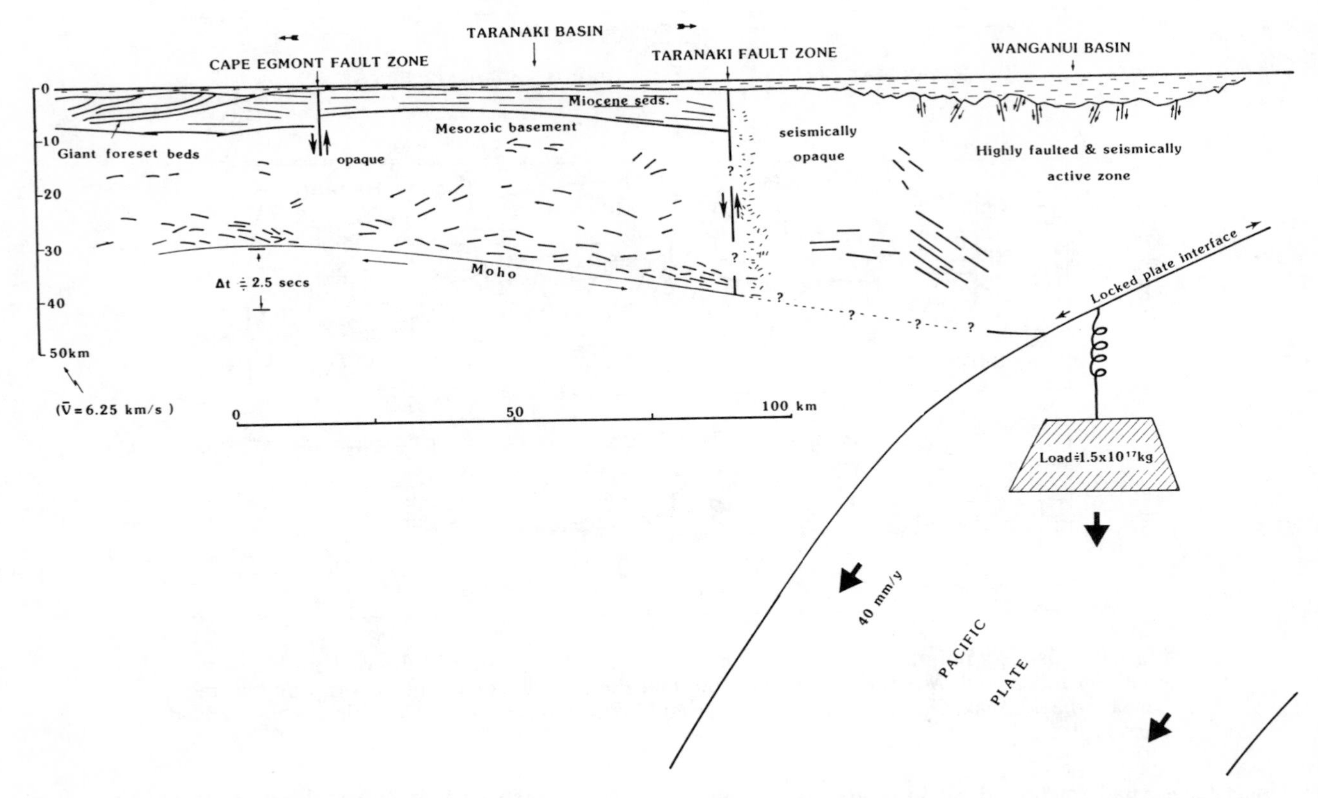

Fig. 11. A conceptual model for the formation of the Wanganui Basin. The driving load is proposed to be the vertical component of the shear coupling between the two plates. The total load required is 1.5×10^{17} kg or 4.3×10^{11} kg/m if considered as an equivalent line load (see text). The depth conversion is on the basis of an average crustal seismic velocity of 6.25 km/s.

ultimately it is the retreating, or roll-back, action of the subducted Pacific plate that is the driving mechanism for the CVR. Similarly, subsidence within the Wanganui Basin appears ro be driven by the excess mass of the subducted Pacific plate, but in a different, more direct way. A conceptual model for the proposed driving force of the Wanganui Basin is shown in Figure 11. Here we speculate that the vertical component of shear, between the subducted Pacific plate and the overriding Australian plate, produces the downwarping action. This model differs somewhat from those outlined by McKenzie [1969] and Davies [1981], as discussed in the introduction. Their models rely on stresses transmitted hydro-dynamically through the mantle wedge to produce the flexural downwarp.

From the flexural model of Figure 5 it is calculated that the equivalent disk load required to form the Wanganui Basin has an area loading of 1.5×10^9 N.m^{-2} or, for a 20 km radius disk, a total load of 1.5×10^{17} kg. As it is contended that the formation of the Wanganui Basin is driven by the excess mass of the subjacent subducted plate, it is useful to compare the disk load with the theoretical magnitude of excess mass in the subducted plate.

Crough and Jurdy [1980] show that the excess mass of a subducted plate per unit length of trench can be estimated by:

$$E d \Delta\rho / 2 \sin \phi$$

where : E = maximum Benioff Zone depth
d = subsidence of lithosphere with respect to ridge prior to subduction
$\Delta\rho$ = asthenospheric density - water density
ϕ = dip of Benioff Zone

For the southern Hikurangi Margin we estimate d = 1700 m. This is a small value for the relative subsidence of the Pacific plate between its spreading ridge and the subduction zone. It represents, however, an effective subsidence as oceanic crust now being subducted at the southern Hikurangi Margin appears to be either unusually thick, or to be capped by a large thickness of sediments [Smith et al., 1988; Robinson, 1986]. Taking $\Delta\rho$ = 2300 kg/m^3, E = 225×10^3 m, and ϕ = 50° [Adams and Ware, 1977] gives an excess mass per unit length of 5.7×10^{11} kg/m for the subducted Pacific plate.

The total load required to produce the Wanganui Basin of 1.5×10^{17} kg, if distributed over the 350 km N.E.-S.W. diameter of the Basin (ie. from Mt Ruapehu in the north to the Wairau hinge (Figure 5) in the south), corresponds to an equivalent line load of 4.3×10^{11} kg/m. Therefore about 75 % of the excess mass for the slab beneath the southern portion of the Hikurangi Margin appears to be compensated

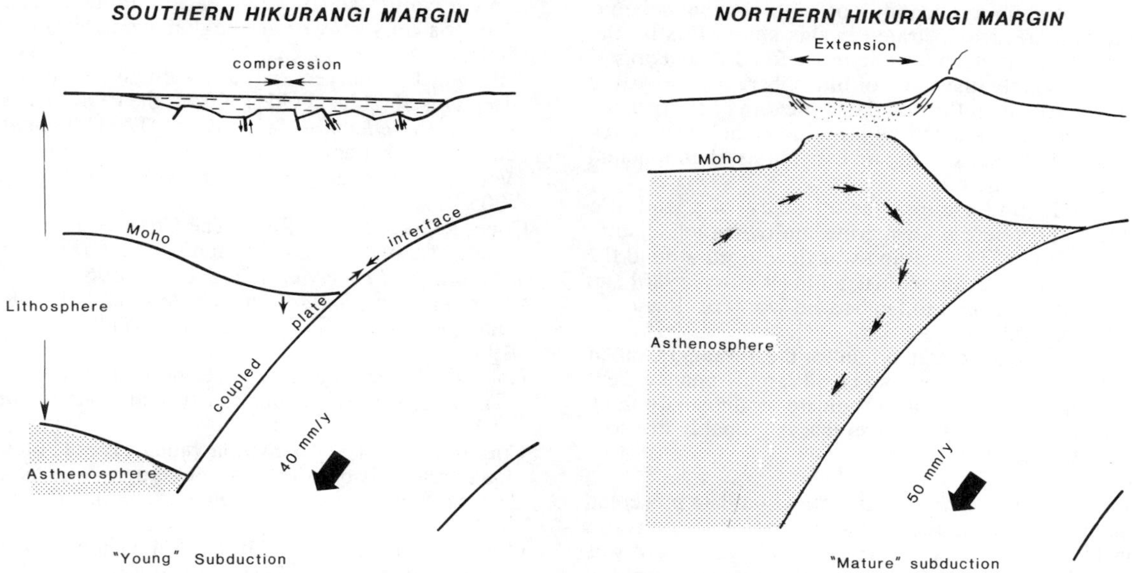

Fig. 12. A cartoon depiction of back-arc basins proposed to form behind "young" and "mature" subduction zones, corresponding to the southern and northern Hikurangi margins respectively.

by lithospheric flexure associated with the Wanganui Basin. The remainder of the load is, presumably, transmitted up the slab to act as part of the driving force for the subduction process [Harper, 1975a].

Time or Space-dependent Back-arc Processes ?

We have described the structure and proposed mechanical formation processes for two different yet coeval forms of back-arc basin that have formed behind the Hikurangi Margin. The question of why there is such a gross difference in structure between these basins remains unanswered. Two broad categories of models that could explain this difference are considered : space-dependent or time-dependent.

Space-dependent. At the south end of the Hikurangi Margin, compared to further north, there are differences in both the subduction process and in the character of the subducted plate. For example, the subducted oceanic crust is thicker to the south which would provide added buoyancy to the subducted plate beneath the southern Hikurangi Margin [Smith et al., 1988]. In addition, subduction is more oblique to the south and because the subduction zone terminates at 41° South (Figure 1), end-effects such as asthenospheric vortices may develop [Jacoby, 1973 ; Harper, 1975b]. Thus special mechanical and dynamical conditions exist at the southern end of the Hikurangi Margin that could partly, or wholly, contribute to the development of the Wanganui Basin.

Time-dependent. In Figure 12 we show a cartoon depiction of the two forms of basin. At the southern end of the Hikurangi Margin, where the Wanganui Basin and compressional tectonics are found, subduction is relatively young. If we were, for example, to numerically "pull back" the subducted plate along its oblique trajectory at the present-day convergence rate, we would find that the leading edge of the subducted Pacific plate, now at a depth of 200 km beneath the northwestern end of the South Island (Figure 1), first penetrated the athenosphere (assumed depth of 100 km) beneath Wanganui about 5 Ma ago. To the north, however, low-potash andesites dating back to 20 Ma, and a more deeply penetrating subducted plate [Adams and Ware, 1976] both attest to the relatively mature subduction regime in this area. Theoretical studies [e.g. Andrews and Sleep, 1974 ; Jurdy and Stefanick, 1983] predict such a time lag (i.e. about 10 Ma) between the onset of subduction and the development of a fully fledged back-arc spreading centre. These studies consider the back-arc basin to be established by induced convection with the asthenosphere which produces tension and then thinning within the overlying lithosphere.

We accordingly attribute an age difference in subduction to be the simplest explanation for the gross differences in structure and tectonics between the Central Volcanic Region and the Wanganui Basin. We can not, nevertheless, completely rule out some of the space-dependent reasons listed above as contributing factors.

Conclusions

The following is concluded from this study of basins formed behind the Hikurangi Margin.

1. The study has highlighted an apparent time-progression for the development of back-arc basins within continental lithosphere. It is proposed that crustal downwarping, that is the hallmark of the Wanganui Basin, represents an early stage in the formation of a back-arc basin. After 5-10 Ma of subduction history a fully fledged, extensional back-arc basin, as exemplified by the Central Volcanic Region, can be expected to develop.

2. Two uncommon applications of multichannel seismic reflection data are demonstrated in this study. Firstly, the joint interpretation of deep seismic reflection data and crustal seismicity. Secondly, the study of lithospheric flexure using multichannel seismic reflection data. Flexure of the oceanic lithosphere has been studied with multichannel techniques [Watts et al., 1985] but seismic investigations of continental lithospheric flexure are rare.

3. This study underscores the efficacy of studying continental lithosphere with marine multichannel seismic reflection methods. The method was developed around the continental shelf of the United Kingdom [Warner, 1986] and has the benefit, compared to land-based work, of low-cost, speed and improved data quality.

4. The problem of a tectonic origin for the Taranaki Graben has been raised by this study, but is left largely unanswered. Our multichannel seismic data indicate an unusually thick crust beneath the Taranaki Graben that is probably due to a form of crustal shortening.

Acknowledgements. The seismic reflection data presented in this paper were acquired under contract to Geophysics Division by Western Geophysical. Initial processing was carried out by GECO(NZ) Ltd. Further processing and analysis of the data were carried out by TAS while on leave at the Institute for the Study of the Continents (INSTOC), Cornell University, New York. We thank Professor J. Oliver and his co-workers for the help and advice on this work. We are indebted to Karen Brown for editing and typing this maunscript.

References

Adams, R. D. and D. E. Ware, Subcrustal earthquakes beneath New Zealand: locations with a laterally inhomogeous velocity model, *N. Z. J. Geol. Geophys.*, 20 : 59-83, 1977.

Allmendinger, R. W., T. A. Hauge, E. C. Hauser, C. J. Potter, S. L. Klemperer, K. D. Nelson, P.Knuepfer and J. Oliver, Overview of the COCORP 40° N transect, western United States : the fabric of an orogenic belt, *Geol. Soc. Am. Bull., 98*, 308-319, 1987.

Anderton, P. W., Structure and evolution of the south Wanganui Basin, New Zealand, *N. Z. J. Geol. Geophys.*, 24, 39-63, 1981.

Andrews, D. J. and N. H. Sleep, Numerical modelling of tectonic flow behind Island arcs, *Geophys. J. R. Astro. Soc., 38* , 237-254, 1974.

Ansell, J. and D. Adams, Unfolding the Wadati-Benioff zone in the Kermadec-New Zealand region, *Phys. Earth & Planet. Interiors*, 44, 274-280, 1986.

Badley, M. E., *Practical seismic interpretation*, 266 pp., International Human Resources Development Corporation, Boston, 1985.

Beaumont, C., Foreland basins, *Geophys. J. R. Astro. Soc.* 65, 291-329, 1981.

Beck, A. C., Sheet 14, Marlborough Sounds (1sted.), *Geological map of New Zealand. 1 : 250,000*. Dept. Sci. & Indus. Res.,Wellington, New Zealand, 1964.

Brotchie, J. F. and R. Silvester, Crustal flexure. *J. Geophys. Res.,74* , 5240-5252, 1969.

Calhaem, I. M., Heat flow measurements under some lakes in the North Island, New Zealand, Ph.D. thesis, 191 pp., Victoria University of Wellington, New Zealand, 1973.

Cheadle, M. J., B. L. Czuchra, C. J. Ando, T. Byrne, L. D. Brown, J. E. Oliver, and S. Kaufman , Geometries of deep crustal faults : evidence from the COCORP Mojave survey, in *Reflection Seismology : The Continental Crust*, edited by M. Barazangi & L. Brown, Geodynamics series vol.14, American Geophysical Union, Washington,D.C., 305-312, 1986.

Cope, R. N. and J. J. Reed, The Cretaceous paleogeology of the Taranaki - Cook Strait area, *Aust. Inst.Mining and Metallurgy, Proceedings*, 222, 63-72, 1967.

Claerbout, J. F., *Imaging the Earth's interior*, 398 pp., Blackwell Scientific Publications, London, Palo Alto, 1985.

Cole, J. W. and K. B. Lewis, Evolution of the Taupo-Hikurangi subduction system, *Tectonophysics, 72* : 1-21, 1981.

Cotton, C. A., The Tuamarina valley : A note on the Quaternary history of the Marlborough Sounds district, *Transactions of the New Zealand Institute, 45* , 316-322, 1913.

Crough, S. T. and D. M. Jurdy, Subducted lithosphere, hot spots and the geoid, *Earth Planet. Sci. Lett.,48* , 15-22, 1980.

Davey, F. J., Seismic reflection measurements behind the Hikurangi convergent margin , southern North Island, New Zealand, *Geophys. J. R. Astro. Soc. 89* , 443-448, 1987.

Davies, G. F., Regional compensation of subducted lithosphere : effects on geoid , gravity and topography from a preliminary model, *Earth Planet. Sci. Lett., 54* , 431-441,1981.

Eiby, G. A., The New Zealand subcrustal rift, *N. Z. J. Geol. Geophys., 7*, 109-133, 1964.

Gallow, S., Linedraw user's guide , 40 pp., Unpublished document of INSTOC, Cornell University, Ithaca, New York, 1986.

Garrick, R. A., A reinterpretation of the Wellington crustal profile (letter), *N. Z. J. Geol. Geophys., 11*, 1280-1294, 1968.

Garrick, R. A. and S. J. Gibowicz, Continuous swarm-like activity : the Wanganui, New Zealand, earthquakes. *Geophys. J . R. Astro. Soc. 75* , 493-512, 1983.

Haines, A. J., Seismic wave velocities in the uppermost mantle beneath New Zealand, *N. Z. J. Geol. Geophys.* , 22, 245-257, 1979.

Hale, L. D. and G. A. Thompson, The seismic reflection character of the continental Mohorovicic discontinuity, *J. Geophys. Res. 87,* 4625-4635, 1982.

Harper, J. F., On the driving forces of plate tectonics, *Geophys. J. R. Astro. Soc. 40*, 465-474, 1975a.

Harper, J. F., Subduction Zone vortices, *Bull. Aust. Soc. Explor. Geophys.*, 6, 79-80, 1975b.

Hatherton, T., The geophysical significance of calc-alkaline andesites in New Zealand , *N. Z. Geol. Geophys., 12* , 436-459, 1969.

Hatherton, T., Upper mantle inhomogeneity beneath New Zealand : surface manifestations, *J. Geophys. Res.,75,* 269-284, 1970.

Haxby, W. F., D. L. Turcotte and J. M. Bird, Thermal and mechanical evolution of the Michigan Basin, *Tectonophysics*, 36, 57-75, 1976.

Hunt, T. M., Basement structure of the Wanganui Basin, onshore, interpreted from gravity data, *N. Z. J. Geol. Geophys. 23* , 1-16, 1980.

Jacoby, W. R., Model experiment of plate movements, *Nature Phys. Sci., 242*, 130-134, 1973.

Jurdy, D. M. and M. Stefanick, Flow models for back-arc spreading,*Tectonophysics, 99* , 191-206, 1983.

Kamp, P. J. J., Late Cretaceous-Cenozoic tectonic development of the southwest Pacific region, *Tectonophysics,122* , 1-27, 1986.

Karner, G. D., M. S. Steckler and J. A. Thorne, Long-term thermo-mechanical properties of the continental lithosphere, *Nature, 304* , 250-253, 1983.

Klemperer, S. L., T. A. Hauge, E. C. Hauser, J. E. Oliver and C. J. Potter, The Moho in the northern Basin and Range Province, Nevada, along the COCORP 40° N seismic reflection transect, *Geol. Soc. Am. Bull.,97*, 603-618, 1986.

Klemperer, S. L. and the BIRPS group, Reflectivity of the crystalline crust : hypotheses and tests, *Geophys. J. R. Astro. Soc. 89*, 217-222, 1987.

Knox, G. J., Taranaki Basin , structural style and tectonic setting, *N. Z. J. Geol. Geophys. 25* , 125-140, 1982.

Lensen, G. J., Sheet 10- Wanganui, *Geological map of New Zealand , 1 : 250 000*, Dept. Scientific and Industrial Research , Wellington, New Zealand, 1959.

Lemiszki, P. J. and L. D. Brown, Variable crustal structure of strike-slip zones as observed on deep seismic reflection profiles, *Geol. Soc. Am. Bull., 100*, 665-678, 1988.

McKenzie, D. P., Some remarks on the development of sedimentary basins, *Earth Planet. Sci. Lett. 40* , 25-32, 1978.

McKenzie, D. P., Speculations on the consequences and causes of plate motions. *Geophys. J. R. Astro. Soc. 18*, 1-32, 1969.

McNutt, M., Implications of regional gravity for the state of stress in the earth's crust and upper mantle, *J. Geophys. Res. 85*, 6377-6396, 1980.

Mooney, W. D., M. C. Andrews, A. Ginzburg, D. A. Peters and R. A. Hamilton, Crustal structure of the northern Mississippi embayment and a comparison with other continental rift zones. *Tectonophysics, 94*, 327-348, 1983.

Otofuji, Y. I. and T. Matsuda, Amount of clockwise rotation of southwest Japan-fan shape opening of the southwestern part of the Japan Sea, *Earth. Planet. Sci. Lett., 85*, 289-301, 1987.

Pilaar,W. F. H. and L. L. Wakefield, Structural and stratigraphic evolution of the Taranaki Basin, offshore North Island, New Zealand, *J. Aust. Petroleum Exploration Ass. 18*, 93-101, 1978.

Reilly, W. I., *Gravity map of New Zealand* , 1 : 4 000 000 isostatic anomalies, 1st edition, Department of Scientific and Industrial Research, Wellington, New Zealand. 1965.

Robertson, E. I. and W. I. Reilly, Bouguer anomaly map of New Zealand, *N. Z. J. Geol. Geophys., 1*, 560-564, 1958.

Robinson, R., Seismicity , structure and tectonics of the Wellington region, New Zealand, *Geophys. J. R. Astro. Soc. 87* , 379-409, 1986.

Sleep, N. H., J. A. Nunn and L. Chou, Platform basins, *Ann. Rev. Earth Planet. Sci. 8* , 17-34, 1980.

Smith, E. G. C., T. A. Stern and M. E. Reyners, Subduction and back-arc activity at the Hikurangi convergent margin, New Zealand, *Pure and Applied Geophysics, 128, no.3 & 4, in press*, 1988.

Steckler, M. S. and U. S. ten Brink, Lithospheric strength variations as a control on new plate boundaries : examples from the northern Red Sea region, *Earth Planet. Sci.Lett. 70* , 120-132, 1986.

Stern, T. A., Asymmetric back-arc spreading,heat flux and structure associated with the Central Volcanic Region of New Zealand, *Earth Planet. Sci. Lett. 85* , 265-276, 1987.

Stern, T. A. and F. J. Davey, A seismic investigation of the crustal and upper mantle structure within the Central Volcanic Region of New Zealand, *N. Z. J. Geol. Geophys., 30*, 217-231, 1987.

Stern, T. A., E. G. C. Smith, F. J. Davey and K. J. Muirhead, Crustal and upper mantle structure of the northwestern North Island, New Zealand, from seismic refraction data, *Geophys. J. R. Astro. Soc., 91*, 913-936, 1987.

Taylor, B. and G. D. Karner, On the evolution of marginal basins, *Rev. Geophys. and Space Phys. 21* , 1727-1742, 1983.

Thompson. B.N., Quaternary volcanism of the Central Volcanic Region,*N. Z. J. Geol. Geophys. 7* , 45-66, 1964.

Walcott, R. I., Geodetic strain and the deformational history of the North Island during the late Cainozoic, *Phil. Trans. R. Soc. Lond., A. 321*, 163-181, 1987.

Walcott, R. I., Gravity, flexure and the growth of sedimentary basins at a continental edge, *Geol.Soc. Am. Bull. 83* , 1845-1848, 1972.

Warner, M. R., Deep seismic reflection profiling the continental crust at sea. In *Reflection Seismology : a global perspective,* edited by M. Barazangi and L. Brown Geodynamics series vol.13, Am. Geophys.Un., Washington, 281-286, 1986.

Watts, A. B., U. S. ten Brink, P. Buhl and T. M. Brocher, A multichannel seismic study of lithosphere flexure across the Hawaiian -Emperor seamount chain, *Nature, 315* , 105-111, 1985.

Watts, A. B., G.D. Karner and M. S. Steckler , Lithospheric flexure and the evolution of sedimentary basins. *Phil. Trans. Roy. Soc. Lond., A 305*, 249-281, 1982.

Watts, A. B. and W. B. F. Ryan, Flexure of the lithosphere and continental margin basins, *Tectonophysics, 36,* 25-44, 1976.

Wilson, C. J. N., A. M. Rogan, I. E. M. Smith, D. J. Northey, I. A. Nairn and B. F. Houghton, Caldera Volcanoes of the Taupo Volcanic Zone, New Zealand, *J. Geophys. Res. 89* , 8463-8484, 1984.

Wright, I. C. and R. I. Walcott, Large tectonic rotation of part of New Zealand in the last 5 Ma, *Earth Planet. Sci. Lett. 80* , 348-352, 1986.

Uyeda, S., Some basic problems in the trench-arc-back arc system. In *Island arcs Deep Sea Trenchs and Back-Arc Basins* , edited by M. Talwani & W. C. Pitman, Maurice Ewing series 1 , American Geophysical Union, Washington, 1-14, 1977.

THE SOUTHERN SAN JOAQUIN VALLEY AS AN EXAMPLE OF CENOZOIC BASIN EVOLUTION IN CALIFORNIA

Emery D. Goodman and Peter E. Malin

Department of Geological Sciences and Institute for Crustal Studies, University of California, Santa Barbara, California 93106

Elizabeth L. Ambos

Center for Earth Sciences, University of Southern California, Los Angeles, California 90089-0741

John C. Crowell

Department of Geological Sciences and Institute for Crustal Studies, University of California, Santa Barbara, California 93106

Abstract. The Tejon Embayment and the crystalline Tehachapi Mountains are located within a rotated crustal block that lies between the San Andreas, White Wolf and Garlock faults. The area's variable tectonic history includes the origin and evolution of the Cenozoic southern San Joaquin Basin and is currently being studied using CALCRUST and industrial seismic reflection and refraction data, borehole data and field observations. The complex structure and the stratigraphy of the Tejon Embayment (Eocene to Recent) provide evidence for a sequence of very different tectonic events, including regional extension and compression.

A graben system of many small blocks was produced in the central Tejon Embayment by presently NE-trending normal faults, episodically active from late Oligocene to latest Miocene time, and by the Springs fault, which shows evidence of strike-slip. Thickness and facies changes across the normal faults reflect the deepening of the Embayment in mid-Miocene time. The normal faults were again reactivated late in the Miocene.

Evidence for late Oligocene-Early Miocene extension in the southern San Joaquin Valley includes normal faults, volcanic extrusives, coarse breccias and basic dikes. The origin of this basin and coincident regional extension may be related to the pre-San Andreas transform history of North American-Pacific plate interaction or, alternatively, to western Basin and Range extension.

At present, the U-shaped Tejon Embayment is closing from three directions. Cross sections and mapping show that a system of mostly buried Pliocene to Recent thrust faults (a) consistently verge basinward along the basin margin, (b) have displaced both the Cretaceous crystalline basement rocks and younger sedimentary rocks, (c) structurally overlie the buried normal faults, and (d) are associated both with petroliferous folds and (e) with vertical and overturned beds south of the exposed Pleito fault. Comanche Point (to the northeast) and Wheeler Ridge (to the northwest) are anticlines whose underlying thrusts are related to the basin-margin thrusts, rather than to the higher-angle White Wolf fault.

During rapid uplift over the past several million years, the thrusts have exhumed normal faults and volcanic rocks along the basin margin. They are part of a regional change in deformation from extension to contraction that can be related to post-Miocene transpression along the nearby San Andreas and Garlock faults, or perhaps to the continual clockwise rotation of the Tehachapi Block.

The White Wolf fault was probably tilted and segmented during this reorganization of stress regimes, accentuating the structural relief across its plane. Latest Miocene to Recent strata are more than 2500 m thicker north of the fault in the 12-km-deep Maricopa subbasin than they are in the Tejon Embayment. This fact and the young compressional features on the southern and western margins of the southern San Joaquin Basin suggest that the Miocene extensional basin has evolved into a foreland-type depression.

Our deep reflection and refraction data and velocity model suggest that the Tehachapi Block is comprised of north-dipping, relatively-high velocity layers with major discontinuities at the White Wolf fault to the north and at the Garlock fault to the south. The apparent continuity of the deep structure suggests that, as the basins formed, the Neogene clockwise rotations of the Tehachapi Block may have been accommodated by detachment faults in the middle crust.

Introduction and Regional Setting

This paper reports on progress on a continuing project aimed at better understanding the development of the Cenozoic southern San Joaquin Basin and other features in the crust of California, as a part of the CALCRUST program of crustal studies in California. Our work has concentrated on unraveling the geological history of the Tejon Embayment, the southernmost subbasin in the San Joaquin Valley, and makes use of structural, stratigraphic, geochemical, paleontological, and geophysical data and methods. We present new seismic reflection and refraction profiles which shed light on the shallow and deep structure of the Tehachapi Block and the regions to its north and south. In addition, we have integrated these data with borehole data and surface geology.

The Tehachapi Block, which includes the Tejon Embayment, is bracketed by the San Andreas, Garlock, and White Wolf faults, and forms the juncture between the deformed and rotated "tail" of the southern Sierran batholith (Tehachapi Mountains) and the southern San Joaquin Valley. Our data base in the Tejon area includes: (1) a grid of 8 proprietary industrial vibroseis lines, (2) a 34 km long CALCRUST deep-reflection vibroseis line, (3) a coincident 115 km long CALCRUST refraction survey, (4) several hundred electrical well logs, velocity logs, cores, drilling histories, biostratigraphic data and other borehole data, and (5) geological maps, including published maps by Dibblee [1973], Bartow and Dibblee [1981] and new, unpublished mapping. From these and other regional data we have begun to construct a model for the origin and evolution of the southern San Joaquin Basin.

Regionally, the origin and evolution of Cenozoic basins in southern California appear to reflect changing patterns of deformation along the North American Plate margin. Basins are viewed as having formed in response to: (1) the subduction of the largely consumed Farallon Plate, (2) crustal extension within the western Basin and Range Province, and (3) dextral shear within the plate boundary zone. Since Late Oligocene time, plate convergence has yielded progressively to a transform margin between the Pacific and North American Plates [Atwater, 1970; Engebretson et al., 1985]. The latter process has been accompanied by major "on land" tectonic events.

Almost 700 km of right-slip and distributed shear [Luyendyk, 1988, written communication] has deformed the terranes along the San Andreas and associated system of faults. With the northward passage of the unstable trench-transform-transform triple junction and lengthening of the San Andreas system came regional thermal events and volcanism [McKenzie and Morgan, 1969; Dickinson and Snyder, 1979]. The dextral shear has manifested itself in clockwise rotations of large fault-bounded blocks of crust [Luyendyk et al., 1980; Atwater, in press,1988].

To summarize and locate these events within the context of our project, we present a map (Figure 1) of the rotated blocks as inferred from paleomagnetic data, and a diagram (Figure 2) of the time-space paths of the plate boundaries, as discussed by Crowell [1987], to which we have added new geologic data from this study.

The active San Andreas and Garlock fault zones are shown as the western and southern boundaries of the Tehachapi Block (Figure 1). The dextral San Andreas transform fault is generally considered to be the boundary between the Pacific and North American Plates. The sinistral Garlock fault has been described as an intracontinental transform fault separating the western Basin and Range and Sierra Nevada to the north from the Mojave Block to the south [Davis and Burchfiel, 1973]. In the Tehachapi Mountains, the Garlock fault forms the boundary between deformed crystalline rocks once buried to depths of greater than 20 km and relatively shallow crustal granites and granodiorites [Sharry, 1981; Saleeby et al., 1987].

As indicated in Figure 2, the Miocene history of the San Andreas system appears to have been one of divergent strike-slip. Within a broad plate boundary composed of discontinuous right-lateral faults, depressions and uplifts formed along releasing and restraining bends (or fault jogs); and pull-apart basins may have evolved [Crowell, 1974 and 1987; Biddle and Christie-Blick, 1985; Sibson, 1985]. Sphenochasms between rotated crustal blocks and intra-block extensional sags may have formed during the block rotations [Luyendyk and Hornafius, 1987].

In central California, the opening of the Gulf of California at the end of the Miocene seems to have produced regional convergence, especially in the Transverse Ranges [Engebretson et al., 1985]. During this same time, much of the 64 km of left slip currently recognized on the Garlock fault took place [Hornafius, 1985; Carter, 1987]. In addition, the local uplift of the Tehachapi Block has taken place near the junction of these two great faults, exposing Cenozoic sedimentary and volcanic rocks along the margin of the San Joaquin basin and contributing to the present 14+ km of structural relief between the mountains and the presumed bottom of the deep Maricopa subbasin to the north (Figure 3). The Maricopa subbasin itself is about 12 km deep, as shown, for example, by a wildcat well, penetrating a structural high, that reached a depth of about 7 km (22,700 ft) without encountering rocks older than Middle Miocene.

The geological history of the relatively shallow Tejon Embayment serves as the key to our understanding of the evolution of the southern San Joaquin Basin, in view of this inaccessibility of the mid to lower Tertiary section in the Maricopa subbasin. Working out the sequence and timing of local expressions of volcanism, subsidence, unconformities, sedimentary facies changes and various styles of faulting is essential to reconstructing the evolution of this portion of California. It will be necessary, for example, to reconcile the timing of the proposed crustal rotations with the relatively undeformed Neogene strata at the Tejon Embayment and with the timing of strike-slip on the Garlock and San Andreas faults. The history and mechanics of the segmented White Wolf fault system, which forms the southern boundary of the Maricopa subbasin, must also be determined.

The Geology of the Tejon Embayment

Geologic Setting

The Tejon Embayment lies within the rotated Tehachapi Block shown in Figure 1. Existing paleomagnetic evidence shows that these rocks of the southernmost Sierra Nevada have been rotated clockwise as much as 60° since Late Cretaceous time [McWilliams and Li, 1985; Kanter and McWilliams, 1982]. Furthermore, the Lower Miocene "Tunis volcanics" (see below) are found to be rotated at least 30° clockwise [Plescia and Calderone, 1986], implying that there have been both Neogene and pre-Neogene rotations. There are no definite constraints on the youngest age of the rotations, although non-rotated 17 Ma volcanic rocks east of Tehachapi Valley (Figure 3) suggested to McWilliams and Li an upper limit to the rotations.

Great uplift and erosion of the region may have taken place during the early rotations, perhaps accounting for the approximately 45 m.y. nonconformity between the exposed lower crustal Cretaceous basement rocks and the overlying Eocene marine sedimentary rocks. Data now in hand suggest that the basement rocks north of the eastern White Wolf fault are not rotated. Thus, the White Wolf fault likely forms the

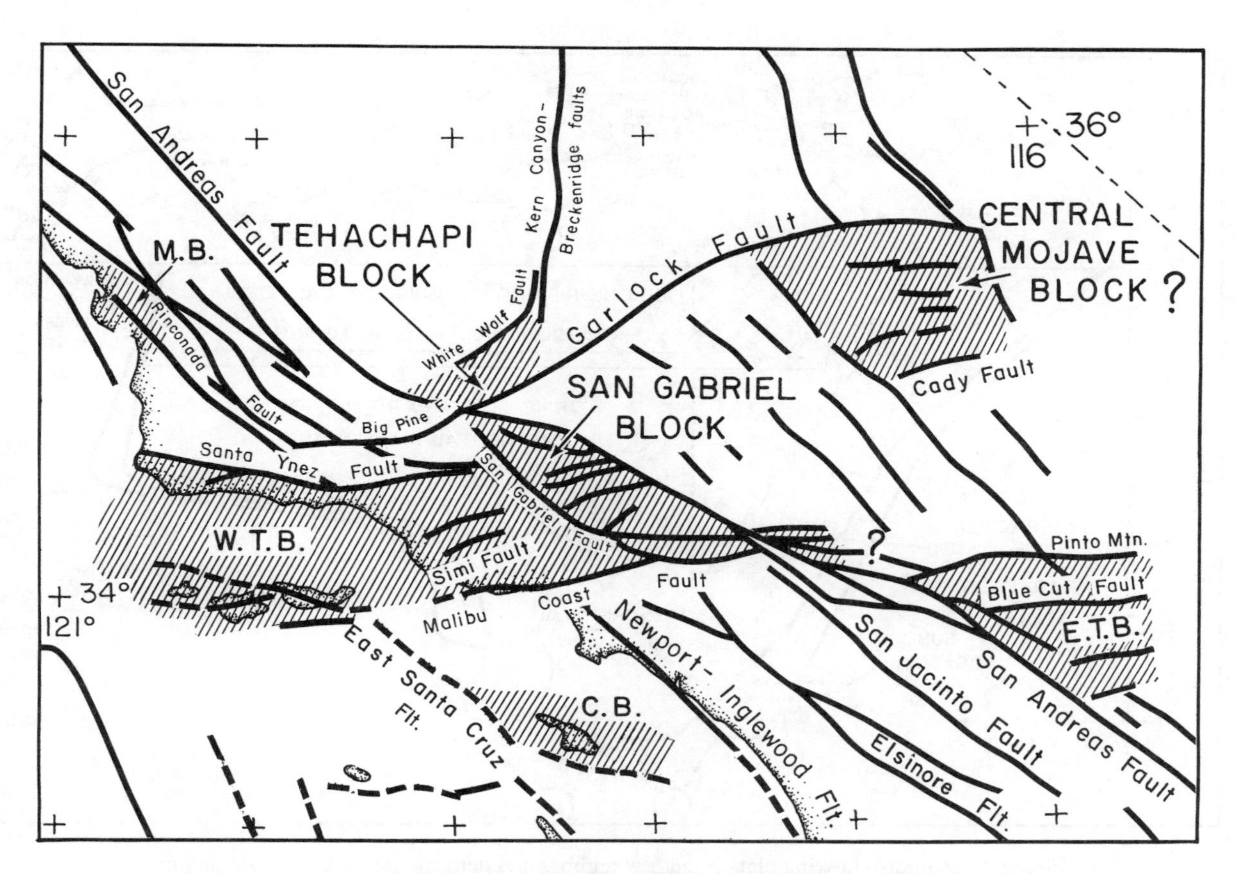

Figure 1. Regional structure map of southern California. The Tehachapi Block (described in this paper) includes the Tejon embayment, the Tehachapi Mountains and the San Emigdio Mountains, and lies between the San Andreas, Garlock and White Wolf faults. Large fault-bounded blocks with paleomagnetic declination anomalies imply clockwise rotation within a right-lateral shear couple [Luyendyk et al., 1980, 1985]. In this model, clockwise-rotated blocks extending to mid-crustal levels are currently bounded on the north and south by E-W- trending, sinistral, strike-slip faults. Paleomagnetic data suggest that crystalline rocks in the Tehachapi Block have been rotated at least 60° clockwise, possibly in more than one occurrence, since Late Cretaceous time. Rocks lying north of the White Wolf fault do not appear to be rotated. M.B. = Morro Block, S.G.B. = San Gabriel Block, E.T.B. = Eastern Transverse Block, W.T.B. = Western Transverse Block, C.B. = Catalina Block. [Figure modified from Luyendyk et al, 1985.]

tectonic boundary separating the terrane to the south from the unrotated terrane to the north. Moreover, undoing the approximately 30° Neogene rotations would align the White Wolf fault with the Kern Canyon-Breckenridge fault system in the Sierra Nevada, with which it may, in part, share a common history [Ross, 1986].

The majority of surface exposures at the Tejon Embayment consist of relatively flat-lying Quaternary alluvium. The gently tilted to overturned Tertiary section is exposed only in a narrow belt along the east, south and western embayment margins; furthermore, field investigations of these rocks are hampered by vegetative cover (Figure 3). These conditions make the use of subsurface data a necessity.

In general, the stratigraphy of the Tejon Embayment is characterized by rapid vertical and lateral facies changes [for stratigraphic columns and more detailed discussion, see, e.g., Blaisdell, 1984; Goodman and Malin, 1988]. The abrupt thickening of both sedimentary and volcanic units across buried normal faults has been reported [Hirst, 1988]. At least three major unconformities can be documented. Middle Eocene marine rocks are nonconformably buttressed against the Cretaceous igneous and metamorphic basement complex along the southern margins of the embayment. A second unconformity occurs near the top of the Late Eocene, and the third within mid-Miocene strata.

The Tejon embayment itself contains Eocene to Recent non-marine and marine sedimentary rocks, and a Saucesian (Lower Miocene) volcanic unit known as the Tunis volcanics. This latter unit consists of basalt and dacite flows, rhyodacitic breccias, tuffs and other volcaniclastic rocks. The volcanic-

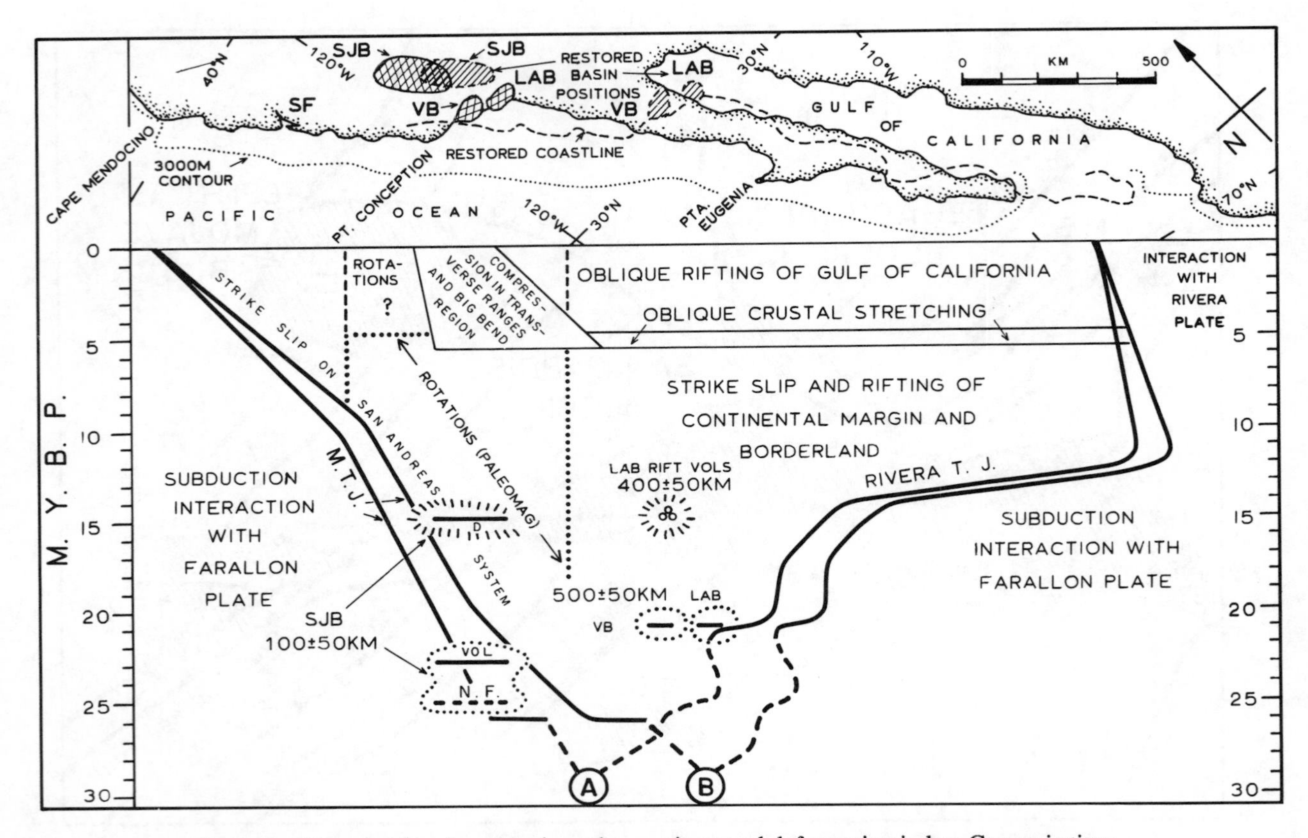

Figure 2. Diagram showing plate-boundary regimes and deformation in late Cenozoic time with respect to the California and Mexican coastlines, restored to their approximate positions about 10 Ma. The restoration is based on sea-floor magnetic anomalies and derived plate-circuit reconstructions [Atwater, 1970, Atwater and Molnar, 1973, and Atwater, in Crowell, 1987]. The present shoreline is shown for approximate geologic reference. The coordinates for the diagram are geography (horizontally) and time (vertically). Two favored time-position curves of the Mendocino and Rivera Triple Junctions show the timing and possible locations for the onset of interaction between the Pacific and North American plates. The 29 Ma date of this event and the locations of the present-day triple junctions are reasonably well established. Path A suggests that this breakup occurred off what is now southern California, whereas path B suggests that this occurred off northern Baja California. This initial interaction between the Pacific and North American Plates was followed by tectonic stretching and subsidence, including basin formation. Only after several millions of years did the San Andreas transform fault begin to lengthen between the northward-migrating Mendocino and southward-migrating Rivera Triple Junctions. The geologic time and geographic locations lying within these paths of plate interaction imply positions within the evolving strike-slip regime at a given time. In the upper diagram, the Ventura, Los Angeles and San Joaquin Basins are shown both as presently located (cross-hatched) and palinspastically restored (shaded). In the lower diagram, the times and locations of the origins of these basins are shown as dotted ellipses based on data described by Crowell [1987] and data presented in this study. The origin of the San Joaquin Basin may have been outside of this boundary or, alternatively, the timing of this event may locate the triple junction and thus favor Path A. The approximate time of deepening for the Tejon embayment is denoted by 'D' within the dashed area, and lies well within the strike-slip regime. See text for discussion. Other symbols: SF=San Francisco, LAB = Los Angeles Basin, VB = Ventura Basin, SJB = San Joaquin basin, M.T.J. = Mendocino Triple Junction, N.F. = onset of normal faulting at Tejon embayment (estimated), VOL = extrusion of Tunis volcanics at the Tejon embayment. [Figure modified from Crowell, 1987.]

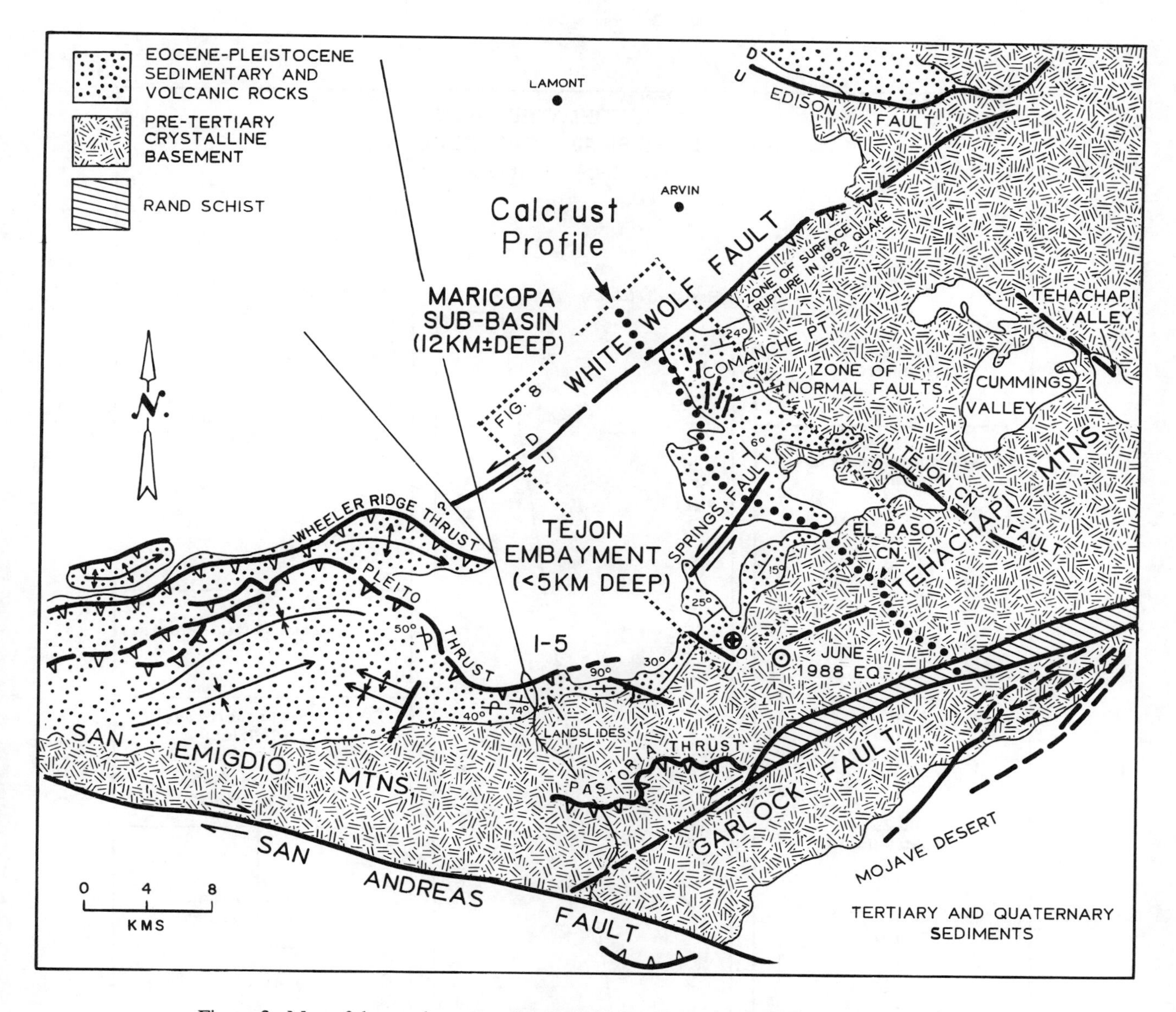

Figure 3. Map of the southernmost San Joaquin Basin, including the Tejon and Maricopa subbasins. showing generalized surface geology, major structures and the location of the CALCRUST reflection survey (dotted line). The non-shaded areas are generally flat-lying alluviated regions where subsurface data have been utilized. The subsurface structure of the area bounded by the dashed rectangle is shown in Figure 8. The 22,700 ft (6880 m) deep well discussed in the text is located within this rectangle; see Figure 8 for a more precise location. The surface expression of the Pleito Thrust appears to die out eastward, but buried thrusts inferred from subsurface data, discussed in this paper, are present to the east and northeast.

flow rocks have calc-alkaline petrologic affinities; their initial 87Sr/86Sr ratios were relatively high; and they appear to be derived both from crustal melts and from more primitive melts that underwent crustal assimilation [Johnson and O'Neil, 1984].

Outline of Embayment History

The oldest sedimentary rocks in the Embayment belong to the fossiliferous Lower and Middle Eocene marine Tejon Formation, which was deposited during a major west-to-east

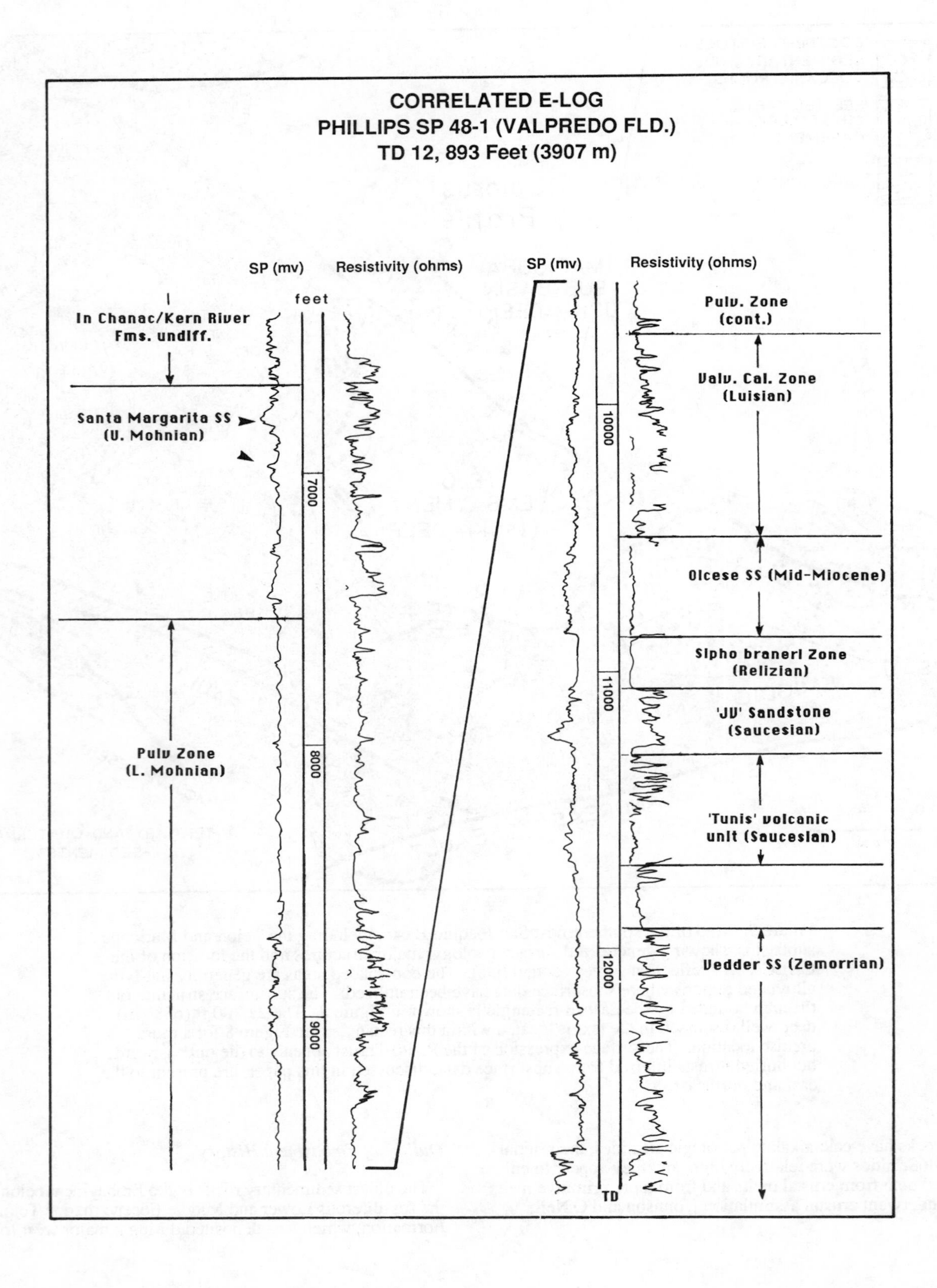
CORRELATED E-LOG
PHILLIPS SP 48-1 (VALPREDO FLD.)
TD 12, 893 Feet (3907 m)
SP (mv)
Resistivity (ohms)
SP (mv)
Resistivity (ohms)
feet
In Chanac/Kern River
Fms. undiff.
Santa Margarita SS
(U. Mohnian)
Pulv Zone
(L. Mohnian)
7000
8000
9000
10000
11000
12000
TD
Pulv. Zone
(cont.)
Valv. Cal. Zone
(Luisian)
Olcese SS (Mid-Miocene)
Sipho braneri Zone
(Relizian)
'JV' Sandstone
(Saucesian)
'Tunis' volcanic
unit (Saucesian)
Vedder SS (Zemorrian)

marine transgression [Nilsen, 1973; Nilsen et al., 1973]. No forearc basin sediments of the Cretaceous to Paleocene Great Valley Sequence, which are preserved to the north and south, have been recognized. If these sediments were not deposited into the Tejon Embayment, then this portion of the San Joaquin Valley was not a part of the extensive, elongate forearc basin which lay to the west of the Sierran magmatic arc [see Nilsen, 1987]. However, this is not a necessary conclusion, since there may have been significant erosion, displacement, or structural attenuation of the Great Valley Sequence in what is now the Tejon Embayment. Significant erosion at the Tejon Embayment in Eocene time is indicated by the presence of Lower Eocene marine rocks and Paleocene non-marine rocks in the Goler Basin, over 100 km to the east [Cox, 1987].

By Early Miocene time, the remnant forearc basin had evolved into a deep Neogene marine basin in the southern San Joaquin, as suggested by paleobathymetry inferred from benthic foraminifera [Bandy and Arnal, 1969]. This transition involved normal faulting and volcanism and apparently coincided with the transformation from the Oligocene broad shelf-slope deposystem to a silled, rugged Early Miocene basin with a narrow shelf and a steep slope [Hirst, 1986]. At this time, and also continuing into the Middle Miocene, the basin was filled by conglomeratic turbidites, whose courses apparently were controlled by the irregular basin topography [Hirst, 1988]. By the early part of Late Miocene time, a broader shelf again existed in the region of the Tejon Embayment, and fan-channel sedimentation began to wane [Macpherson, 1978].

In the Late Miocene, the modern San Andreas fault began to truncate the southern San Joaquin basin and to displace the southwestern portion of the basin. The Tejon and Maricopa subbasins began closing during the Pliocene, and were eventually isolated from marine waters [Addicott, 1968; Crowell, 1987]. The Plio-Pleistocene to Recent history of the Tejon Embayment is characterized by reverse and oblique-slip faulting, rapid uplift, alluvial and fluvial sedimentation, and the reworking of exhumed Tertiary units. Today there is pronounced relief along the embayment margins, as rapid uplift and erosion continues [e.g., Seaver et al., 1986]. South of the eastern Pleito Thrust, there is a 14% topographic slope difference between the top and bottom of the Tertiary section, which are folded and overturned (Figure 3).

Cross Sections and Embayment Structure

Our interpretation of the seismic reflection data at the Tejon embayment was facilitated by dense well control. More than one hundred logs and well histories from wells as deep as 4000 m were used in preparing nine seismic-based structural cross sections of this area. (For reasons of brevity and clarity, the areal distributions of wells are not presented in this paper; they are shown in California Division of Oil and Gas Maps W 4-2, 430 and 432.) The major correlation markers are: (1) the top of the Upper Miocene Santa Margarita Formation sandstone, (2) Middle Miocene shale markers, (3) the top of the Lower Miocene Tunis volcanics, and (4) the top of the Cretaceous crystalline basement. The markers were then related to the reflection sections using available velocity surveys. No "typical" log exists for this area, but a sample of a correlated electric log is shown in Figure 4. As Figure 4 shows, the California benthic foraminiferal stages correlate well with distinctive rock sequences [Kleinpell, 1938, in Blaisdell, 1984; Hirst, 1988], and other paleontological data are useful in recognizing and correlating lithostratigraphic markers.

Crossing seismic profiles show that the relatively thin Paleogene section is overlain by a Neogene section that thickens significantly to the west of Comanche Point and in general thickens to the northwest (e.g., Figures 5 and 6). The Tunis volcanics appear in all profiles, and are characterized on logs by an inhibited spontaneous potential response, with high resistivities in flows and volcaniclastic facies. Low resistivities characterize the interbedded tuffaceous shales and clays (Figure 4) [Hirst, 1988].

The structure of the subsurface is shown by our cross sections (Figures 5-7) and derived structure map (Figure 8; which also shows the locations of the cross sections). The latter shows that within a relatively small area, normal, thrust and oblique-slip faults all occur. This is very different from the gentle surface topography and geology of today's alluviated central embayment. Going west to east and north to south, we find that the central embayment is broken into a northeast-trending central graben and associated normal faults. The graben system is truncated by the White Wolf fault to the north, and may have been displaced both by strike-slip and by dip-slip. The largest normal faults in this system have vertical separations exceeding 1200 m. Most of the Tertiary section and the crystalline basement are displaced by these faults. Though the faults can be traced down into the basement, it is difficult to ascertain how deep they extend.

These same fault relationships, with equivalent amounts of separation and the age of the section offset, are seen in outcrop on the southern rim of the basin (Figure 3). Both these faults and those in the subsurface show normal separation, but they may also have a component of strike slip.

On their eastern and southern ends, near the basin margins, the reflection profiles show basinward thrusting and associated anticlinal folds. These blind thrusts occur immediately adjacent to the normal faults and appear to overlie them structurally. The crystalline basement is also offset by the thrusts. The basinward transition from thrust to normal faults is observed on all of the relevant seismic profiles and in the

Figure 4. (Opposite) Correlation of electric log from the lower portion of the Phillips SP 48-1 well, Valpredo Field (see Figure 8). Borehole depths are shown in feet. To the left of the borehole is the spontaneous potential curve; to the right of the borehole is the resistivity curve. The California benthic foram stages of Kleinpell [1938, in Blaisdell, 1984] are shown in parentheses. The ages of these stages are as follows: Mohnian = Middle and Late Miocene; Luisian and Relizian = Middle Miocene; Saucesian = Early Miocene; Zemorrian = Oligocene and earliest Miocene (where base Miocene is about 23 Ma). The Chanac Formation, lying immediately above the Santa Margarita Formation, is Late Miocene in age [Bartow and McDougall, 1984]. Note the inhibited S-P/high resistivity responses of the volcanic unit. Normal faults have affected the "thicknesses" in this well.

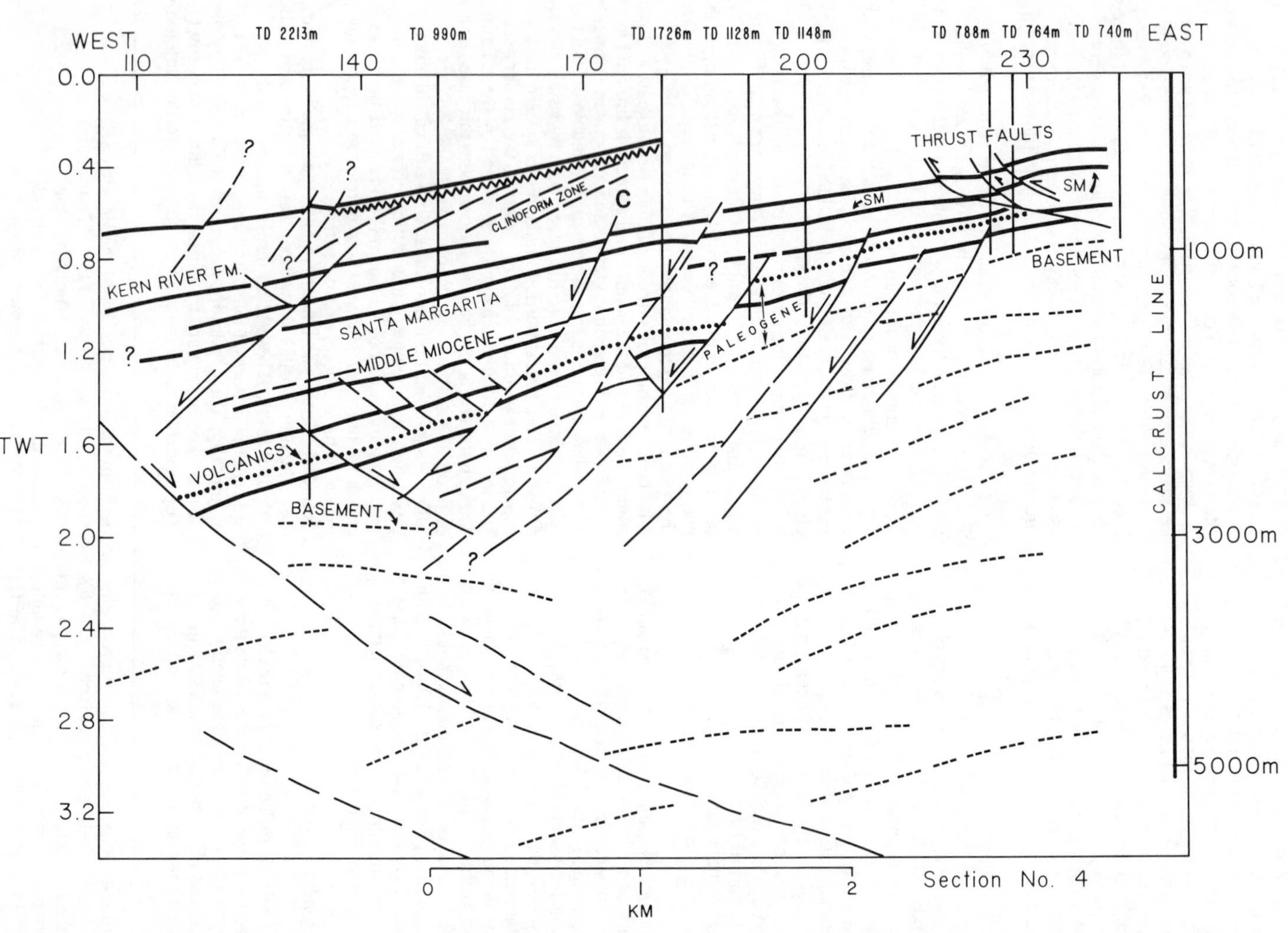

Figure 5. Cross section based on an interpretation of seismic line 4 in Figure 8. Buried normal faults exist from the central embayment to the west to the basin margin on the east, where they are structurally overlain by younger thrust faults. Top of basement and underlying crystalline rock reflectors are indicated by short dashed lines. The depth to top of basement is poorly known in the center of the graben. The Santa Margarita (SM) Formation is Late Miocene in age. The age of the Kern River Formation is poorly known, but may range from Late Miocene to Pleistocene [Bartow and McDougall, 1984]. The Lower Miocene volcanic unit is indicated by a dotted line. Depths shown on the side are generalized from velocity surveys. Some of the wells utilized are shown with their total depths. Normal fault "C", discussed in text, is interpreted as a growth fault.

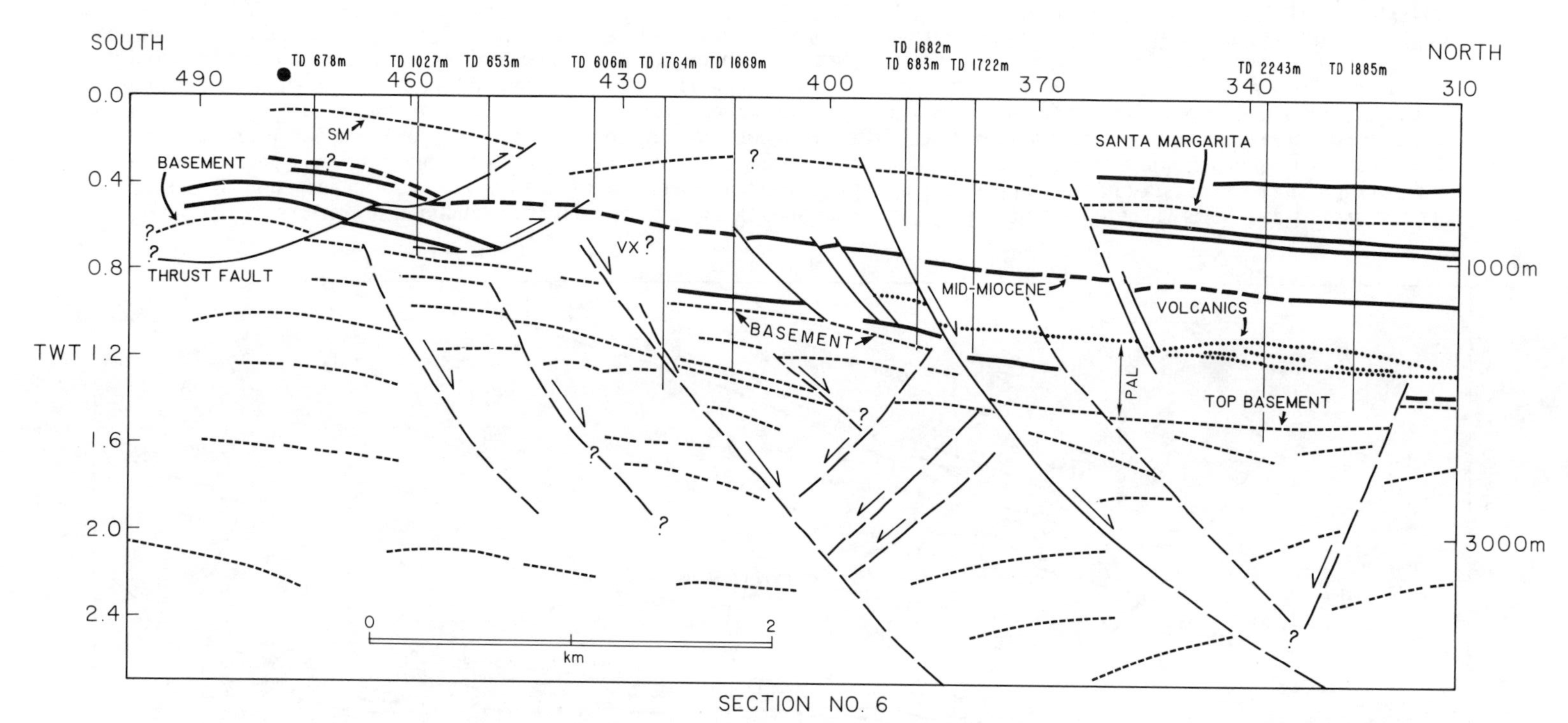

Figure 6. Cross section as in Figure 5, based on the southern portion of line 6 in Figure 8. The updip limit of the dotted volcanic unit ("vx") is not well defined. The top of the Santa Margarita Formation ("SM") is shown as a dashed line. Younger thrust faults at the south end of the profile structurally overlie the normal faults. The dot represents oil production at Tejon S.E. Field (from a fold on a hanging wall block). PAL = Paleogene.

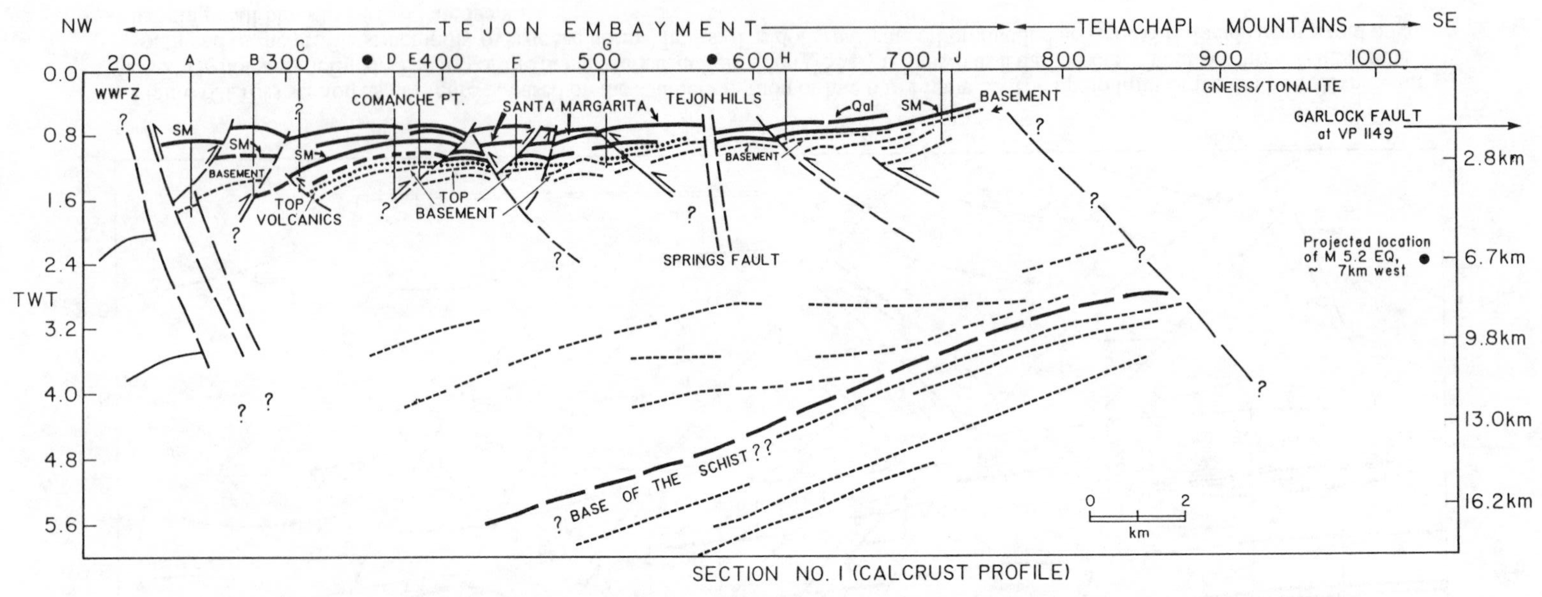

Figure 7. Cross section as in Figure 5, based on the middle portion of the CALCRUST profile, preliminary stack (line 1 in Fig. 8). The Tertiary section along the embayment margins is extremely thin, and is broken by high and low angle reverse faults. Dots indicate the location of the Comanche Point and Tejon Hills oil fields. The dotted line represents the Lower Miocene volcanic unit. Basement reflectors are shown by dashed lines, the uppermost line being top of basement. The heavy dashed line in the middle crust may represent the base of the Rand Schist and/or a detachment surface. Alternatively, this and other north-dipping deep reflectors are truncated by an unmapped, range-bounding fault (queried). The preliminary location of the M 5.2 "Tunis Canyon" earthquake of June 10, 1988, is projected onto this section and suggests a north-dipping, oblique-slip structure (see text). Note faults with reverse separation which branch south from the White Wolf fault. Depths of wells utilized are as follows: (A) 1418 m (4680 ft), (B) 991 m (3270 ft), (C) 987 m (3257 ft), (D) 1212 m (4000 ft), (E) 696 m (2296 ft), (F) 670 m (2210 ft), (G) 712 m (2350 ft), (H) 304 m (1003 ft), (I) 487 m (1608 ft), and (J) 149 m (492 ft). The depths shown on the side label are based on the interval velocities of crystalline rocks. SM = Santa Margarita, Qal = Quaternary alluvium, VP = vibrator point.

borehole data. A series of contractional features are seen in the cross section shown in Figure 7; the trace of this section crosses the eastern embayment margin between vibrator points (VPs) 220-740.

The blind thrusts discussed above are in marked contrast to the pronounced scarps of the Pleito fault to the west and of the eastern half of the White Wolf fault. The buried thrusts may also be active, but because the recurrence intervals of reverse faults are an order of magnitude higher than the nearby San Andreas fault, slip may only occur on the Tejon thrusts on the order of every 10,000 years [Yeats and Berryman, 1987]. Furthermore, in a broad sense, the folds on the hanging wall blocks of the buried thrusts explain the accumulations of hydrocarbons at Tejon Southeast Field, Comanche Point Field and Tejon Hills Field (Figures 6, 7 and 8), although these fields have complex trapping configurations beyond the scope of this paper.

One puzzling aspect of the fault relationships in this area has been the apparent sandwiching of the thrust faults between the central graben system, which is made up of subsurface normal faults, and a second system of normal faults exposed along the southern margin of the embayment and at Comanche Point (Figure 8). The exposed normal faults do not appear to be active and have normal separations similar to those seen in the buried central embayment fault system. For example, at exposed faults where the north-dipping Oligocene and Lower Miocene sections are exposed along the southern embayment margin, the horizontal fault-separation of the Oligocene-top of basement unconformity exceeds 1400 m. We assume that this horizontal fault-separation is primarily due to normal slip. Where the younger, less lithologically distinctive Miocene section is exposed in scattered outcrops at Comanche Point, vertical separation greater than 25 m cannot be demonstrated across normal faults ("b" in Figure 7). The relatively small amount of separation of these Comanche Point faults is consistent with the up-section decrease in vertical separation seen in the cross sections.

Thus, the exposed normal faults at the basin margins are probably exhumed members of the buried central Embayment faults. Seismic data show that the Comanche Point normal faults are located on the hanging-wall block of a basin-verging thrust. In addition, where the profile shown in Figure 7 hugs the western margin of Comanche Point, this blind thrust was encountered in the form of a compressional wedge at about VP 380. Beneath this wedge lies the inferred thrust (marked "a" in Figure 7), which we infer dips to the east under Comanche Point.

The relief, rapid dissection, and folding of the exposures of Tertiary strata at Comanche Point are apparently due to thrust-generated uplift, as is Wheeler Ridge at the western end of the Embayment (Figure 3). Wheeler Ridge anticline is a fault-bend fold which formed at the tip of an active, north-verging wedge above the Wheeler Ridge Thrust system [Medwedeff, 1988]. Recent quarrying at Wheeler Ridge has exposed several of these "blind" thrusts, whose existence was predicted many years ago on the basis of subsurface data [Davis, 1986].

The Cenozoic section and the underlying basement are offset at the prominent sub-vertical discontinuity known as the Springs fault (Figure 7; VP 575). This fault is presumed to be active based on a 6 m-high fault scarp at Tejon Hills Field (Figure 8). The Springs fault is sub-parallel to both the White Wolf and Garlock faults, and in fact has also been described as an oblique-slip fault [Bartow, 1984]. Indeed, several cross sections indicate the presence of "palm tree" structures branching out from the Springs fault (lines 2 and 9 in Figure 8; Goodman and Malin, 1988). Hirst [1986] has suggested that the Springs fault may be a reactivated normal fault.

The White Wolf fault zone appears to consist of high-angle, multiple strands west of Comanche Point, where it is associated with branching faults showing reverse separation and back thrusts to the southeast (Figure 7). North of Comanche Point, the White Wolf fault forms a prominent escarpment that reflects surface rupture during the 1952 Arvin-Tehachapi and earlier earthquakes [Buwalda and St. Amand, 1955; Bartow and Dibblee, 1981].

Interpretation of CALCRUST Deep Reflection and Refraction Data

Several preliminary results from the CALCRUST deep-reflection and refraction profiles are relevant to our discussion of the Tejon Embayment and its connection to the crustal blocks to the north and south [Goodman and Malin, 1988; Ambos and Malin, 1987]. The location of the reflection profile is shown in Figures 3 and 8, and an interpreted brute stack section is shown in Figure 7. The location of the refraction survey is shown in Figure 9, with the corresponding forward model and interpretation shown in Figures 10 and 11.

The CALCRUST reflection profile was designed to cross more or less normal to the strikes of the White Wolf, Springs, and Garlock faults and to image deep crustal structures. This line was shot with 6 heavy vibrators and recorded on 33 m station spacings with a 400 channel optical fiber recorder. Vertical stacks of 4 to 8 rolled sweeps, from 8 to 32 hz over 32 seconds, were used with a 12 s listen. The number of channels recorded and the VP interval were allowed to vary, resulting in a CMP-stacking fold of 40 to 80.

The most important mid-crustal reflection horizon seen so far in the CMP data lies in the central portion of the profile, and dips north between depths of 10 to 15 km. This horizon may correlate with a highly reflective horizon in the Rand Schist, a section of which is exposed between the north and south branches of the oblique-slip Garlock fault (Figures 3 and 7). The north branch of the Garlock fault, which probably dips at a moderate angle to the north, separates the Rand Schist from the overlying tonalites and gneisses, with mylonites mapped along the fault zone [Sharry, 1981]. Therefore, these field relations may persist into the mid-crust. The preliminary epicenter of the M 5.2 "Tunis Canyon" earthquake of June 10, 1988, has been placed some 4 km north of the North Branch of the Garlock fault and 6-7 km west of the CALCRUST profile ("bull's eye" in Figure 3) [D. Given, U.S.G.S., personal communication, June 17, 1988]. The focal mechanism is apparently left-lateral with a component of thrusting (rake=35°). The trend of the preliminary focal plane solution is N 81° E, dipping 51° north, and the locations of aftershocks apparently define a north-dipping plane. Thus, the structure associated with this earthquake may correlate with one of the north-dipping structures identified in the mid-crust (Figure 7). Alternatively, the main reflector may be a tilted mid-crustal detachment surface, where rocks of disparate densities and velocities have been juxtaposed.

The point here—supported by the more regional scale refraction data—is that high velocity materials are rapidly coming up to shallow crustal depths south of the White Wolf fault. Although signal processing of the lower crustal data

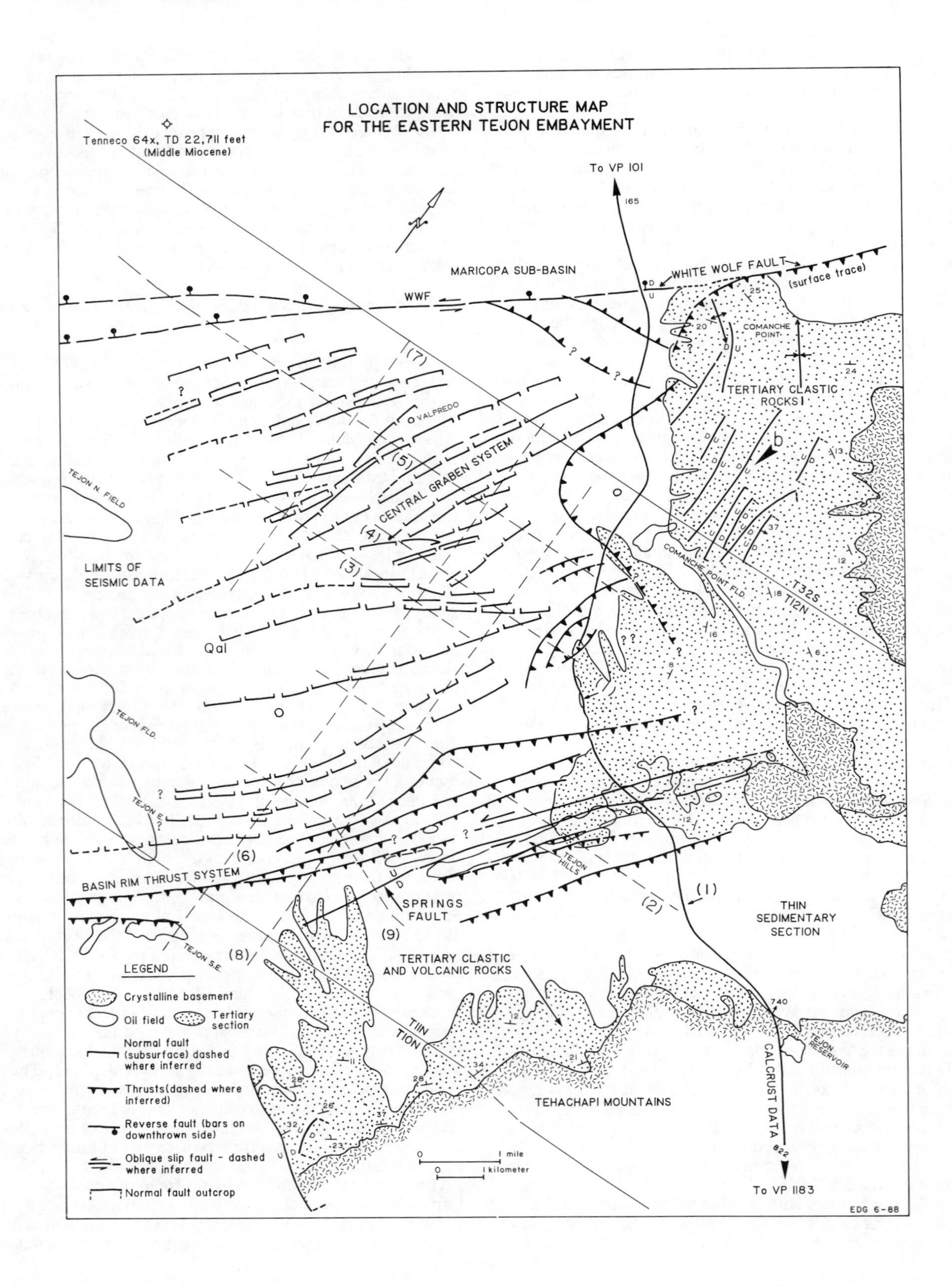
LOCATION AND STRUCTURE MAP
FOR THE EASTERN TEJON EMBAYMENT
Tenneco 64x, TD 22,711 feet
(Middle Miocene)
To VP 101
MARICOPA SUB-BASIN
WHITE WOLF FAULT
(surface trace)
WWF
COMANCHE POINT
TERTIARY CLASTIC ROCKS
VALPREDO
CENTRAL GRABEN SYSTEM
TEJON N. FIELD
COMANCHE POINT FLD.
T32S
T12N
LIMITS OF SEISMIC DATA
Qal
TEJON FLD.
TEJON E.
BASIN RIM THRUST SYSTEM
TEJON HILLS
SPRINGS FAULT
THIN SEDIMENTARY SECTION
TEJON S.E.
TERTIARY CLASTIC AND VOLCANIC ROCKS
T11N
T10N
TEHACHAPI MOUNTAINS
TEJON RESERVOIR
CALCRUST DATA
To VP 1183
LEGEND
Crystalline basement
Oil field
Tertiary section
Normal fault (subsurface) dashed where inferred
Thrusts (dashed where inferred)
Reverse fault (bars on downthrown side)
Oblique slip fault - dashed where inferred
Normal fault outcrop
1 mile
1 kilometer
EDG 6-88

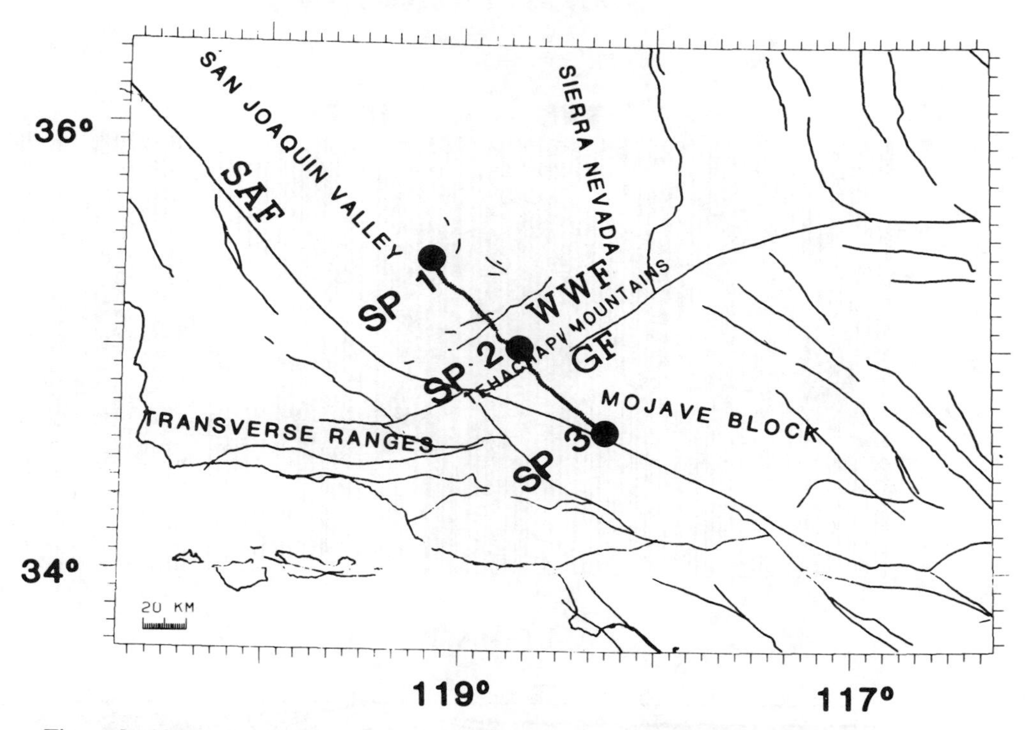

Figure 9. Shot point ("SP") and receiver station map of the CALCRUST refraction profile connecting the southern San Joaquin Valley to the western Mojave Desert. The end point shots were located west of Bakersfield and west of Quartz Hill. The central shot was within the Tehachapi Mountains. The total profile length was 115 km, which was covered with an average station spacing of 1.0 ± 0.2 km. Because of this density of coverage, the triangular station location symbols have overlapped to plot as a single broad line. SAF, WWF, GF = San Andreas, White Wolf and Garlock faults.

remains incomplete, the discontinuous reflections there appear to echo the northward-dipping mid-crustal structure. Earthquake-sourced refraction data across the San Andreas, Garlock, and White Wolf faults suggest that the upper crust of the Tehachapi block is composed of grossly faster geologic materials than the crust to the north and south [Hearn and Clayton, 1986a].

Our dynamite-sourced refraction data were collected with the 120 recording-unit system developed by the United States Geological Survey, Menlo Park [Blank et al., 1979]. Three

Figure 8. (Opposite) This is both a location map of the seismic grid used in this study (dashed lines) and a structure map constructed at the Lower Miocene level. The outcrop patterns of older crystalline rocks and Tertiary sedimentary and volcanic rocks are indicated; unshaded areas are alluvium. The well locations and vibrator-point numbers are not shown on the map for reasons of clarity (selected wells are located on the cross sections [Figs. 5-7] which were derived from lines 4, 6 and 1 respectively). Numbers in parentheses refer to reflection line numbers. The Tertiary outcrop pattern is modified from Bartow and Dibblee [1981] and Dibblee [1973]. The limits of oil fields are indicated: Tejon Field, Tejon Hills Field, North Tejon Field and Comanche Point Field, along with Wheeler Ridge Field (located to the west of the area shown in this map) have collectively produced over 110 MMBO. The borehole data and velocity surveys used in this study are from the wells drilled within and around these fields. The heaves of the buried normal faults are shown where applicable. The normal faults which crop out at the basin margins (see legend) are interpreted as having been exhumed by the thrust faults rimming the embayment (see text for discussion). The White Wolf fault (WWF) is segmented near Comanche Point.

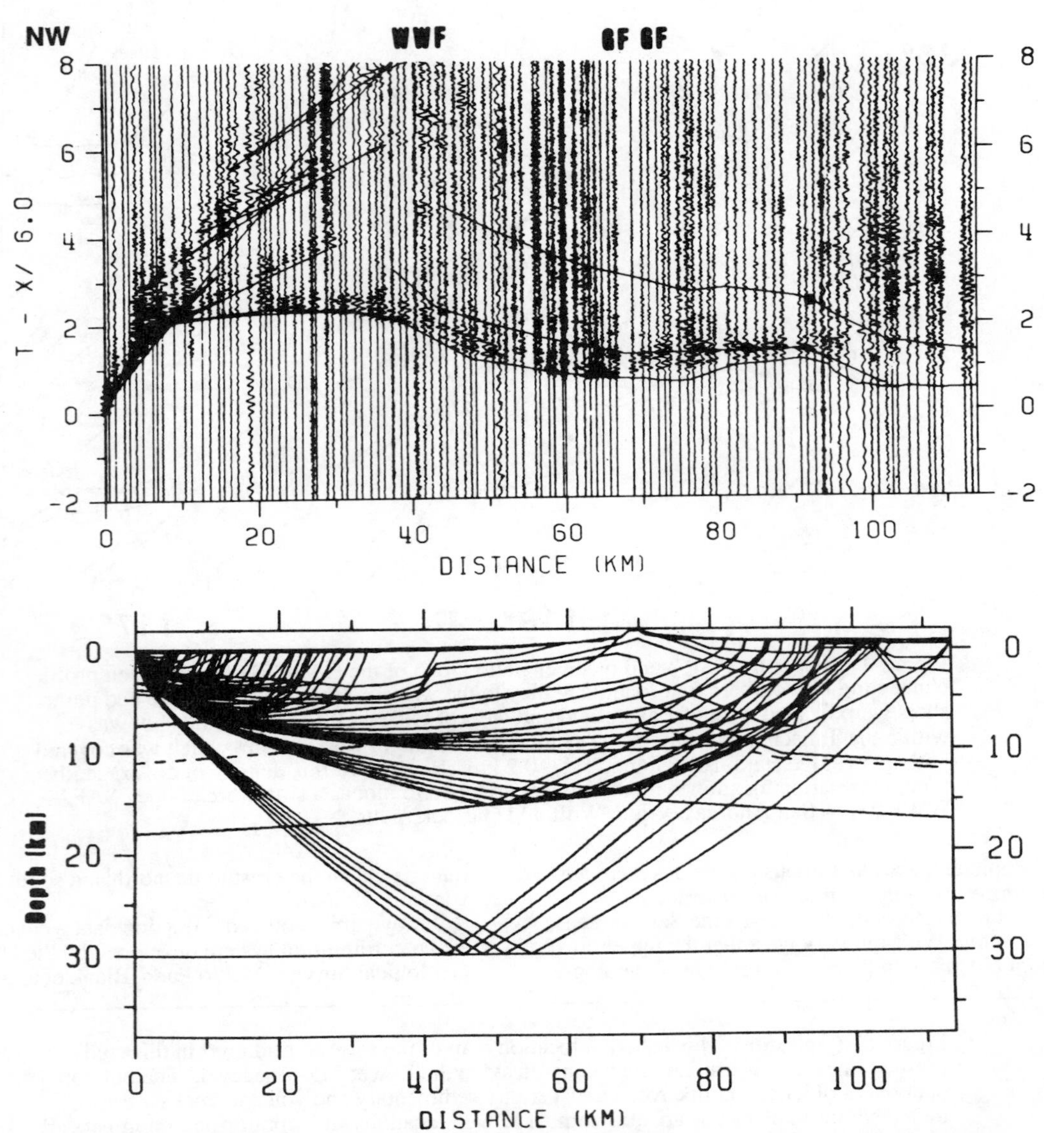

Figure 10. Shown on the top is a reduced travel-time record section of seismograms from shot point 1 on the north end of the refraction profile. The reduction velocity was set equal to 6.0 km/sec. The surface locations of the White Wolf fault and the north and south branches of the Garlock fault are shown at the top of the section. The thin lines on the section show theoretical travel times, which are based on our current best fit forward model of the data. Below the record section is a cross sectional model which shows the velocity structure and ray paths corresponding to the theoretical travel times in the record section. The profile design resulted in relatively good sampling of the upper half of the crust. On the far north end of the profile, several branches of reflected refractions in the low velocity basin sediments can be seen in the data and model. In the Tehachapi Mountains portion of the line, a strong increase in apparent velocity exists. South of the Garlock fault, this increase is reversed.

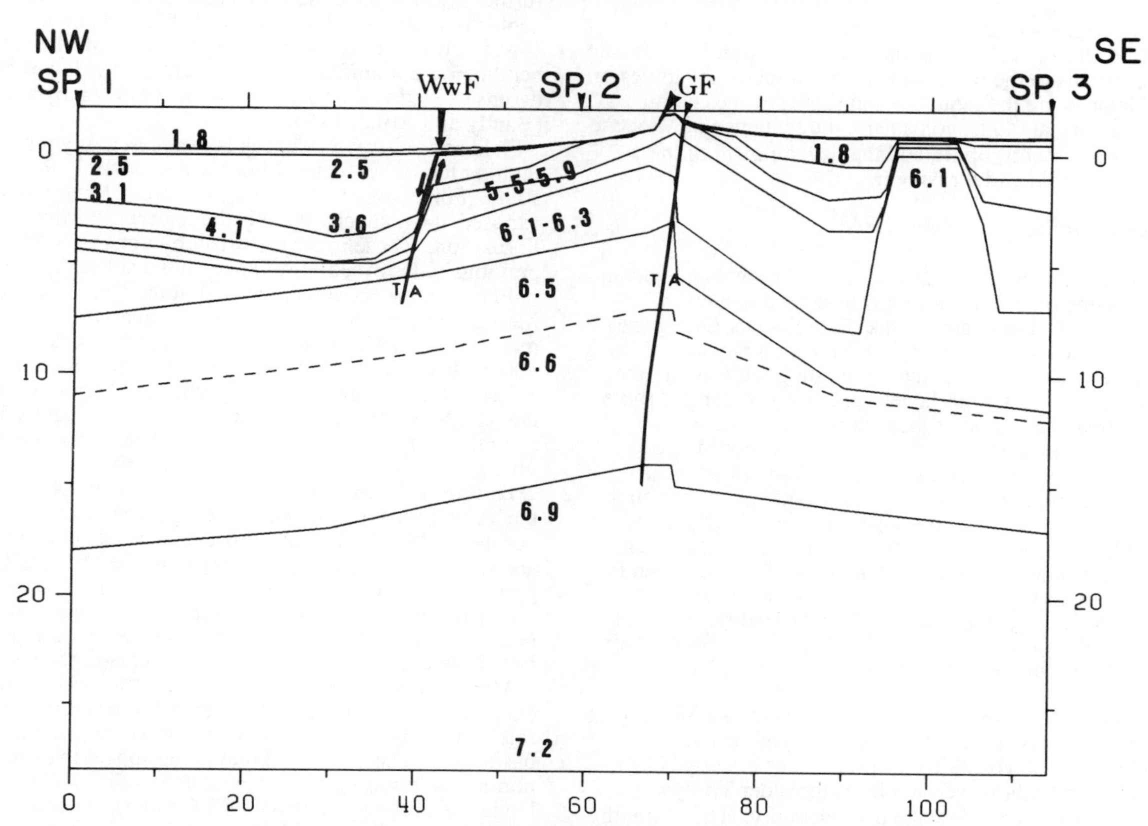

Figure 11. Velocity model for the 115 km Tehachapi refraction profile. Interval velocities are in km/sec and depths are shown along the side in kms. Note high velocities at shallow depths in the Tehachapi/Tejon Block between the White Wolf and Garlock faults (WWF and GF), where the velocity structures dip to the north. North of the White Wolf fault is the Maricopa subbasin, with a thicker sedimentary section and a thinner upper basement section. Assigning a northward-dip to the White Wolf fault west of Comanche Point may explain the pronounced kink in the velocity profile, but reflection data suggest that this fault dips southeast at shallow depths (see text). Note the isolated basement block in the Mojave Desert, south of the Garlock fault. (This block is currently not well constrained by the data.)

shot points were recorded into a single layout of the spread at slightly less than 1.0 km station spacings. Over the route of the reflection profile, the refraction stations were set up on the VP survey points. The forward modeling described below made use of the ray-tracing code of Cervený et al. [1977].

The first-order travel time "fits" to our data show that the refraction profile design resulted in good sampling of the upper half of the crust (Figure 10). The data show the apparent velocity increase resulting from the difference in the thickness of the Maricopa and Tejon subbasins. Furthermore, north-dipping high velocity layers are also seen. In fact, the apparent velocity continues to increase until the Garlock fault is encountered, but south of this fault the crust is significantly slower. Our current crustal velocity model accounts for these features and is shown in Figure 11.

The velocity contrast across the Garlock fault does not appear to extend into the lower crust [Hearn and Clayton, 1986b]. As suggested by Hearn and Clayton [1986a, b], the velocity model shows that the profound discontinuity at this fault extends to mid-crustal depths. Somewhat surprisingly, on the north side of the Tehachapi Block the refraction data require steep north-dipping velocity contours at relatively shallow depths north of the surface trace of the White Wolf fault. In the reflection shot gathers acquired north of the White Wolf fault, there exists a strong back reflection which is not seen for equivalent source positions to the south. This implies a northward-dipping velocity structure. The CMP data from this section of the line (Figure 8) is highly confused, as one would expect from the locally complicated structure. As can be seen from the travel time fits south of the Garlock fault, the current model is too fast there and the large plug of high velocity material northwest of shot point 3 is not well constrained. This particular aspect of our model requires further work.

Thoughts on the Evolution of the Southern San Joaquin Basin

Our observations in the Tejon Embayment have direct implications for understanding the adjacent Maricopa subbasin and other basins to the north and south. Some of the topics of especial interest are the sequence and timing of the central graben and normal faults, volcanism, the sequence of Miocene deep water sediments, uplift, and the development of the young, active, fold and thrust belt.

The Ages of Faulting and Subsidence

Normal faulting at the Tejon Embayment probably began in Late Oligocene time. This timing is best documented by outcrop relations. For example, the (Late Eocene-Oligocene) Tecuya Formation clastic rocks are thicker on the downthrown side of several exposed, northwest-trending faults along the southern Embayment, with facies generally coarser and more angular across the largest of these faults (circled cross in Figure 3). Across this fault, the younger, Lower Miocene Tunis volcanic section, dated at 22.7 ± .4 Ma [J. Plescia, written communication, 1987], is almost 300 m thicker on the downthrown side than it is on the upthrown side [Dibblee, 1973]. Paleomagnetic data, based on sampling of the volcanic flows, suggest that only normally polarized fields exist on the upthrown side of this same fault, whereas the thicker, downthrown section shows reversely polarized fields which are overlain by normally polarized fields [Hirst, 1986]. This implies that faulting was synchronous with the volcanic extrusions.

This same fault is overlapped by by a laterally extensive mid-Miocene (post-Saucesian, pre-Delmontian) erosional unconformity. The volcanic unit usually lies directly below this unconformity; however, in places the older Tecuya Formation clastics lie below the unconformity. Therefore, the inference that the reversely-polarized volcanic rocks were laid down before the normally-polarized rocks is critical, because it answers the argument that the volcanic section on the upthrown block is thinner solely due to erosion associated with this unconformity. That is, the reversely-polarized sections could not have been eroded from the upthrown block without also stripping away the overlying normally-polarized section.

Thus, we conclude that normal faulting was synchronous with the deposition of both Oligocene and Lower Miocene rocks. This normal growth-fault interpretation assumes little or no early right-lateral slip.

Based on the seismic sections, it appears that the major subsidence in the Tejon area associated with the normal faulting began after the extrusion of the Tunic volcanics, that is, in late Early and Middle Miocene time. For example, in Figure 5 an abrupt thickening of Miocene section occurs across the fault denoted by the letter "c", which we interpret as primarily the result of greater deposition on the downthrown block rather than as a later oblique-slip that juxtaposed unlike sequences. In general, the greatest thickening occurs in the section between the top of the Lower Miocene volcanic reflection and the Middle Miocene reflection. This increased rate of subsidence may also have been the result of a short period of post-volcanic cooling that ended with the next episode of faulting. Similarly, large Middle Miocene subsidence, occurring roughly between 16 and 14 Ma, has also been documented both in the Maricopa subbasin and also further north in the central San Joaquin Basin, using geohistory analysis [Olson et al., 1986; Graham and Williams, 1985]. Furthermore, paleontologic studies of Miocene benthonic foraminifera suggest that middle to lower bathyal depths existed in the southern San Joaquin Valley at this time [Bandy and Arnal, 1969].

Field and seismic evidence also suggest that younger normal faulting has offset Upper Miocene and possibly Pliocene strata. For example, the youngest pre-Quaternary unit exposed at Comanche Point is the uppermost Miocene Chanac Formation. These rocks are offset by normal faults, as previously described. Similarly, in the subsurface, as shown in Figure 5, the section composed of the Upper Miocene Santa Margarita Formation and some of the overlying younger non-marine section are offset. Borehole data show that greater than 500 meters of vertical separation across the normal faults exists as high as the Santa Margarita level, with expansion of the post-Santa Margarita section. The normal faults in the central embayment do not reach the surface, and the age of the youngest strata offset has not been determined.

To summarize, we recognize three episodes of extension at the Tejon Embayment: the Late Oligocene-Early Miocene normal faulting, volcanic extrusives, coarse volcanic breccias and igneous dikes; the late Early to Middle Miocene time of great subsidence, with expansion of the sedimentary section across normal faults and deep-water marine sedimentation; and relatively minor normal faulting which began during or after Late Miocene time and reactivated the existing faults.

After Late Miocene time, there was a transition from extension to contraction at the Tejon Embayment. Thrusts and folds currently rim the Embayment and consistently verge basinward. The younger thrust faults appear to truncate the normal faults along the embayment margins, and have offset Upper Miocene strata (Figures 5,6 and 7). Therefore, they are no older than very late Miocene, and in fact appear to be Pliocene to Recent in age. The precise timing of this transition is important and is a focus of our continuing research. Fission-track studies suggest that the uplift of the Tejon Block with respect to the Maricopa subbasin began after 5 Ma [Naeser, 1981; Briggs et al., 1981]. Furthermore, in the San Emigdio Mountains on the southwest side of the Embayment, there is a larger scale Pliocene-Recent fold-thrust belt (Figure 3) [Namson and Davis, 1988; Seaver et al., 1986]. The loading of these thrust sheets is probably related to the great thickening of the similarly-aged sedimentary section seen north of the White Wolf fault [Bartow, 1984, section G-G']. Latest Miocene to Recent strata are more than 2500 m thicker in the Maricopa subbasin than they are in the Tejon Embayment. This implies a foreland depression in the Maricopa subbasin similar to that modeled for the central San Joaquin basin [Rentschler and Bloch, 1988].

The White Wolf Fault Zone

Our data from the seismic reflection lines and the refraction study suggest that the history of the White Wolf fault may be as complex as that of the faults in the Tejon Embayment, with episodes of both normal and reverse faulting and with a poorly known component of strike slip.

Well and seismic data have shown that the White Wolf fault zone offsets the upper surface of the basement complex at least 4.5 km down to the northwest [Davis, 1986]. Based on subsurface data, the total (or net) left-lateral displacement may be as small as 600 m [Dibblee, 1955] or greater than 15 km [Hill, 1955]. Basement-rock correlations have suggested net right-lateral displacement on the White Wolf-Breckenridge-Kern Canyon fault zone [Ross, 1986]. Aftershocks from the 1952 M 7.2-7.7 Arvin-Tehachapi earthquake have defined a fault plane dipping 50°-60° southeast above 15 km, and suggest both thrust and lateral shear at the surface [Buwalda and St. Amand, 1955; Cisternas, 1963]. Reflection profiles across the fault zone suggest multiple strands which dip to the south in the shallow section.

However, retrodeformable geologic cross sections across the presumed trace of the White Wolf fault near Wheeler Ridge require that, beginning at a depth greater than 3 km, the fault dips to the north [Davis and Lagoe, 1984; Davis, 1986]. This implies that the south dip and apparent reverse separation in the shallower section are not necessarily due to reverse faulting, but result instead from the rotation of the upper surface of the White Wolf fault due to deep thrusting and folding. By this interpretation, the White Wolf fault had a long history as a north-dipping normal growth fault. There is some evidence cited for this: for example, mid-Upper Miocene marine strata on the north side of the fault are some 3700 ft. (1120 m) thicker than they are on the south [Davis, 1986]. Such an interpretation may in part explain the north-dipping velocity contours north of the mapped trace of the fault (Figure 11).

Another problem is the absence of a scarp along the surface trace of the White Wolf fault west of the very prominent scarp at Comanche Point due to ground rupture from the 1952 and earlier earthquakes. Aftershock and geodetic studies suggest that at shallow crustal depths, the fault is probably segmented in the vicinity of Comanche Point [Stein and Thatcher, 1981]. The zone of surface rupture northeast of Comanche Point may slip along a south-dipping, low-angle fault segment, whereas the fault zone bounding the Tejon Embayment east of Wheeler Ridge is a more steeply-dipping fault segment. The active, thrust segment of the fault system, trending into the embayment from the northeast, may curve and continue southeast around Comanche Point as its southwest-facing escarpment. Thus, the lower-angle segment of the White Wolf fault can be related to the system of buried thrust faults that we mapped using our subsurface methods (Figure 8).

Relationship to Regional Events

The timing of these local events is related to the regional compression and uplift throughout the central and western Transverse Ranges. Transpression has occurred along the San Andreas fault system since the Pliocene, particularly in the "Big Bend" area, which evolved from a dominantly releasing-bend to a more restraining-bend geometry [Crowell, 1979]. This was probably a result of the opening of the Gulf of California about 5.5 million years ago (Figure 2) [Atwater and Molnar, 1973; Hornafius, 1985].

The U-shaped Tejon embayment is closing from three directions (Figures 3 and 8). The basinward displacement of older extensional features along the margin of the current basin and the presence of Miocene nearshore marine facies lapping onto basement along the eastern margin of the embayment suggest that both the structural and depositional limits of the Tejon Embayment were much larger in Miocene time than today. Outcrop evidence along the east-west trending Edison fault suggests that Late Oligocene/Early Miocene normal faulting extended northeastward into the southern Sierra Nevada (Figure 3) [Dibblee and Chesterman, 1953; Olson et al., 1986]. The normal faulting seen in cross sections through many of the oil fields in the Bakersfield area suggests that Miocene extension took place in the Maricopa subbasin as well.

Assuming a similar origin for the subbasin to the north, our data from the Tejon Embayment suggests that the Cenozoic southern San Joaquin Basin originated with normal faulting in the Late Oligocene, was volcanically active in the Early Miocene (22-23 Ma), and subsided rapidly later in the Miocene. Palinspastic restoration of strike-slip suggests that the San Joaquin, Los Angeles, and Ventura Basins all formed in Early Miocene time, although separated by some 500 km (Figure 2). The age and distribution of volcanic centers in southern and central California, including the Tunis volcanics, suggest similarly widespread volcanism in Late Oligocene and Early Miocene time [Stanley, 1987].

There are two hypotheses for placing this early extensional event, as inferred from basin formation and volcanism, into a regional context. This extension was too widespread to be explained by the triple junction/transtensional model. This model depicts deformation as proceeding northward through time, associated with the passing of the unstable Mendocino Triple Junction [e.g., Blake et al., 1978; Dickinson and Snyder, 1979]. The Triple Junction is inferred to have migrated northward since the first interaction between the North American and Pacific Plates, about 29 Ma. However, about 7 m.y. seems to have been required for the plate readjustments to reach inland from their margins and produce a well organized transform boundary [Atwater, 1988, in press]. Therefore, basin origin and volcanism in Late Oligocene/Early Miocene time may be related to this early history of plate interaction, before the development of the San Andreas system [Goodman and Malin, 1988].

As an alternate hypothesis, it is possible that this Late Oligocene-Early Miocene period of extension in southern California may represent the limits of western Basin and Range extension, from which it appears indistinguishable. For example, recent work in the Mojave Desert suggests the presence of a large rift which opened between 22 and 17 Ma (Early Miocene) [Dokka, 1988; Serpa, 1988]. Furthermore, undoing the 30° of Neogene rotation on the northeast-trending normal faults in the Tejon embayment would align many of them with the common, generally north-trending, early Miocene extensional structures of the southwestern U.S.

In contrast, the distribution of the younger, post-20-Ma volcanic centers implies that these have migrated northward with time [Fox et al., 1985] and can be correlated with the northward path of the unstable triple junction and the related lengthening of the San Andreas fault system [e.g., Stanley, 1987; Goodman and Malin, 1988]. It is possible that the late Early and Middle Miocene deepening of the southern San Joaquin Valley, discussed above, occurred under this transtensional regime (Figure 2). However, due to the subsequent (or perhaps) coincident rotation and translation of this region, the inferred framework strike-slip faults bounding the subsiding Miocene basin have not been identified. For example, the White Wolf may have been such a fault boundary for the subsiding Maricopa basin during the Miocene (Figure 12).

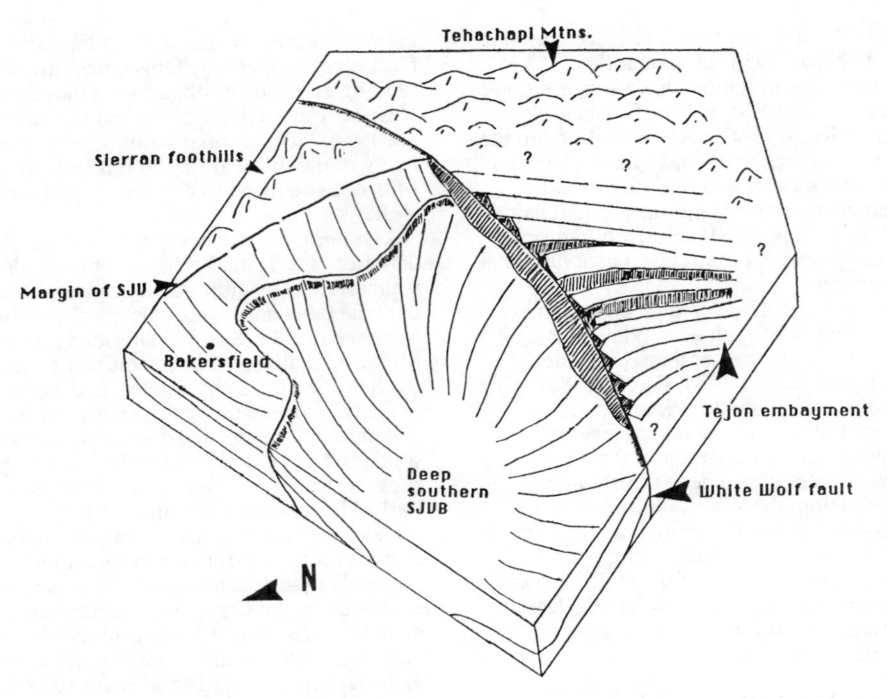

Figure 12. Block diagram cartoon viewed looking southeast, showing the inferred truncation of the Tejon Embayment normal faults by the younger, oblique-slip White Wolf fault, indicated with large normal separation in late Miocene time. At this time, the White Wolf fault is the inferred southeastern boundary of the deep, extensional Maricopa subbasin to the north. Since this time, the White Wolf fault has been tilted and segmented. A generalized graben system is indicated for the Tejon Embayment. Highlands bounding these basins are exaggerated for late Miocene time. SJVB = San Joaquin Valley Basin.

Conclusions

A sequence of disparate tectonic events has resulted from a unique geological setting between the San Andreas, Garlock and White Wolf faults. The complex structure of the Tejon Embayment consists mainly of buried, high-angle normal faults, which were active in at least three episodes from the Late Oligocene to the Late Miocene. The first episode was synchronous with the deposition of non-marine clastics, the extrusion of volcanic flows, and the deposition of coarse volcanic breccias. The second facilitated the deepening of the embayment, and the third apparently reactivated existing faults.

Pliocene to Recent thrust faults, which have offset the basement and overlying section, structurally overlie the normal faults at the margins of the embayment and verge basinward. These younger thrust faults are local manifestations of a Pliocene change in regional tectonics which coincided with transpression along the San Andreas and Garlock faults. During this time, the U-shaped Tejon Embayment has closed from three directions and is in the early stages of a fold-thrust belt. The much larger Late Oligocene-Early Miocene extensional basin formed too early and too far inland to have originated as a pull-apart basin in the evolving San Andreas strike-slip regime. However, later in Miocene time, the basin subsided within the transtensional plate margin.

On the upper plates of the thrust faults are petroliferous folds as well as the older normal faults which have been exhumed along the basin margins by the thrusts. Oblique-slip faults have also been a part of this structural overprinting. The overprinting is surprisingly mild given the presumed 30+° of Neogene clockwise rotations observed for the Tehachapi Block. This observation, and the evidence for the continuity of the deep structure beneath roughly 8 km provided by the refraction study, suggests that the Neogene rotations may have been accommodated by detachment structures in the middle crust, with the White Wolf fault serving as the rotated northern boundary.

The White Wolf fault itself reflects this complex history, and was probably tilted and segmented during the Pliocene or younger transition from extension to contraction. During its early history, the fault was an extensional expression of the Miocene-basin forming process, trending close to north-south, and later may have been an oblique-slip boundary to the deepening Maricopa subbasin. Then, according to our present interpretation, during the reorganization of the regional stress regimes the fault zone north of the central Tejon Embayment became decoupled from the segment to the east, upon which the northeastern Tehachapi Mountains are now being uplifted.

Furthermore, as the crustal block south of the White Wolf fault was strongly tilted and rotated, the ancestral fault itself was rotated and brought towards the vertical, thus accentuating the structural relief across the fault zone.

In our model, both Comanche Point and Wheeler Ridge are fault-generated anticlines, and they are more plausibly linked by the segmented system of thrust faults along the Tejon Embayment margins than by the more linear, deformed high-angle White Wolf fault segment at the northern embayment margin, as traditionally mapped.

Thus, the use of seismic reflection and refraction data, borehole data, and field observations allows us to see the evolution of the relatively shallow Tejon Embayment as an example of multi-cyclic Cenozoic basin development in California.

Acknowledgements. We gratefully acknowledge Tenneco, Texaco, Arco, Cities Service and Rancho Energy for contributing data to our study and for many useful discussions with their geoscientists. Jack Hunt and Mildred Wiebe have allowed us access to the Tejon Ranch for field geology and geophysics on many occasions. Walter Mooney of the USGS, Menlo Park, was instrumental in providing equipment for the refraction survey, which Beth Ambos organized, presided over and processed. Western Geophysical was the contractor for the CALCRUST seismic surveys. Thomas Henyey is the overall Principal Investigator for CALCRUST. This paper was substantially improved by critical reviews by Gerard Bond, Raymond Price and an anonymous reviewer. Pamela Morgan edited the manuscript. We have benefited from discussions with Steve Graham, Tanya Atwater and Thom Davis. We also thank Dave Okaya, Dave Crouch and Claudia Martin, as well as the California Division Oil and Gas. This research was supported by NSF Grant EAR 86-09347. ICS Contribution No. 0003-02CS 02TC.

References

Addicott, W. O., Mid-Tertiary zoogeographic and paleogeographic discontinuities across the San Andreas fault, California, *Stanford Univ. Publ. Geol. Sci., 11*, 144-165, 1968.

Ambos, E. L., and P. E. Malin, Combined seismic reflection/refraction investigations in the Tehachapi Mountains, southern California: Results from the 1986 CALCRUST experiment (abstract), *EOS Trans. AGU, 68* , 1360, 1987.

Atwater, T. M., Implications of plate tectonics for the Cenozoic tectonic evolution of western North America, *Geol. Soc. Am. Bull, 81*, 3513-36, 1970.

Atwater, T. M., Plate tectonic history of the northeast Pacific and western North America, in *Geol. Soc. Am. DNAG Vol. N: The Eastern Pacific and Hawaii*, edited by E. L. Winterer, D. M. Hussong, and R. W. Decker, (in press) 1988.

Atwater, T. M., and P. Molnar, Relative motion of the Pacific and North American plates deduced from sea-floor spreading in the North Atlantic, Indian and South Pacific oceans, *Stanford Univ. Publ. Geol. Sci, 13*, 136-148, 1973.

Bandy, O. L., and R. E. Arnal, Middle Tertiary basin development, San Joaquin Valley, California, *Geol. Soc. Am. Bull., 80*, 783-820, 1969.

Bartow, J. A., Geologic map and cross sections of the southeastern margin of the San Joaquin Valley, California, scale 1:125,000, *Geol. Misc. Invest. Map I-1496*, U.S. Geol. Surv., Reston, Va., 1984.

Bartow, J. A., and T. W. Dibblee, Jr., Geologic map of the Arvin Quadrangle, Kern County, California, scale 1:24,000, *Open File Report 81-297*, U.S. Geol. Surv., Reston, Va., 1981.

Bartow, J. A., and K. McDougall, Tertiary Stratigraphy of the Southeastern San Joaquin Valley, California, *U.S. Geol. Surv. Bull. 1529-J*, 41 pp., 1984.

Biddle, K. T., and N. Christie-Blick, Glossary—Strike-slip deformation, basin formation, and sedimentation, in *Strike-slip Deformation, Basin Formation, and Sedimentation,* edited by K. T. Biddle and N. Christie-Blick, pp. 375-386, SEPM Spec. Publ. 37, 1985.

Blaisdell, R. C., COSUNA: Correlation of stratigraphic units of North America project, central California region, *AAPG Chart #745*, columns 25 and 26, Am. Assoc. of Petrol. Geol., Tulsa, Oklahoma, 1984.

Blake, M. C., Jr., R. H. Campbell, T. W. Dibblee, Jr., D. G. Howell, T. H. Nilsen, W. R. Normark, J. C. Vedder, and E. A. Silver, Neogene basin formation in relation to plate-tectonic evolution of San Andreas fault system, California, *AAPG Bull., 62*, 344-372, 1978.

Blank, H. R., J. H. Healey, J. C. Roller, R. Lamson, F. Fischer, R. McClearn, and S. Allen, Seismic refraction profile, Kingdom of Saudi Arabia—Field operations, instrumentation and initial results, *Saudi Arabian Mission Project Report 259*, 49 pp., U.S. Geol. Surv., Reston, Va., 1979.

Briggs, N. D., C. W. Naeser, and T. H. McCulloh, Thermal history of sedimentary basins by fission track dating, *Nuclear Tracks, 5*, 235-237, 1981.

Buwalda, J. P., and P. St. Amand, Geological effects of the Arvin-Tehachapi earthquake, *Calif. Div. Mines and Geol. Bull, 170*, 131-142, 1955.

Carter, B., Quaternary fault-line features of the central Garlock fault, Kern County, California, in *Geol. Soc. Am. (Cordilleran Sect.) Centennial Field Guidebook 29*, 133-135, 1987.

Červený, V., I. A. Molotkov, and I. Pšenčik, *Ray Method in Seismology,* University Karlova, Prague, 214 pp., 1977.

Cisternas, A., Precision determination of focal depths and epicenters of local shocks in California, *Bull. Seismol. Soc. Am., 53*, 1075-1083, 1963.

Cox, B. F., ed., *Basin Analysis and Paleontology of the Paleocene and Eocene Goler Formation, El Paso Mountains, California*, SEPM, Pacific Section, Field Trip Guidebook 57, 67 pp., 1987.

Crowell, J. C., Origin of late Cenozoic sedimentary basins in California, *SEPM Spec. Publ., 22*, 190-204, 1974.

Crowell, J. C., The San Andreas fault system through time, *J. Geol. Soc. London, 136*, 293-302, 1979.

Crowell, J. C., Late Cenozoic basins of onshore southern California, in *Cenozoic Basin Development of Coastal California*, edited by W. G. Ernst and R. V. Ingersoll, UCLA Rubey Volume 6, pp. 207-241, Prentice-Hall, Englewood Cliffs, N.J., 1987.

Davis, T. L., A structural outline of the San Emigdio Mountains, in *Geologic Transect across the western Transverse Ranges*, edited by T. L. Davis and J. S.

Namson, Soc. Econ. Paleont. Mineral., Pacific Section, 1986 Annual Meeting Field Trip Guidebook 48, pp. 23-32, 1986.

Davis, T. L., and M. B. Lagoe, Cenozoic structural development of the north-central Transverse Ranges and southern margin of the San Joaquin Valley (abstract), Geol. Soc. Am. 1984 Ann. Meet., *Absttr. with programs, 16*, 484, 1984.

Dibblee, T. W., Jr., Geology of the southeastern margin of the San Joaquin Valley, California, in *Earthquakes in Kern County, California, during 1952*, edited by G. B. Oakeshott, *Calif. Div. Mines and Geol. Bull., 171*, 23-24, 1955.

Dibblee, T. W., Jr., Geologic map of the Pastoria Creek Quadrangle, California, scale 1:24,000, *Open File Map 73-57,* U.S. Geol. Surv., Reston, Va., 1973.

Dibblee, T. W., Jr., and C. W. Chesterman, Geology of the Breckenridge Mountain Quadrangle, *Calif. Div. Mines and Geol. Bull., 168*, 56 pp., 1953.

Dickinson, W. R., and W. Snyder, Geometry of triple junctions related to San Andreas transform, *J. Geophys. Res., 84*, 561-572, 1979.

Dokka, R., Synthesis of Middle and Late Cenozoic tectonic history of the Mojave Desert block , Geol. Soc. Am., Cordilleran Sect., 1988 Ann. Meet., *Abstr.with programs,* p. 156, 1988.

Engebretson, D. C., A. Cox, and R. G. Gordon, Relative motions between oceanic and continental plates in the Pacific basin, *Geol. Soc. Am. Spec. Pap. 206*, 59 pp., 1985.

Fox, K. F., Jr., R. J. Fleck, G. H. Curtis, and C. E. Meyer, Implications of the northwestwardly younger age of the volcanic rocks of west-central California, *Geol. Soc. Am. Bull., 96*, 647-654, 1985.

Goodman, E. D., and P. E. Malin, Comments on the Geology of the Tejon Embayment from seismic reflection, borehole and subsurface data, in *Studies of the Geology of the San Joaquin Basin*, edited by S. A. Graham, pp. 89-108, SEPM Pacific Section Book 60, 1988.

Graham, S. A., and L. A. Williams, Tectonic, depositional and diagenetic history of Monterey Formation (Miocene), central San Joaquin basin, California, *AAPG Bull, 69*, 385-411, 1985.

Hearn, T. M., and R. W. Clayton, Lateral velocity variations in southern California, part I. Results for the upper crust from Pg waves, *Bull. Seismol. Soc. Am., 76*, 495-509, 1986a.

Hearn, T. M., and R. W. Clayton, Lateral velocity variations in southern California, part II: Results for the lower crust from Pn waves, *Bull. Seismol. Soc. Am., 76*, 511-520, 1986b.

Hill, M. L., Nature of movements on active faults in southern California, in *Earthquakes in Kern County, California, during 1952,* edited by G. B. Oakeshott, *Calif. Div. Mines and Geol. Bull., 171,* 37-40, 1955.

Hirst, B. M., Tectonic development of the Tejon and adjacent areas, Kern County, California, in *Southeast San Joaquin Valley Field Trip, Kern County, California, Part II: Structure and stratigraphy*, edited by P. Bell, pp. 1-8, Am. Assoc. Petrol. Geol., Pacific Sect., and San Joaquin Geol. Soc. Field Trip Guidebook, Bakersfield, Ca., 1986.

Hirst, B. M., Early Miocene tectonism and associated turbidite deposystems of the Tejon area, Kern County, California, in *Studies of the Geology of the San Joaquin Basin*, edited by S. A. Graham, pp. 207-222, SEPM Pacific Section Book 60, 1988.

Hornafius, J. S., Neogene tectonic rotation of the Santa Ynez Range, Western Transverse Ranges, California, suggested by paleomagnetic investigations of the Monterey Formation, *J. Geophys. Res., 90 , B-14,* 12503-12522, 1985.

Johnson, C. M., and J. R. O'Neil, Triple junction magmatism: A geochemical study of Neogene volcanic rocks in western California, *Earth Planet. Sci. Lett., 71*, 241-262, 1984.

Kanter, L. R., and M. O. McWilliams, Rotation of the southernmost Sierra Nevada, California, *J. Geophys. Res., 87*, 3819-3830, 1982.

Luyendyk, B. P., M. J. Kamerling, and R. R. Terres, Geometric model for Neogene crustal rotations in southern California, *Geol. Soc. Am. Bull, 91*, 211-217, 1980.

Luyendyk, B. P., M. J. Kamerling, R. R. Terres, and J. S. Hornafius, Simple shear of southern California during Neogene time, suggested by paleomagnetic declinations, *J. Geophys. Res., 90, B-14,* 12454-12466, 1985.

Luyendyk, B. P., and J. S. Hornafius, Neogene crustal rotations, fault slip and basin development in southern California, in *Cenozoic Basin Development of Coastal California*, edited by W. G. Ernst and R. V. Ingersoll, pp. 259-283, UCLA Rubey Volume 6, Prentice-Hall, Englewood Cliffs, N.J., 1987.

Macpherson, B. A., Sedimentation and trapping mechanisms in Upper Miocene Stevens and older turbidite fans of southeastern San Joaquin Valley, California, *AAPG Bull, 62*, 2243-2274, 1978.

McKenzie, D. P., and W. J. Morgan, The evolution of triple junctions, *Nature, 225*, 125-133, 1969.

McWilliams, M., and Y. Li, Tectonic oroclinal bending of the southern Sierra Nevada batholith, *Science, 230*, 172-175, 1985.

Medwedeff, D. A., Structural analysis and tectonic significance of Late Tertiary and Quaternary, compressive-growth folding, San Joaquin Valley, California, Ph.D. dissertation, 184 pp., Princeton University, 1987.

Naeser, C. W., The fading of fission tracks in the geologic environment—data from deep drill holes, *Nuclear Tracks, 5*, 248-250, 1981.

Namson, J. S., and T. L. Davis, Seismically active fold and thrust belt in the San Joaquin Valley, Central California, *Geol. Soc. Am. Bull., 100*, 257-273, 1988.

Nilsen, T. H., Facies relations in the Eocene Tejon Formation of the San Emigdio and western Tehachapi Mountains, CA, in *Sedimentary Facies Changes in Tertiary Rocks—California Transverse Ranges and Coast Ranges,* 1973 Annual Meeting, Am. Assoc. Petrol. Geol. and Soc. Econ. Paleont. Mineral., Soc. Econ. Paleont. Mineral. Trip 2 Guidebook, pp. 7-23, 1973.

Nilsen, T. H., Paleogene tectonics and sedimentation of coastal California, in *Cenozoic Basin Development of Coastal California*, edited by W. G. Ernst and R. V. Ingersoll, pp. 81-123, UCLA Rubey Volume 6, Prentice-Hall, Englewood Cliffs, N.J., 1987.

Nilsen, T. H., T. W. Dibblee, Jr., and W. O. Addicott, Lower and middle Tertiary stratigraphic units of the San Emigdio and western Tehachapi Mountains, California, *U.S. Geol. Surv. Bull 1372-H*, pp. H1-H23, 1973.

Olson, H. C., G. E. Miller, and J. A. Bartow, Stratigraphy,

paleoenvironmental depositional setting of Tertiary sediments, southeastern San Joaquin Basin, in *Southeast San Joaquin Valley Field Trip, Kern County, California, Part II: Structure and stratigraphy*, edited by P. Bell, pp. 18-55, Am. Assoc. Petrol. Geol., Pacific Sect., and San Joaquin Geol. Soc. Field Trip Guidebook, Bakersfield, Ca., 1986.

Plescia, J. B., and G. J. Calderone, Paleomagnetic constraints on the timing of rotation of the Tehachapi Mountains, California (abstract), *Geol. Soc. Am. Abstr. with Programs, 18*, 171, 1986.

Rentschler, M., and R. Bloch, Flexural modeling of the central San Joaquin Basin, California, in *Studies of the Geology of the San Joaquin Basin*, edited by S. A. Graham, pp. 29-52, SEPM Pacific Section Book 60, 1988.

Ross, D. C., Basement-Rock Correlations across the White Wolf-Breckenridge-Southern Kern Canyon Fault Zone, Southern Sierra Nevada, California, *Geol. Surv. Bull. 1651*, 1-25, 1986.

Saleeby, J. B., D. B. Sams, and R. W. Kistler, U/Pb zircon, strontium, and oxygen isotopic and geochronological study of the southernmost Sierra Nevada batholith, California, *J. Geophy. Res., 92, B-10* , 10443-10466, 1987.

Seaver, D. B., R. L. Zepeda, E. A. Keller, D. M. Laduzinsky, D. L. Johnson, and T. K. Rockwell, Active folding: southern San Joaquin Valley, California (abstract), *EOS Trans. AGU, 67*, 1223, 1986.

Serpa, L. and R. K. Dokka, Reinterpretation of Mojave COCORP data: implications for the structure of the Mojave rift, , Geol. Soc. Am., Cordilleran Sect., 1988 Ann. Meet., *Abstr.with programs,* p. 230, 1988.

Sharry, J., The geology of the western Tehachapi mountains, California, Ph.D. dissertation, 215 pp., Massachusetts Institute of Technology, 1981.

Sibson, R. H., Stopping of earthquake ruptures at dilational fault jogs, *Nature, 316*, 248-251, 1985.

Stanley, R. G., Implications of the northwestwardly younger age of the volcanic rocks of west-central California: Alternative interpretation, *Geol. Soc. Am. Bull., 98*, 612-614, 1987.

Stein, R. S., and W. Thatcher, Seismic and aseismic deformation associated with the 1952 Kern County, California earthquake and relationship to the Quaternary history of the White Wolf Fault, *J. Geophy. Res., 86, B-6*, 4913-4928, 1981.

Yeats, R. S., and K. R. Berryman, South Island, New Zealand and Transverse Ranges, California: A seismotectonic comparison, *Tectonics, 6*, 363-376, 1987.

THE MECHANISM OF FORMATION OF THE NORTH SEA BASIN

Eugene V. Artyushkov

Institute of Physics of the Earth, Academy of Sciences, Moscow, USSR

Michael A. Baer

Ministry of Geology USSR, Moscow, USSR

Abstract. The North Sea basin is filled by a thick sequence of sediments which are underlain by strongly thinned continental crust. This is commonly attributed to lithospheric stretching ($\beta \sim 1.5$-2.0). An intense stretching is, however, possible only under specific deformation types that do not occur in the North Sea. The geometry of the normal faults ensures extension of only <10-20%.

During the interval from Permian to Cenozoic three rapid phases of subsidence without intense stretching occurred in the North Sea; in each case deeper-water basins were formed (~0.5 km). The subsidences were caused by the destruction of the lowermost crust by asthenospheric upwelling. Under regional tensile stresses the thinned crust broke into wide blocks (>10 km), these subsided isostatically along normal faults. This was associated with a moderate extension of the basin.

Introduction

The North Sea is a deep sedimentary basin underlain by thinned continental crust. Its subsidence was accompanied by the development of a major graben system characterized by numerous normal faults [Ziegler, 1982a]. Under some of the grabens the thickness of continental crust has been attenuated to about half the normal thickness [Ziegler, 1982b; Barton and Wood, 1984; Beach, 1986]. It has been postulated that the North Sea Basin was formed by lithospheric stretching [Sclater and Christie, 1980; Wood and Barton, 1983; and others]. Back-stripped subsidence curves, including correction for depositional water depths, were used, which gives a stretching factor β of about 1.5-2.0 [Barton and Wood, 1984].

Normal faults in the North Sea indicate that the subsidence occurred under tensile stresses. This does not, however, prove that the extension was sufficient to produce the observed crustal thinning and subsidence. Using multichannel reflection seismic data, Ziegler [1983] estimated relative extension in the North Sea as $\varepsilon = \beta - 1 \sim 0.10$-0.15 for grabens and $\varepsilon \sim 0.4$-0.06 for the total cross-section. This is quite insufficient to produce the observed subsidence and two-fold thinning of the crust in the grabens. Ziegler's estimates were, however, not accepted by the proponents of crustal stretching as the sole subsidence mechanism [Barton and Wood, 1983].

Our analyses revealed that most deep basins on continental crust formed without significant stretching [Artyushkov and Baer, 1983, 1984a, b, 1986a, b]. The subsidence was commonly associated only with a moderate amount of extension $\varepsilon \sim 3$-5%. Typical rift valleys produced by an intense crustal stretching ($\varepsilon > 50\%$) occur only in a few per cent of the cases.

Many deep basins developed in response to very rapid water loaded subsidence (~0.5 km/Ma) even in the absence of significant crustal stretching. In the main hydrocarbon provinces of the world, large hydrocarbon fields occur only in areas of rapid subsidence or just near them.

The objective of this paper is to estimate the magnitude of crustal stretching in the North Sea using the available data and to discuss possible causes of attenuation and subsidence of the crust in this area.

Main Modes of Stretching

The main objection against Ziegler's estimates of upper crustal extension at the base of the pre-rift Zechstein salt level is that the reflection seismic profiling data are unable to resolve offsets on presumably numerous small faults [Barton and Wood, 1983]. Reconstruction of the fault configurations is not quite reliable and the faults can flatten at great depth which increases the extension. Thus higher resolution seismic data might give evidence for larger β-factor.

On the other hand it should be kept in mind that the amount of extension can be estimated from the geometry of normal faults and the surface of tilted blocks even in the absence of high resolution seismic data [Artyushkov, 1987]. Two simple models for normal faulting extension in deep sedimentary basins can be analyzed geometrically: S-type in which the fault blocks are not tilted during the faulting, or are tilted toward the basin axis (Fig. 1a) and A-type in which the blocks are tilted toward the down-thrown side of the boundary normal faults (Fig. 1b).

In some basins S-type faulting is predominant; this is typical, for example, of the Baikal Graben [Artyushkov et al., 1987] or the Horn and Glückstadt Grabens (see Fig. 3). There are also some basins, such as those in the Basin and Range Province or in the Afar Rift where A-type faulting with strongly tilted blocks is mostly observed. In many basins

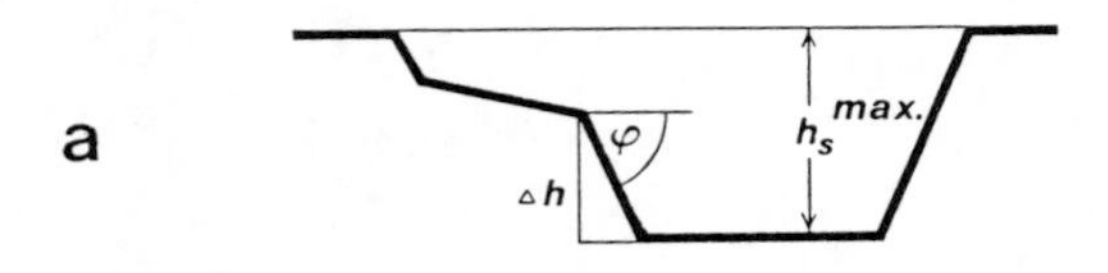

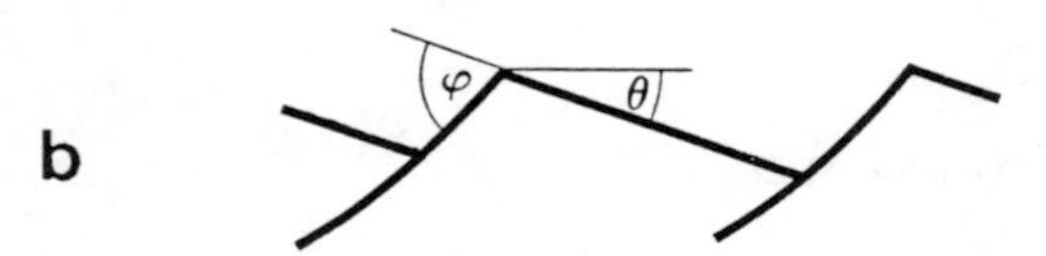

Fig. 1. Main types of extensional deformations in deep sedimentary basins. a. S-type faulting with fault planes and the surface of individual fault blocks both tilted to the axis of the basin, S-type basin. b. A-type faulting with the fault planes and fault blocks tilted in the opposite directions, A-type basin.

S-type and A-type faulting exist in different parts of the profiles (see Figs. 4-6). In some strongly deformed regions a combination of both S-type and A-type faulting can occur [Gibbs, 1984]. This has also been demonstrated by the laboratory experiments of McClay and Ellis [1987].

If we assume that the faults are perfectly planar, then under a small angle of block tilting $\theta < 20°$ extension in a basin with normal faults can be estimated as

$$\Delta L = (\Sigma \Delta_i) \mathrm{ctg}\, \psi \qquad (1)$$

where $\Sigma \Delta h_i$ is a sum of vertical offsets along the normal faults and ψ is their average dip angle. This expression is valid for both S-type and A-type faulting at small θ.

Many of the normal faults are listric. In order to determine extension on curved faults ψ should be measured as a deep angle of the straight line that connects the intersections of the fault with the top of the blocks separated by the fault. Then the average value of ψ defined in this way should be used in (1).

The amount of extension at the top of the basement or at a sedimentary bed of a certain age is determined only by the geometry of the surface, which includes both the tops of the blocks and those parts of the faults that connect them (see, Fig. 1b). Below this surface curved listric faults can have considerably lower dip angles. This cannot, however, increase extension at the higher level which is determined by the average dip angle of the faults between the tops of the blocks and not below the tops of the blocks.

Consider the basins with S-type faulting. In such basins $\Sigma \Delta h_i$ cannot exceed twice the maximum depth (h_s^{max}) of the basin. Hence according to (1), for these basins

$$\Delta L \leq 2 h_s^{max} \mathrm{ctg}\, \psi \qquad (2)$$

This extension does not depend on the width of the basin (L) hence relative extension

$$\varepsilon = \beta - 1 = \Delta L / (L - \Delta L) \qquad (3)$$

decreases as the width of the basin increases. On the other hand extension ΔL_{st} that is necessary to produce the same basin by stretching is proportional to L. It is easy to show (Artyushkov, 1987) that for $\psi > 50\text{-}60°$ the ratio

$$\Delta L / \Delta L_{st} < h_c^0 / L \qquad (4)$$

where h_c^0 is the pre-stretched thickness of the crust. In basins produced by stretching the observed extension (ΔL) and that produced by stretching (ΔL_{st}) are, by definition, equal: $\Delta L = \Delta L_{st}$. According to (4) this is possible only for narrow basins with S-type faulting ($L \sim h_c^0$). In wide basins with S-type faulting ($L > 100$ km) stretching can account only for a portion $\sim h_c^0/L$ of the subsidence.

In basins with A-type faulting (Fig. 1b) the stretching factor is [Le Pichon and Sibuet, 1981]:

$$\beta st \approx \sin \psi / \sin(\psi - \theta) \qquad (5)$$

where θ is the angle of block tilting and ψ is the initial dip angle of normal faults. Intense stretching can occur in A-type basins of any width. This is, however, possible only under a large angle of block tilting. For instance, at typical value of $\psi \sim 50\text{-}60°$ stretching by $\beta \sim 1.5\text{-}2.0$ requires block tilting by

$$\theta \sim 20\text{-}35° \qquad (6)$$

For most basins where tilted blocks occur, θ is considerably smaller ($\sim 3\text{-}5°$). This corresponds to a rather small extension ($\beta \sim 1.03\text{-}1.08$) that can form only shallow sedimentary basins ($h_s \sim 0.7\text{-}1.7$ km).

The above estimated make it possible to determine the role of stretching by normal faulting in the formation of deep basins on continental crust. The subsidence in very narrow basins ($L < h_c^0$) can result from stretching by both S-type and A-type faulting. It should be, however, demonstrated that the displacement of the fault blocks along the fault ensure the extension sufficient for the subsidence. Wide basins can be formed by stretching only if they include strongly tilted blocks over most of the region. Consider from this point of view typical profiles in the North Sea.

Horn Graben

The Horn Graben is located in the eastern part of the North Sea (see Fig. 2). In this basin the sedimentary strata are well resolved only to the base of the Late Permian Zechstein Salt (Fig. 3a). Hence only the history of the subsidence since that time can be discussed using the profile of Figure 3a.

All the faults in the Horn Graben dip toward the basin axis. In all the individual blocks the base of the Zechstein is horizontal or also tilted toward the axis. Hence this is a typical S-type basin. The width of the graben in Fig. 3a ($L \sim 60$ km) exceeds the pre-stretched thickness of the crust in cratonic areas ($h_c^0 \sim 30\text{-}40$ km). Hence the graben could not have been produced solely by stretching. In addition at the margins A and F of the graben the base of Zechstein is not at the surface but a depth $h_{s1} = 3$ km and $h_{s2} = 4.5$ km respectively. Hence the total subsidence along the normal faults is less than twice the depth of the basin:

$$\Sigma \Delta h_i \leq 2 h_s^{max} - h_{s1} - h_{s2} = 10.5 \text{ km} < 2 h_s^{max} = 18 \text{ km} \qquad (7)$$

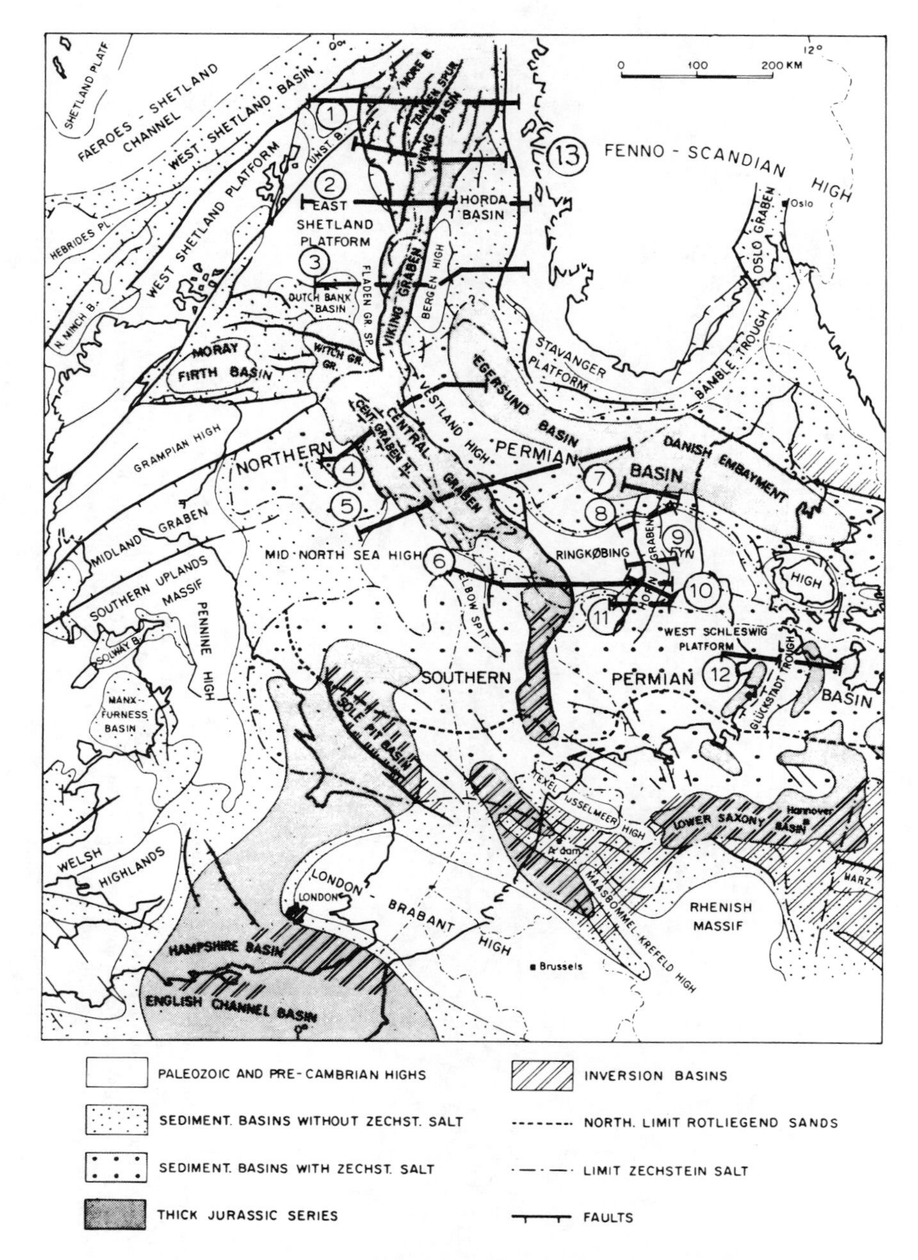

Fig. 2. North Sea – Permian and Mesozoic tectonic units (after Ziegler, 1980).

Furthermore, within the graben a considerable portion of the subsidence is accomodated by tilting of the base of the Zechstein in some blocks toward the axis of the graben. A direct measurement of the sum $\Sigma\Delta h_i$ from point A to point F gives $\Sigma\Delta h_i \approx 6$ km. Taking $\psi \sim 50\text{-}60°$ we find from (1) that ,

$$\Delta L \sim 4\text{-}5 \text{ km} \qquad (8)$$

The same value (~5 km) can be obtained by a direct measurement of the sum of horizontal offsets along the normal faults in segment AF. The width of segment AF in profile (a) of Fig. 3 is about 60 km. The profile is, however, not normal to the strike of the graben, but inclined by about 67° to its axis. This gives the real width of the graben $L \sim 55$ km. Then the average relative extensions is

$$\varepsilon = \beta - 1 = \Delta L/(L - \Delta L) \sim 0.08\text{-}0.10 \qquad (9)$$

In order to form a sedimentary basin by crustal stretching with subsequent thermal contraction and subsidence of the crust and mantle, relative extension should be

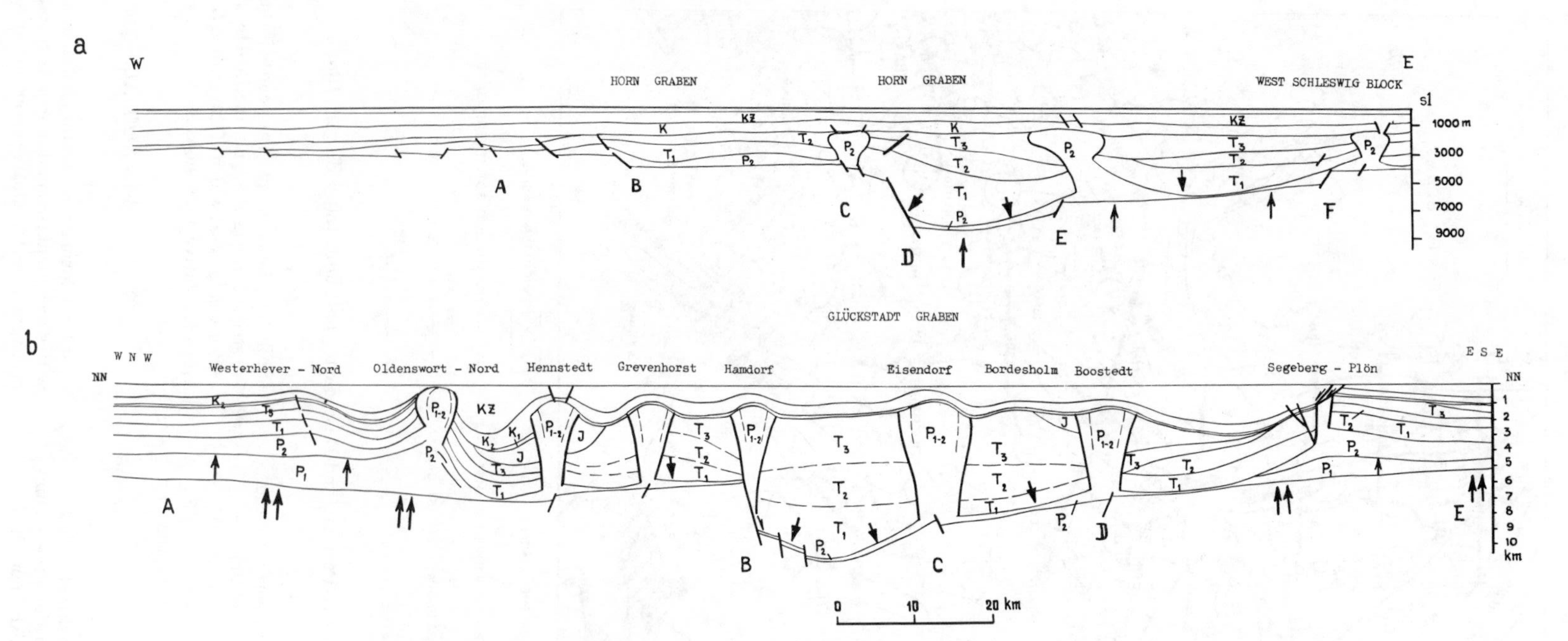

Fig. 3. Seismic reflection profiles. a. Across the Horn Graben (line 11 in Fig. 2) (after Best et al., 1983). No large stretching can be seen at the base of the Upper Permian (P_2) and Lower Triassic (T_1). These levels are marked by thin and solid arrows respectively. b. Across the Glückstadt Graben (line 12 in Fig. 2) (Best et al., 1983). There is no large stretching at the base of the Lower Permian Rotliegend (P_1) (double thin arrows), Upper Permian Zechstein (P_2) (single thin arrows) and Lower Triassic (T_1) (solid arrows).

$$\varepsilon_{st} = \beta_{st}\text{-}1[(h_c^0/h_s)(\rho_m\text{-}\rho_c)/(\rho_m\text{-}\rho_c)/(\rho_m\text{-}\rho_s)\text{-}1]^{-1} \quad (10)$$

where ρ_m is the density of the mantle lithosphere, ρ_c is the density of the crust and ρ_s is the density of the sediments. Let us take $\rho_m = 3.35$ g/cm^3, $\rho_c = 2.85$ g/cm^3, $\rho s = 2.45$ g/cm^3 $h_c^0 = 35$ km. Then (10) reduces to:

$$\varepsilon_{st} = \{19/[(h_s)\text{km}]\text{-}1\}^{-1} \quad (11)$$

In segment AF of Fig. 3a the average depth of the base of the Zechstein is $h_s = 6.1$ km. Taking this value in (11) we have

$$\varepsilon_{st} = 0.47 \quad (12)$$

which is five times larger than the observed value (9). In the inner part of the graben (segment DE) the average depth of the Zechstein floor is $h_s = 8.6$ km. According to (11) for this region $\varepsilon_{st} = 0.83$ (β_{st}-1.83). The real extension (9) is thus quite insufficient to produce the Horn Graben by crustal stretching only.

Glückstadt Graben

This sedimentary basin is located onshore in northern Germany, in the southeastern part of the area (see Fig. 2). The lowest bed resolved by seismic profiling (Fig. 3b) is the base of the Early Permian Rotliegend sands that disconformably overlay Carboniferous strata. In segment BD of the Glückstadt Graben all the fault except for a small fault near point C dip forward the axis of the basin. The base of the Rotliegend also dips toward this axis. Hence the Glückstadt Graben is a S-type basin.

The structure of the Glückstadt Graben is rather similar to that of the Horn Graben. The base of the Rotliegend at the margins of the Glückstadt Graben at points B and D is at a considerable depth ($h_{s1} = 5.5$ km and $h_{s2} = 7$ km).

A portion of the subsidence is accomodated by tilting of the base of the Rotliegend toward the basin axis (in segments BC and CD). As a result the sum of vertical offsets along the normal faults in the graben appears to be rather small: $\Sigma\Delta h_i \approx 4.5$ km. Taking $\psi \sim 50\text{-}60°$ we find from (1) that ΔL is about 2.6-3.8 km. For a width of the graben $L = 47$ km this gives relative extension

$$\varepsilon \sim 0.09 \quad (13)$$

The average depth of the graben is $h_s = 9.3$ km. According to (11) in order to produce such a basin by stretching ε_{st} should be ~ 0.96 ($\beta_{st} \sim 1.96$), which is much larger than the observed extension (13).

Beyond the Glückstadt Graben in segments AB and DE the sediment thickness above the base of the Rotliegend is $h_s \sim 6.2\text{-}6.8$ km. According to (11) in a basin of this depth produced by crustal stretching, $\varepsilon_{st} = \beta_{st}\text{-}I \sim 0.48\text{-}0.56$. The base of the Rotliegend in the above regions is practically undisturbed. There are no normal faults in segment DE, and the normal faults in segment AB involve an extension of less than 1%.

Viking Graben

Three schematic profiles across the Viking Graben are shown in Fig. 4. The consolidated crust under the graben is thinned by about two times [Ziegler, 1982b; Beach, 1986]. Profiles 1 to 3 are plotted with 5:1 vertical exaggeration. This exaggerates by 5 times block rotation so, to that extent, gives a false impression of the amount of extension.

It should also be taken into account that the present seismic data reliably resolve the structure only to the depth of the base of the Triassic in the Viking Graben and pre-Zechstein in the Central Graben. There is no Zechstein salt in profiles 1 and 2 in Figure 4 and a rather small volume of salt is present in profile 3. Hence we can use the geological structure to estimate the amount of extension at the top of the pre-Zechstein (Rotliegend and Devonian in Fig. 4).

Profile 2

First consider profile 2 where the basin has the simplest structure. This includes the Viking Graben segment AB and the Horda Basin segment BC. Except for one block, the top of Rotliegend is everywhere tilted in the same direction as the normal faults, i.e., the basins are S-type. The top of Rotliegend (or the basement where the Rotliegend or older sedimentary rocks are not present) is at $h_{s1} = h_{s2} = 1.5$ km at the left and right margins of the basin (A and C). The maximum depth of the base of Zechstein is $h_{s3} = 8$ km. The relative high, near point B, has an amplitude $\Delta h_s \sim 1$ km with respect to the deepest part of the Horda Basin. Hence the total subsidence along the normal faults is

$$\Sigma\Delta h_i < 2h_{s3}\text{-}h_{s1}\text{-}h_{s2} + 2\Delta h_s = 15 \text{ km} \quad (16)$$

In the schematic profiles of Fig. 4 with 5:1 vertical exaggeration the normal faults are shown as characterized by very low values of the initial dip angle ψ. As can be seen in the real (non-schematic) profiles [see Fig. 5, in this paper, or Fig. 2 in Badley et al., 1984] for most normal faults in the North Sea $\psi \sim 50\text{-}60°$. Substituting this value together with $\Sigma\Delta h_i$ determined by (16) into (1) we have:

$$\Delta L < 9\text{-}13 \text{ km} \quad (17)$$

The present width of the basin segment AC is $L = 178$ km, which gives

$$\varepsilon \sim 0.05\text{-}0.08 \quad (18)$$

The average depth of the system of the Viking Graben and Horda Basin in profile 2 is $h_s \sim 6$ km. According to (11) in order to produce a basin of this depth by stretching ε_{st} should be ~ 0.5.

Profile 3

At the boundaries of the graben segment AB the basement is at $h_{s1} = h_{s2} = 2$ km, while the maximum depth of the top of Rotliegend is $h_{s3} = 7$ km. Within the graben the basement and the normal faults are tilted slightly in opposite directions, which ensures an additional subsidence along the faults $\delta h_s \sim 3$ km. The total subsidence of the basement along the faults is

$$\Sigma\Delta h_i \approx 2h_{s3}\text{-}h_{s1}\text{-}h_{s2} + \delta h_s = 13 \text{ km} \quad (19)$$

Substituting this quantity into (7) with $\psi \sim 50\text{-}60°$ we obtain

$$\Delta L \sim 8\text{-}11 \text{ km} \quad (20)$$

Due to the small width of the graben in this profile (64 km) the average relative extension is comparatively high:

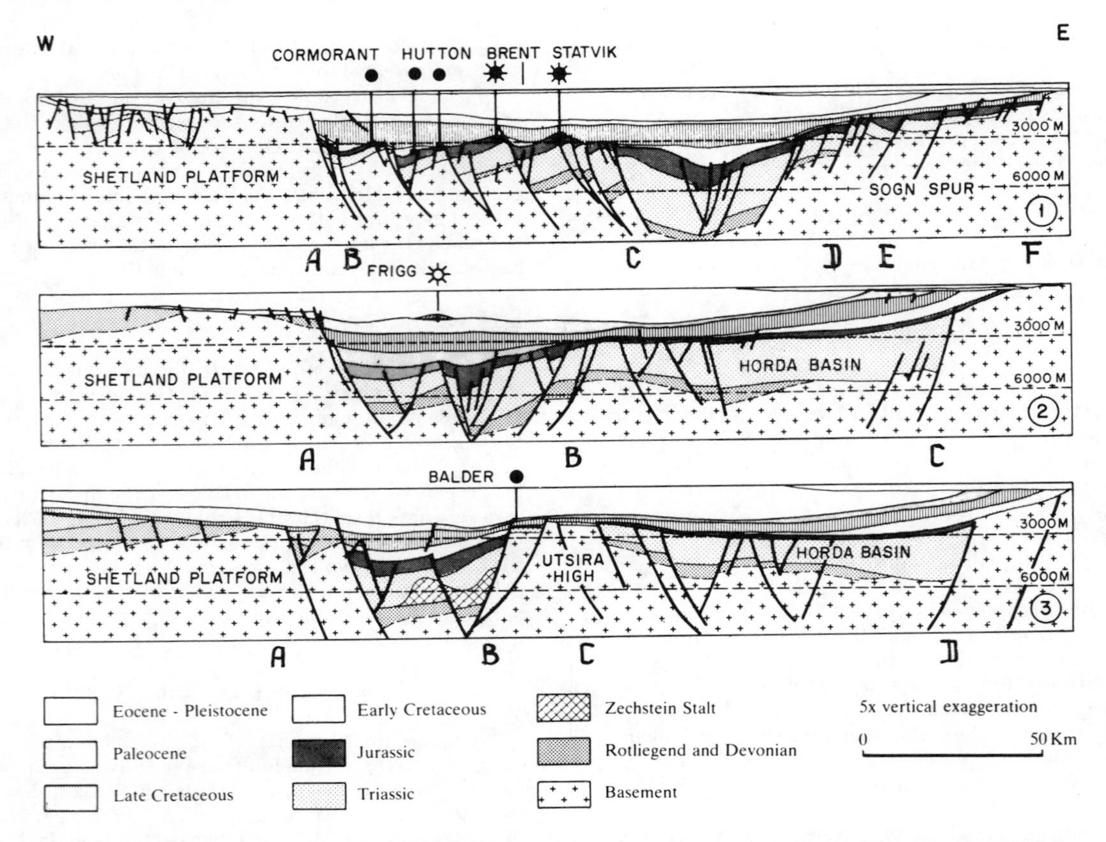

Fig. 4. Structural cross-sections of the northern North Sea (see Fig. 2 for locations) (after Ziegler, 1982a). Vertical exaggeration is 5:1.

$$\varepsilon \sim 0.12\text{-}0.19 \tag{21}$$

This is, however, several times smaller than the extension ($\varepsilon_{st} \sim 0.5$), that is necessary to produce a basin of the same depth by stretching.

In the Horda Basin CD ΔL can be evaluated as ~4-6 km and ε as ~0.4-0.06.

Profile 1

This profile includes the segments where the basement and faults are tilted both in the direction of the basin axis (segments AB, CD, EF) and in the opposite directions (segments BC, DE). In a first approximation the extension in profile I can be represented as a sum of extension due to S-type faulting of the top of Rotliegend ΔL_1, and extension ΔL_2 due to A-type faulting in segments BC and DE. Extension ΔL_1 can be evaluated as

$$\Delta L_1 = (2h_s^{max} - h_{s1} - h_{s2}) \operatorname{ctg} \psi \tag{22}$$

where $h_s^{max} = 9$ km is the maximum depth of the Rotliegend, $h_{s1} = h_{s2} = 1.5$ km are the depths of the basement at the margins of the graben A and F. Taking $\psi \sim 50°\text{-}60°$ we have

$$\Delta L_1 \sim 9\text{-}13 \text{ km} \tag{23}$$

Extension ΔL_2 is

$$\Delta L_2 = \varepsilon_{st} L_{20} = (\beta_{st} - 1) L_{20} = [\sin\psi/\sin(\psi-\theta) - 1] L_{20} \tag{24}$$

where L_{20} is the total initial length of segments BC and DE. The average angle of block tilting in segments BC and DE is $\theta \sim 4\text{-}5°$. At $\psi \sim 50\text{-}60°$ this gives the relative extension

$$\varepsilon_{st} = \sin\psi/\sin(\psi-\theta) - 1 \sim 0.04\text{-}0.08 \tag{25}$$

The present length of the segments BC and DE is $L_{21} = (\varepsilon_{st} + 1) L_{20} = 95$ km. Hence $L_{20} \sim 88\text{-}91$ km. Then extension (24) due to the block tilting is

$$\Delta L_2 \sim 4\text{-}7 \text{ km} \tag{26}$$

The total extension is

$$\Delta L = \Delta L_1 + \Delta L_2 \sim 13\text{-}20 \text{ km} \tag{27}$$

The present width of the graben segment AE is 195 km which gives the average relative extension

$$\varepsilon \sim 0.07\text{-}0.11 \tag{28}$$

The average depth of the Viking Graben in profile 1 is $h_s \sim 5$ km.

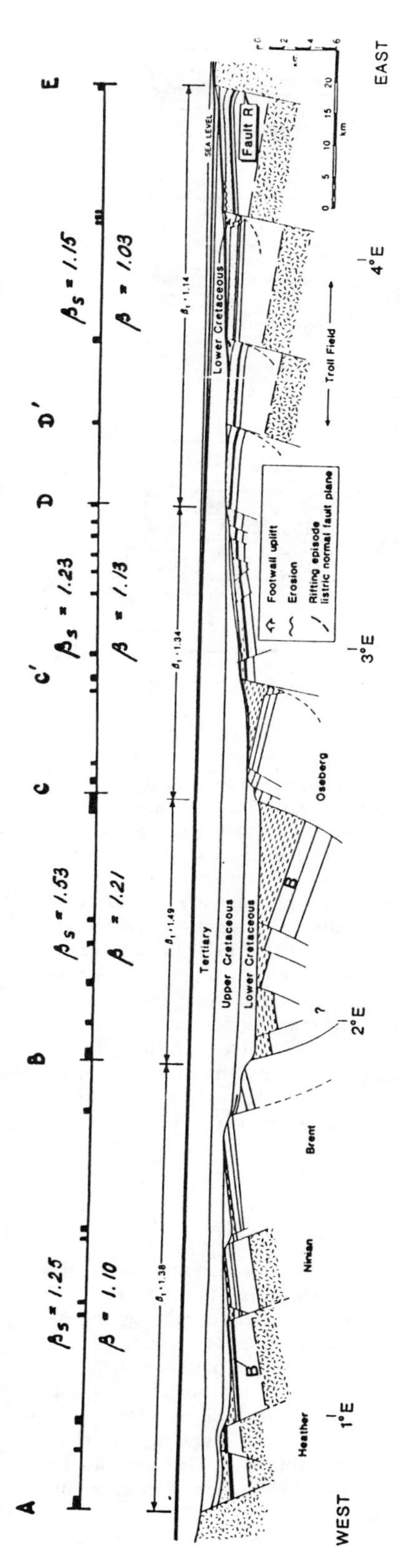

Fig. 5. Structural cross-section across the Viking Graben (Badley et al., 1988). Vertical exaggeration is 2:1. β_t – the values of the stretching factor at the top Brent (B) for the segments AB, BC, CD and DE that were calculated by the authors of the profile; solid bars on the upper line above the profile are projections of the displacements of the fault blocks along the normal faults. Extension in each segment is calculated as a sum of these projections. β is the stretching factor calculated as a ratio of the extension in each segment to the length of the segment minus the extension. β_s is the stretching factor that would be necessary to produce the subsidence by crustal stretching; it is calculated as $\beta_s = \varepsilon_s + 1$ where ε_s is determined by (11).

According to (11) in a basin of this depth that was produced by stretching of the crust ε_{st} would be about 0.4.

Figure 5 illustrates non-schematic profile across the Viking Graben that has been released recently [Badley et al., 1988]. This profile, is ~200 km long, and crosses the graben between profiles 1 and 2 (see Fig. 2).

Two main epochs of stretching were identified by Badley et al. [1988]. The first one is related to the Late Permian and/or Triassic. At that time broad tilted blocks were formed in segment DE. The second epoch of crustal stretching occurred during the mid Middle Jurassic-early Early Cretaceous (Bathonian-Ryazanian) interval; this was associated with significant block tilting in segments AB, BC and in the western part CC' of block CD. During the mid Middle Jurassic epoch of extension continuous syn-rift sedimentary beds of the Brent series B were disrupted. Badley et al. [1988] calculated the stretching factor β_t for the total thermal (post-rifting) subsidence (see lower line in Fig. 5). The maximum value of β_t was evaluated as 1.49 for the deepest part of the basin segment BC. The average value of β_t for the total section AE can be calculated as $\beta_t = 1.32$

Most of the profile of Figure 5 is covered by tilted blocks, which indicate crustal stretching. The profile is plotted with approximately 2:1 vertical exaggeration. This exaggerates by about two times the angle of block tilting in Fig. 5. The maximum tilting occurs in segments AB and BC where at the top of Brent the average value of θ is $\sim 5°$ and $\sim 8°$ respectively.

Taking into account the vertical exaggeration 2:1 the angle between the top Brent and straight lines that approximate the faults between the top of the adjacent fault blocks can be estimated as $\psi \sim 50\text{-}75°$ for different blocks. Then taking $\psi \sim 60°$ on the average, and $\theta \sim 5°$, $8°$ for segments AB and BC, we obtain from (5) $\beta_{st} = 1.06$ and 1.10.

In addition to block tilting the top of Brent series subsided along the normal faults to a considerable average depth $h_s \sim 6\text{-}7$ km in the axial parts of the graben. As in the other sections of the graben this requires the additional relative extension $\varepsilon \sim 0.5\text{-}0.10$. Together with $\varepsilon_{st} \sim 0.06\text{-}0.10$ this gives the net extension in segments AB and CD of about 0.11 and 0.20.

A net extension can be calculated as a sum, ΔL, of horizontal offsets along the normal faults (shortening due to tilting of the surface of the blocks can be neglected at small θ). These offsets are projected onto the upper line above the profile in Figure 5. The corresponding values of $\beta = L/(L\text{-}\Delta L)$ for different segments are shown below the upper line. They are considerably smaller than those calculated by the authors of the profile (see the lower line).

The values of β_{st} that are necessary to produce the subsidence after the deposition of the Brent series are plotted above the upper line. The corresponding relative extension $\varepsilon_{st} = \beta_{st}\text{-}1$ is several times larger than the observed extension $\varepsilon = \beta\text{-}1$.

The average value of β for the Viking Graben for profile of Fig. 5 is

$$\beta = 1.11 \tag{29}$$

This is of the same order as the values of extension (18), (21) and (28) for the other profiles across the graben. This confirms that even schematic profiles can be used for the estimates of extension.

In segment D'E the basement (which can be followed only tentatively) is tilted considerably more intensely than the top of Brent. For this segment the stretching factor at the top of

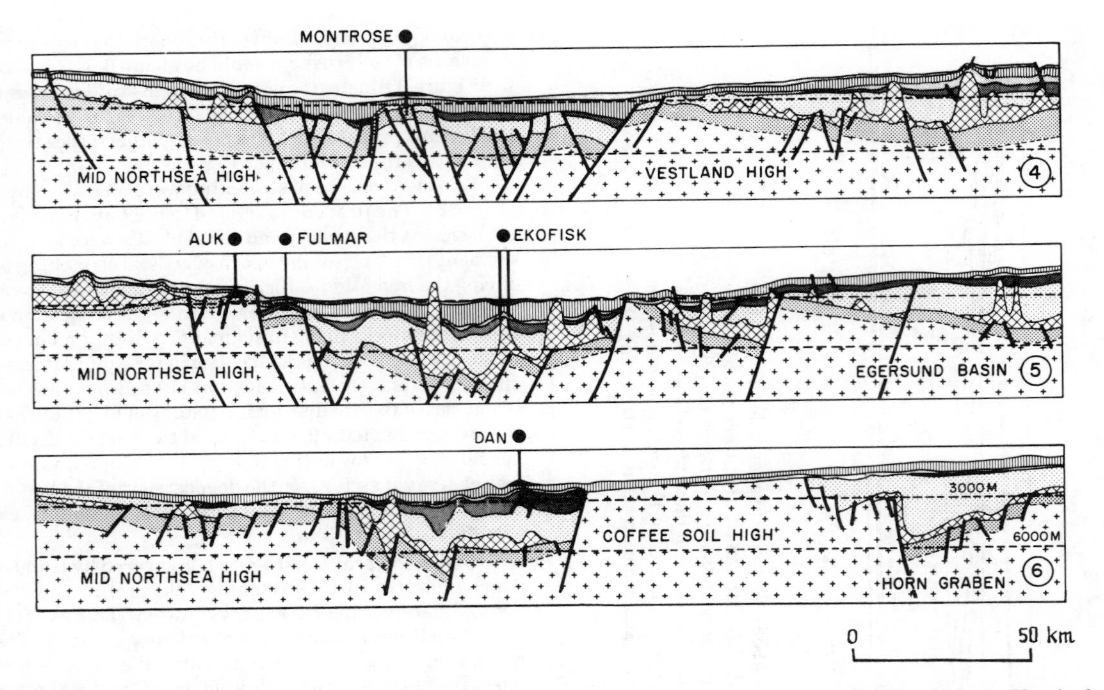

Fig. 6. Structural cross-sections of the central North Sea (see Fig. 2 for locations) (after Ziegler, 1982a). Vertical exaggeration is 5:1. Legend same as in Figure 4.

the basement can be estimated as $\beta = 1.03$, which is a rather low value.

A deep seismic reflection profile was shot in a NW-SE direction [Beach, 1986; Beach et al., 1987] near the region crossed by the profile in Figure 5. Major fault blocks were defined; most of them are 15-20 km wide and tilted by $\theta < 10°$. In the Viking Graben and the adjacent deep basins the crust under the Mesozoic sediments is thinned by 1.5-2 times as compared to the adjacent Horda Platform to the east. The maximum thinning of the crust (by 2.3 times) is observed under the deepest part of the basin.

It was suggested [Beach et al., 1987] that the initial thinning of the crust by about 50%, which is inferred from the subsidence curves, occurred in the Early Mesozoic due to extension along a gently dipping detachment shear zone passing eastwards down through the crust. This is an example of a qualitative approach to the problem, when extensional features are considered as a direct indicatiom on the extensional origin of crustal thinning. The profiles in the above two papers are not as detailed as that by Badley et al. [1988]; however, the extension can be measured in them in the same way as that used for the above profile. For instance, for the profile in Figure 5 of Beach et al. [1987] this gives $\varepsilon \sim 0.1$ which is five times smaller than the presumed value (0.5).

The above authors suggested that the rest of the thinning of the crust under the basin occurred due to extension along the Viking Graben in the Late Jurassic-Early Cretaceous. They, however, gave no evidence on extension in that direction. Moreover, the maps of the faults [see, Badley et al., 1984] show that most extensional normal faults are oriented along and not across the basin.

Thus, extension in the Viking Graben and the adjacent deep basins was not large: about 10% on the average and up to 20% in some narrow zones. This is much smaller than thinning of the crust under the region (1.5-2.3 times). It cannot be precluded that the extension took place in the hanging wall above the detachment shear zone; however, this was not the main cause of thining of the crust.

Numerous gently dipping shear zones were revealed by seismic reflection profiling in different regions [Barazangi and Brown, 1986a, b; Matthews and Smith, 1987]. In many cases they were formed due to the emplacement of large nappes during an intense shortening of the crust. Such shear zones can probably be reactivated and transformed into low-angle detachments during the extensional phases. The fact of extension along the detachment does not, however, immediately mean that this was a cause of crustal thinning under the region. The amount of extension should be measured and compared with that of thethinning of the crust.

Central Graben

Three profiles across the Central Graben are shown in Figure 6. Considering the low level of resolution of the basement, we estimate the maximum possible stretching at the basement level as shown in Figure 6. The estimates, which are similar to those for the profiles in Figure 4, give $\varepsilon \sim 0.1$. The depth of the basement in this graben is $h_s \sim 5$-7 km which would require $\varepsilon_{st} \sim 0.4$-0.6 if the subsidence resulted from crustal stretching alone. A deep seismic sounding line [Barton and Wood, 1984] gives a two-fold thinning of the consolidated crust under the graben. If the basin was formed by stretching this would correspond to $\varepsilon_{st} \sim 1$.

The Mid North Sea High to the west of the graben is a sedimentary basin, 3-6 km deep. In order to produce this basin by stretching ε_{st} should be about 0.2-0.5. The real extension in this region does not exceed a few per cent. At $\psi \sim 50$-$60°$ the same extension should be typical of the Vestland High in profile 4 and Egersund Basin in profile 5 because they are sedimentary basins that are 3-6 km deep.

A profile across the Central Graben and the adjacent deep basins was also published by Gibbs [1984]. At a distance of 80 km this includes a set of tilted fault blocks. Their tops are at a depth of 4-6 sec. (two-way travel time), which probably corresponds to the Late Jurassic. A direct measurement of extension in this profile gives $\varepsilon \sim 17\%$.

Intensity of Stretching

In order to produce the observed subsidence and thinning of continental crust in the North Sea the magnitude of stretching should be $\varepsilon \sim 0.5$-1.0 ($\beta \sim 1.5$-2) in the grabens and $\varepsilon \sim 0.2$-0.5 ($\beta \sim 1.2$-1.5) outside them. The above considerations show that the actual magnitude of stretching is several times or more smaller: $\varepsilon \sim 0.1$-0.2 in the grabens and a few per cent (~ 0.02) in most regions outside them. This has already been shown by Ziegler [1982a, 1983], but the above analysis provides an easy check on the estimates of stretching.

Our estimates of extension in schematic profiles are based on fault dip angles of $\psi \sim 50$-$60°$. Low-angle normal faulting with large detachment of the sedimentary strata could considerably increase the magnitude of extension. As was mentioned above, a gently dipping detachment passing down through the crust had been supposed for the northern North Sea. Such detachments, if they exist, do not, however, produce large disruptions in the sedimentary cover which would be required by a large stretching.

Crustal uplift in the Middle Jurassic resulted in erosion of the Early Mesozoic and in some cases of the Paleozoic deposits on regional highs. Yet in some profiles the Devonian and Lower Permian (Rotliegend) strata can be traced over virtually all of the basin. For instance, in profiles 4 and 5 of Figure 6 the net detachment of these strata on a number of normal faults is only about 20 km. In Figure 3b the Rotliegend sands can be traced almost continuously for 180 km. In the other profiles moderately disrupted Triassic, Permian (Rotliegend) and Devonian strata are preserved in the deepest parts of the basin (see profiles 1 to 3 in Figure 4 and profile 6 in Figure 6). They would be disrupted most intensely there if the subsidence resulted from low-angle normal faulting.

The absence of large crustal stretching can easily be proved for the Mesozoic and Cenozoic deposits where normal faults are well resolved by reflection seismic data. The accuracy of identification and location of the normal faults in the Paleozoic deposits is much less certain. The normal faults may flatten with depth which would increase the amount of extension.

Large stretching in wide basins requires tilting of the blocks through a large angle θ. This produces an unconformity with the same angle θ between the strata deposited before and after stretching [Artyushkov and Baer, 1983, 1984a]. The main subsidence during the Paleozoic occurred in Permian time when the Zechstein deeper-water sea was formed (see below). Numerous wells show a conformable relation between the Upper Permian Zechstein strata and the underlying Rotliegend sands, which is considered as a part of the pre-rift sedimentary sequence. This precludes significant stretching during the Paleozoic. Thus the bulk of the observed crustal stretching evident in the North Sea was achieved during Triassic to Cretaceous.

Possible Causes of Crustal Thinning

Despite a large sediment thickness in many regions under the margins of the North Sea, the crustal thickness is the same there as under the adjacent areas. It is strongly reduced under the Central and Viking Grabens. In the presence of only limited upper crustal extension this indicates thinning or even the destruction of the lower crust. Two possible causes of this phenomenon can be discussed. Firstly, the lower crust could have been eroded or stretched by convection flow in the underlying mantle. Secondly, contraction of rocks with an increase in seismic velocities could have occurred in the lower crust from the gabbro-eclogite phase transformation.

The former mechanism was applied for the North Sea by Ziegler [1983] and other authors. It can be shown that subcrustal erosion or stretching of the lower crust are unable to produce a considerable crustal thinning; however, a correct proof of this statement requires a special paper, which will be published elsewhere. The main arguments are the following.

Thinning of the lower crust under the basin produces an additional compressive force in the lithosphere [Artyushkov, 1973]. This force should be overcome by viscous drag at the base of the crust. The drag results from the stresses produced by density inhomogeneities in the mantle under the crust.

The flow in the mantle can reach the base of the crust only in regions with an increased heat flow and high temperature gradient in the mantle. Due to a strong temperature dependence, the viscosity in the uppermost mantle in such regions decreases with depth by one order of magnitude per each 10 km.

Under a strong decrease in the viscosity with depth, density inhomogeneities creating the flow under the crust can be concentrated only in the uppermost layer of the mantle of a thickness $\ell \sim 10$ km. This produces shear stresses at the base of the crust $\tau \sim \Delta\rho g \ell/2$ where $\Delta\rho$ is the density contrast and g is the gravity. For $\Delta\rho \sim 0.5$ g/cm^3 and $\ell \sim 10$ km the stresses τ are of ~ 25 bar.

A decrease in the crustal thickness by about 20 km requires the additional stresses in the crust of the order of 1 kbar. They cannot be produced by viscous friction $\tau \sim 25$ bar at the base of the crust in a narrow region. These low stresses are able to reduce the crustal thickness only by about 1 km.

Thus, subcrustal erosion or stretching of only the lower crust cannot explain the large observed thinning of the crust under the North Sea. The only remaining possibility is a compaction of the lower crust under the region due to the transformation of gabbro in the lower crust into dense garnet granulites or eclogite. This mechanism is very speculative from the petrological point of view; however, among all the known ones it seems to fit better the subsidence without large stretching at the surface [Artyushkov and Baer, 1983, 1984a, 1986a].

In many deep basins on continental crust subsidence was very rapid and occurred during one or a few million years. This is probably associated with migration into the lower crust from the mantle of a water-containing fluid, which strongly increases the transformation rate. The subsidence in narrow linear zones such as the grabens of the North Sea can probably be explained by the ascent of fluid along large fault zones dipping into the asthenosphere. For instance, the Baikal Graben formed during the last 3-4 Ma [Artyushkov et al.,

1987]. The subsidence occurred along a huge boundary fault zone of the Siberian Platform which reaches the asthenospheric upwelling at a depth of ~100 km.

The phase transformation produces basic rocks that can be both lighter or denser than the mantle. The former rocks – garnet granulites with P-wave velocities Vp~7.2-8.2 km/sec [Manghani et al., 1974] should rest under the residual crust. It is quite probable that they still exist under the seismically determined crust in some places in the North Sea. Then they are a part of the earth's crust from the petrological point of view.

Dense eclogites with high P-wave velocities can sink into the mantle, when the asthenospheric upwelling reaches their base. New seismic studies are necessary to determine what happened to the dense basic rocks under the North Sea.

Thinning of the crust and its subsidence due to contraction of basic rocks are possible only where the lower crust includes a thick layer of gabbro. The seismic profile across the Central Graben [Barton and Wood, 1984] shows that basic rocks with low density (Vp~6.6-7.0 km/sec) can be present in the lower crust outside the graben and under its flanks. They almost disappear from the section in the region of the largest subsidence, i.e., under the axial parts of the Central Graben.

Probable Rapid Subisdences in the North Sea

As was mentioned in the last section, contraction of basic rocks in the lower crust in many cases occurred very rapidly, which produced a rapid crustal subsidence. Rapid subsidence is commonly inferred from an abrupt replacement of shallow-water deposits by deep-water or deeper-water sediments. In most cases the thickness of the transition is ~1.10 m, which corresponds to a time interval of ~0.1-1 Ma. In some rare cases changes in the water circulation may also produce an abrupt replacement of shallow water sands or carbonates by bituminous deposits that look like deeper-water sediments. Rapid flooding of pre-existing depression may result in abrupt replacement of terrestrial deposits by deeper-water sediments. Hence in order to relate an abrupt facies change to rapid subsidence it is necessary to have some independent proof of the phenomenon. This commonly is provided by clinoforms or reefs that rapidly (in ~1 Ma) form near the margins of the basin and after backstripping indicate its depth with respect to the adjacent shelf. The data on the North Sea do not substantiate rapid subsidence as reliably as that for West Siberia, Timan-Pechora, Pre-Caspian Basin, and many other regions. They do, however, permit the suggestion that such subsidence was a rather higher probability during three separate epochs.

Late Permian

At the transition from Carboniferous to Permian large crustal uplift and volcanism took place in the North Sea and adjacent regions [Ziegler, 1982a]. In the late Early Permian a slow sediment-loaded subsidence formed the North and South Permian sedimentary basins (Fig. 7). At the initial stage terrestrial clastics deposited on the basin slopes due to erosion of the adjacent highs. By the end of the Early Permian deposition of clastics ceased, which indicates that the relief became smooth [Füchtbauer, 1968; Maximov, 1975; Van Adrichem Boogaert and Burger, 1983].

In the early Late Permian terrestrial deposits and evaporites in the Permian basins were abruptly overlain by deeper-water bituminous Zechstein shales and limestones. Since deeper-water deposits overlaid terrestrial ones, and not shallow-water marine sediments, it cannot be precluded that the transgression resulted from flooding of the pre-existing depression. This was suggested by Ziegler [1982a]; he explained the above facies change by the break-down of topographic barriers and flooding of the area that thermally subsided below the level of the world ocean after the crustal uplift between the Carboniferous and Permian. The transgression produced the wide basin of the Zechstein Sea (Fig. 7). Rapid flooding is indicated by the generally small thickness (measured in meter) of the transitional layer from the terrestrial to marine deposits.

No large-scale stretching has been suggested for the Zechstein Sea area in the Late Paleozoic. In the absence of large-scale stretching the crustal surface after thermal relaxation must subside to the level that it occupied before heating of the crust and mantle, i.e., before the uplift. In the Zechstein Basin area before the uplift at the transition from the Carboniferous to Permian the crust was above sea level. Considerable erosion occurred during the uplift. This thinning of the crust was most probably compensated or overcompensated by the deposition of 1.5-2.5 km of clastics and evaporites in the South Permian Basin before the transgression in the early Late Permian. Therefore it is very improbable that the crust which was above sea level before the uplift could subside below sea level from thermal relaxation in 20 Ma by the Late Permian.

After the transgression 2-3 km of sedimentary deposits accumulated in the basins in the Late Permian and Triassic. The sediment-loaded subsidence of at least 2-3 km below the initial level was possible only because of contraction of crustal rocks from phase transformations.

The Zechstein Sea included a shallow-water shelf and a deeper-water inner part (see Fig. 7). After the transgression, evaporitic clinoforms 200-300 m high, rapidly formed in the marginal parts, 20-50 km wide, of this inner depression (Fig. 8) [Grachevsky, 1974; Ziegler, 1982a]. Isostatic adjustment does not commonly occur in regions of such a small width. Hence the height of the clinoforms should equal the minimum depth of the basin with respect to the adjacent shelf. Thus the outer shelf and inner deeper-water part of the Zechstein Sea were separated by a rather steep slope. As was mentioned above this slope did not exist before the transgression. Hence it is most probable that it was formed concomitantly with the transgression, i.e., the transgression resulted mainly from crustal movements.

Exactly the same situation was typical of the West Siberian Basin in the late Late Jurassic [Artyushkov and Baer, 1986b]. During ~1 Ma the area subsided by ~0.5 km with the formation of a steep slope that separated the deeper-water basin from the adjacent shelf. Clinoforms, 200-300 m high, formed rapidly near this slope.

It is also easy to see from Fig. 7 that the boundary of the deeper-water part of the basin does not coincide with the boundary of the Permian basins filled by terrestrial deposits. For instance, the deeper-water basin spread over the highs to the east of the Northern Permian Basin and to the west of the Belorussian High. It is very improbable that no deposition could occur in these regions just before the transgression, if at that time they really were 200-300 m below the adjacent areas. The discrepancy of the contours of the terrestrial and subsequent deeper-water deposition most probably indicates a rapid crustal subsidence in the Zechstein Basin. According to the studies of varved sediments the rate of subsidence was

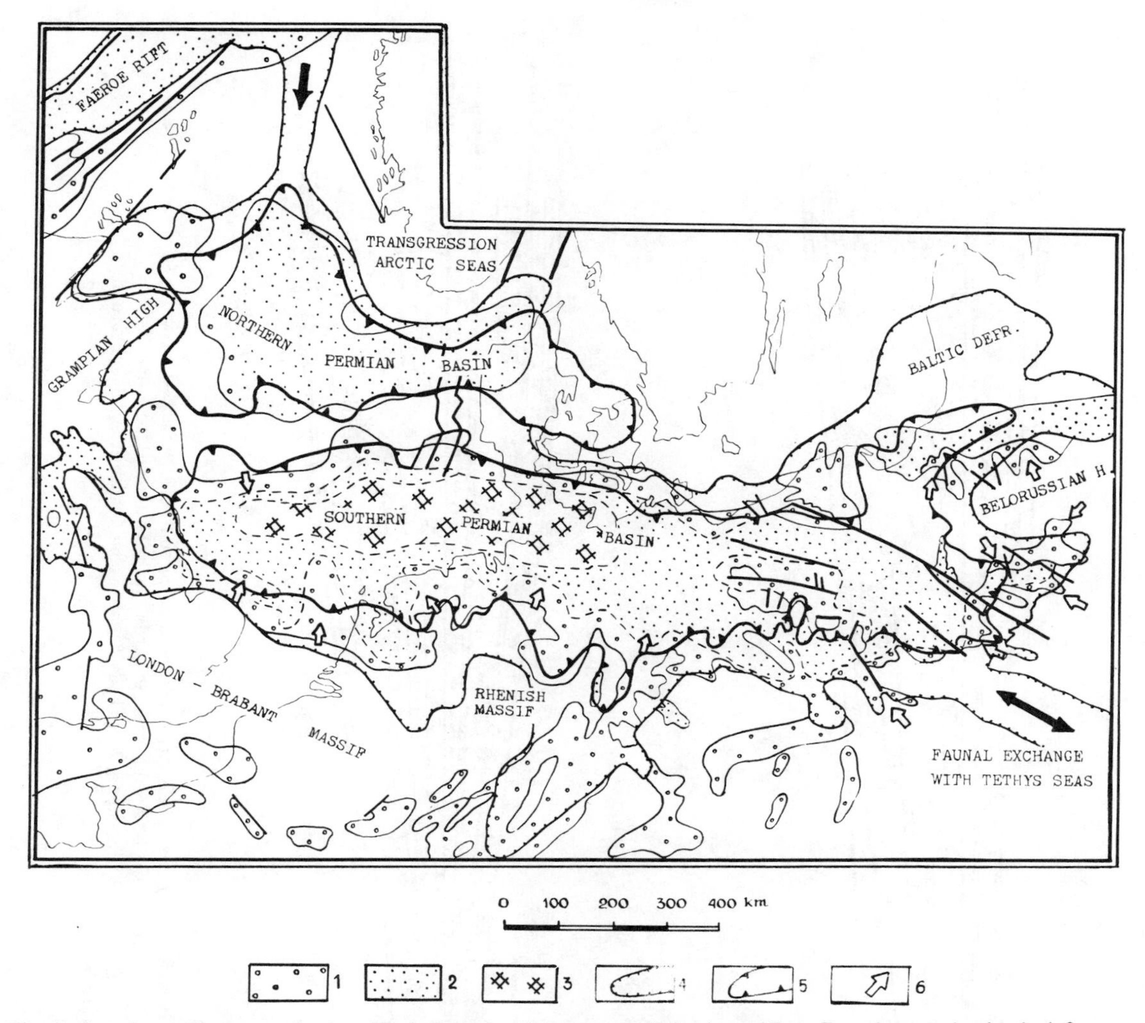

Fig. 7. Permian sedimentary basins: Early Permian intramountain basin and Late Permian marine basin (after Ziegler, 1982a). 1-3 Early Permian (Rotliegend) Basin (1. sands and conglomerates, 2. sands, shales and sulphates, 3. shales and halite), 4-5 Late Permian (Zechstein) Basin (4. boundary of marine basin, 5. boundary of deeper-water part of the basin), 6. Direction of clastic influx.

~0.5 km/Ma [Füchtbauer, 1968]. The initial depth of water was 400-500 m [Grachevsky, 1974; Ziegler, 1982a].

No nappes were emplaced near the Zechstein Basin in the Late Permian, which precludes subsidence due to thrust loading. We suggest that rapid subsidence in this basin (if it did occur) resulted from the contraction of the lowermost crust due to gabbro-ecologite transformation under an increase in the inflow of fluid. Then the break-down of the topographic barriers, which resulted in flooding of the basin should also be attributed to the same mechanism.

As follows from the condition of isostatic balance, destruction of a layer of lower crust of density ρ_{lc} and thickness Δh_{lc} produces, after cooling of the crust and mantle, a sedimentary basin of the depth

$$h_s = [(\rho_m - \rho_{lc})/(\rho_m - \rho_s)]\Delta h_{lc} \qquad (30)$$

From two to three kilometers of sediments were deposited in the Zechstein basins in the late Late Permian and Triassic. Substituting $h_s = 2\text{-}3$ km into (30) and taking $\rho_m = 3.35$ g/cm^3, $\rho_{lc} = 2.95$ g/cm^3 and $\rho_s = 2.45$ g/cm^3 we find

$$\Delta h_{lc} = 5\text{-}7 \text{ km} \qquad (31)$$

Late Jurassic

In the Early Jurassic the North Sea represented a shelf region. In the Middle Jurassic a considerable crustal uplift and volcanism took place for ~10 Ma in the central part of the area [Ziegler, 1982a]. These phenomena indicate asthenospheric upwelling. The intensity of vertical movements was nonuniform in space. From one to three kilometres of the Paleozoic-Triassic deposits were eroded from the highs. Only minor erosion took place in the grabens of the Central North Sea: the Central Graben, the Horda, Egersund and Norwegian-Danian Basins. This indicates relative

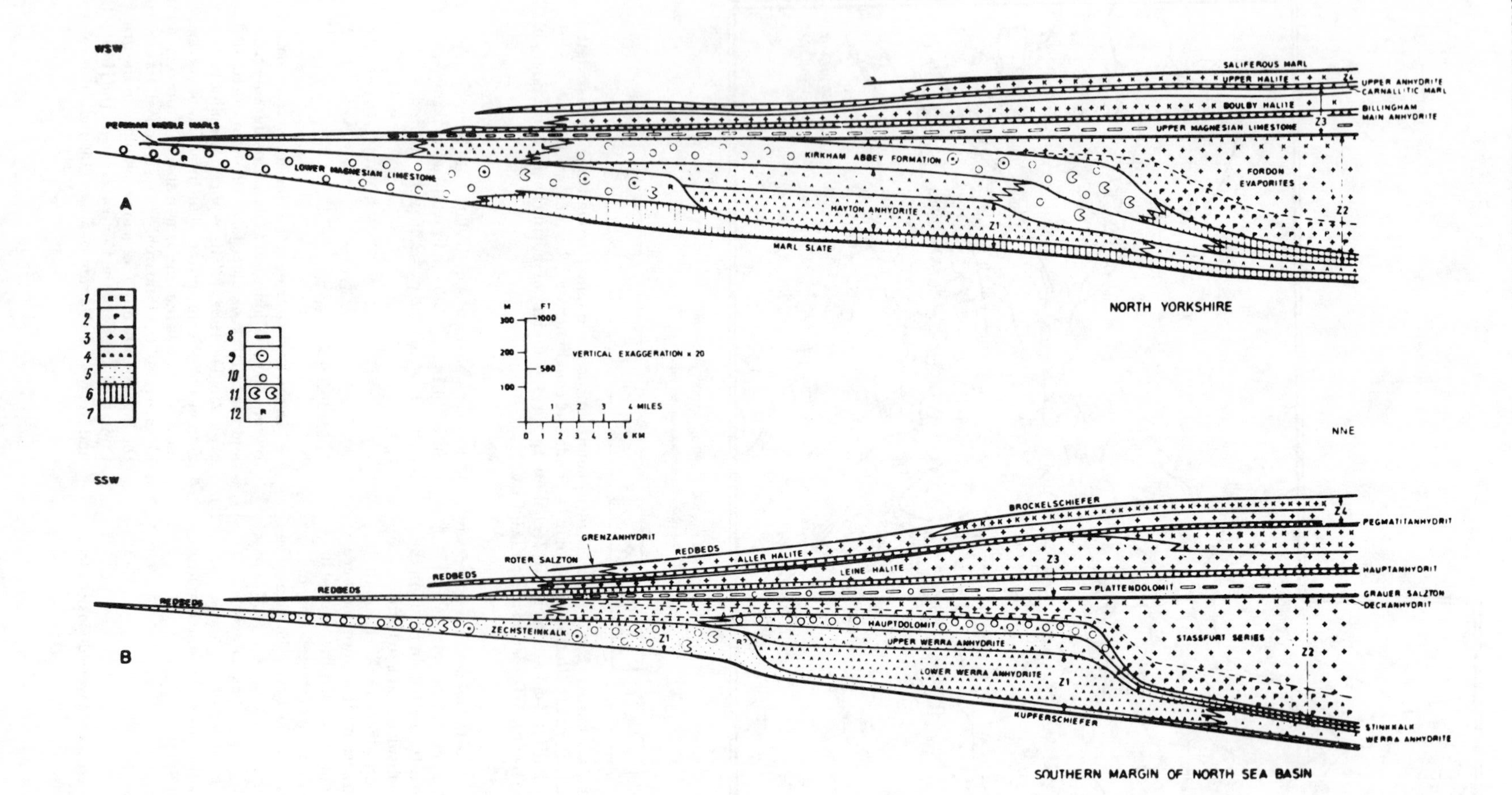

Fig. 8. Generalized cross-sections into the Upper Permain Zechstein deposits of the southern North Sea Basin (after Taylor and Colter, 1975). a. North Yorkshire, b. Norfolk coast towards Viking Graben. 1. potash salts, 2. polyhalite, 3. halite, 4. anhydrite, 5. dolomite, 6. limestone, 7. red beds, 8. calcinema, 9. pisoliths, 10. oolitic and pelletoid sands, 11. skeletal grains, 12. reef.

subsidence of the grabens by 1-2 km with respect to the adjacent highs. The stretching in the grabens by $\beta = 1.10$-1.15 times could result in subsidence of 2-3 km due to sediment loading (see below). Stretching took place in the Middle Jurassic-Early Cretaceous, and only a small part occurred during the Middle Jurassic uplift. Hence relative subsidence by 1-2 km in the grabens of the central North Sea most probably required an increase in crustal density.

In the Viking Graben a considerable subsidence (up to 3 km of the sediments) began earlier – in the Triassic. Hence contraction of the crust should have begun there earlier than in the central North Sea.

After the crustal uplift in the Middle Jurassic a gradual marine transgression took place in the North Sea for ~20 Ma in the Late Jurassic (Callovian-Kimmeridgian). During that time interval there is supposed to have been a relative rise of sea level up to 300 m [Vail et al., 1977]; however, another mechanism of sea level change has also been suggested [Cloetingh et al., 1985]. During this period of time three abrupt facies changes took place in different parts of the North Sea. As a result grey shallow-water sands and clays containing minor carbonates were overlain by black and dark-grey organic shales of euxinic type. The thickness of the transition is several meters or less [Hay, 1978, and others].

The first facies change occurred in the Northern Viking Graben in the Early Callovian [Ziegler, 1982a]. The second facies change took place in the Early Oxfordian in the Central Graben and in the Southern Viking Graben. Deposition of turbidites began in the marginal parts of the Central and Viking Grabens after this facies change [Ziegler, 1982a; Hamar et al., 1983]. The third facies change took place between the Oxzfordian and Kimmeridgian and covered the residual part of the North Sea: the Moray-Firth and Norwegian-Danian Basins and the Eastern England Shelf. After the facies change, deposition of turbidities began in the Moray-Firth Basin. On the Eastern England Shelf the rate of sediment-loaded subsidence increased up to 100-500 m/Ma which proceeded for ~1.5 Ma [Penn et al., 1986].

The occurrence of deeper-water sandy fan deposits and turbidites in the Central and Viking Grabens and in the Moray-Firth Basin indicates that their depth reached ~0.5 km with respect to the adjacent regions [Michelsen, 1982; Ziegler, 1982a; Hamar et al., 1983]. At the same time some large faults were formed on the margins of the grabens. The relative depth of the marginal intrashelf basins – the Norwegian-Danian, Horda on the north-east and Sole-Pit, Midland on the south-west (see Fig. 2) was not very large, ~100-200 m.

The abrupt facies changes in the Late Jurassic are described in all the publications as a replacement of shallow-water deposits by deeper-water ones. Deposition of turbidites began concomitantly with the facies changes. These data most probably indicate several episodes of rapid subsidence in the Late Jurassic which was more intense in the grabens.

Other explanations are of course possible. For instance, deeper-water basins could have been formed gradually by slow subsidence; then the facies changes and deposition of turbidites could be attributed to abrupt changes in the water circulation and intensity of erosion of the adjacent regions. Concomitant occurrence of these changes seem, however, less probable than rapid crustal subsidence.

Late Paleocene

The third level of an abrupt change from shallow-water to deeper-water facies is in the Upper Paleocene. Shallow-water deposition took place in the North Sea in the Late Cretaceous-Middle Paleocene, except in the Viking Graben and some regions in the Central Graben [Ziegler, 1982a; Michelsen, 1982]. Some chalks, particularly in the Central Graben were deposited in considerable water depth. Between the Middle and Late Paleocene shallow-water marls and chalks were abruptly replaced by deeper-water carbonate free clays. The thickness of the transition is ~1 m. After the facies change sandy turbidites and fans were deposited in the Central and Viking Grabens and in the Moray-Firth Basin. The depth of water in the grabens was ~0.5 km [Ziegler, 1982a]. At the adjacent highs it was probably 0.2-0.3 km.

It cannot be precluded that the North Sea already had a considerable depth before the above facies change. The abrupt facies change together with a concomitant appearance of turbidites and fans, however, indicates a rapid deepening of the basin. In some places deposition of turbidites could also result from the uplift of the basin margins (e.g., thermal uplift on the Shetland Shelf).

The presumed subsidence occurred long after the preceding thermal events in the North Sea: 110 Ma after the crustal uplift in the Middle Jurassic, 90 Ma after the rapid subsidence in the Late Jurassic and 70 Ma after the termination of moderate stretching in the Middle Jurassic-early Early Cretaceous [Ziegler, 1982a; Badley et al., 1988]. From two to three kilometres of sediments were deposited in the North Sea after the subsidence in the Paleocene. No stretching has been inferred for this epoch. Hence a considerable increase of density in the lower crust or the desctruction of the lowermost crust is necessary to account for the subsidence in the Cenozoic. The distribution of isopachs (Fig. 9) is isometric and not directly related to the grabens.

The Amount of Crustal Thinning

The total sediment thickness in the inner part of the North Sea is h_s~6-10 km. In this region the crust was stretched by

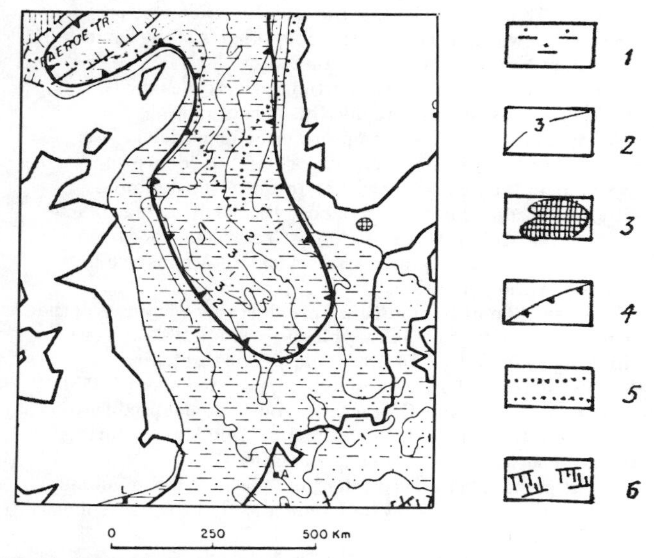

Fig. 9. Isopach map of Tertiary deposits in the North Sea (after Ziegler, 1978, 1982a) 1. Tertiary deposits, 2. Tertiary isopachs in 1000 m, 3. Lower Tertiary volcanics, 4. boundary of the Paleocene-Eocene deeper-water deposits, 5. Viking and Faeroe grabens, 6. faults.

$\beta = 1.10$-1.15 times on the average. This could attenuate crust of prestretched thickness $h_c^0 = 35$ km by 3.2-4.6 km. Crustal stretching produces (after cooling of the crust and mantle) a sedimentary basin of a depth

$$(h_s)_{st} = [(\rho_m - \rho_c)/(\rho_m - \rho_s)](1 - 1/\beta)h_c^0 \qquad (32)$$

Taking the above values of h_c^0 and β, $\rho_m = 3.35$ g/cm^3 and the densities of the crust and sediments as $\rho_m = 2.85$ g/cm^3, $\rho_s = 2.45$ g/cm^3 we obtain:

$$(h_s)_{st} = 1.8\text{-}2.5 \text{ km} \qquad (33)$$

This can only account for a minor part of the subsidence. Deposition of the remaining 4-8 km of rocks must be due to contraction or destruction of the lower crust. Assume, for example, that a layer of the lower crust, Δh_{lc} thick, was destroyed, or acquired a density equal to that of the mantle and remained under the crust. Then as follows from (30):

$$\Delta h_{lc} = 9\text{-}18 \text{ km} \qquad (34)$$

In regions of lithospheric stretching the uppermost crust is broken into narrow tilted blocks [Barberi et al., 1975; Proffett, 1977, and others]. As can be seen from Figures 3 to 6 most of the North Sea is underlain by wide blocks (>10-15 km). Within the blocks the strata are continuous or almost continuous. In the presence of such deformation (apparent) crustal thinning may be independent of crustal stretching, and it could arise in the following way. The phase transformation produces contraction of rocks in the lower crust. Under regional tensile stresses the upper crust breakes into wide blocks which subside isostatically in the upwelled asthenosphere. In this case the (apparent) crustal thinning in the regions of no significant block tilting is entirely produced by the phase transformation, and crustal extension occurs as a passive response (isostatic subsidence along the normal fault) of the attenuated crust and lithosphere to regional tensile stresses. Then the initial thickness of the contracted or destroyed lower crust under the deepest parts of the basin can be estimated from (30) as $\Delta h_{lc} \sim 14$-23 km.

Asthenospheric upwelling strongly decreases the thickness of the lithosphere and this permits the easy development of tensile or compressive deformations. This occurred in many basins formed by rapid subsidence, for example, in West Siberia and Timan-Pechora. A similar situation occurred in the North Sea. Block tilting took place there in the Middle Jurrassic-Early Cretaceous; moderate compression occurred during the latest Cretaceous and Late Paleocene [Ziegler, 1987].

These phenomena indicate a considerable weakness of the lithosphere in the North Sea which is incompatible with its thinning by $\beta = 1.10$-1.15 times solely from stretching. Furthermore, moderate stretching terminated about 130 Ma ago, in the early Early Cretaceous, and could not produce a significant thinning of the lithosphere 55-60 Ma ago during the last epoch of compressive deformations.

The present heat flow in the North Sea is high: $q \sim 60$-80 mWt/m^2 [Čermák, 1979] which indicates that the lithospheric thickness is still considerably reduced.

Conclusions

The above considerations show that S-type normal faulting is typical of most of the North Sea Basin. Under this condition only very narrow basins can be produced by stretching. The North Sea and the grabens within it have a large width. Hence most of the subsidence was not produced by stretching.

In some regions of the North Sea A-type normal faulting with block tilting is observed. The angle of tilting is, however, small (<5-8°) which accounts for only minor stretching ($\beta \sim 1.05$-1.10). Together with the S-type normal faulting this gives extension 10-20% in the grabens and a few per cent outside them.

The amount of upper crustal extension is at least several times smaller than the observed amount of lower crustal attenuation and does not fit the amount of well documented subsidence in the North Sea Basin. Subcrustal erosion or stretching of the lower crust by convection flows in the mantle are able to attenuate the crust only by about 1 km. Hence, large thinning of the crust under the North Sea without an adequate stretching at the surface can be attributed to contraction of the lower crust due to gabbro-garnet granulite-eclogite phase transformation.

The dense basic material with high P-wave velocities can still remain under the seismically determined crust or it could have sunk into the underlying mantle. The thinned crust broke into wide blocks (>10 km) and subsided isostatically under regional tensile stresses. Thinning of the lithosphere after the asthenospheric upwelling ensured the development of moderate block tilting and compressive deformations.

In many other basins studied earlier, contraction of rocks in the lower crust from phase transformation took place at a high rate and rapidly produced deep-water or deeper-water basins. Three abrupt transitions from shallow-water or terrestrial facies to deeper-water facies are observed in the North Sea. They occurred in the Late Permian, Late Jurassic and Paleocene. The transitions were associated with the formation of clinoforms or deposition of turbidites. In many other basins similar features were connected with rapid crustal subsidence (West Siberia, Timan-Pechora, Volga-Urals, Pre-Caspian Basin, Permian Basin of Texas and many others). Hence it can be concluded that rapid crustal subsidence occurred in the North Sea at the above three epochs.

Acknowledgement. The authors are very indebted to P.A. Ziegler, Shell International Petroleum Mij.Bv. for valuable comments and discussions.

References

Artyushkov, E.V., Stresses in the lithosphere caused by crustal thickness inhomogeneities, J. Geophys. Res., 78, 7675-7708, 1973.

Artyushkov, E.V., Rifts and grabens, Tectonophysics, 133, 321-331, 1987.

Artyushkov, E.V., and M.A. Baer, Mechanism of continental crust subsidence in fold belts: the Urals, Appalachians and Scandinavian Caledonides, Tectonophysics, 100, 5-42, 1983.

Artyushkov, E.V., and M.A. Baer, On the mechanism of continental crust subsidence in the Alpine Fold belt, Tectonophysics, 107, 193-228, 1984a.

Artyushkov, E.V., and M.A. Baer, Mechanism of continental crust subsidence around the Northern Pacific. II. North American Cordilleras, Tikhookean. Geol., 5, 3-15, 1984b.

Artyushkov, E.V., and M.A. Baer, Mechanism of formation of deep basins on continental crust in the Verkhoyansk fold belt: miogeosynclines and cratonic basins, Tectonophysics, 122, 217-245, 1986a.

Artyushkov, E.V., and M.A. Baer, Mechanism of formation of hydrocarbon basins: the West Siberia, Volga-Urals, Timan-Pechora basins and the Permian Basin of Texas, Tectonophysics, 122, 247-281, 1986b.

Artyushkov, E.V., F.A., Letnikov, and V.V. Ruzhich, On the possible mechanisms of the Baikal origin (abstract), in Intracontinental Mountainous Terrains: Geological and Geophysical Aspects Symposium,Irkutsk, pp. 238-284, 1987.

Badley, M.E., T. Egeberg, and O. Nipen, Development of rift basins illustrated by the structural evolution of Oseberg feature, Block 30/6, offshore Norway, J. Geol. Soc. London, 141, 639-649, 1984.

Badley, M.E., J.D. Price, C. Rambech Dahl, and T. Agdestein, The structural evolution of the Northern Viking Graben and its bearing upon extensional modes of basin formation, J. Geol. Soc. London, 145, 455-472, 1988.

Barazangi, M., and L.D., Brown (Eds.), Reflection Seismology: A Global Perspective, Geodynamics Series, 13, 311 pp., American Geophysical Union, Washington, D.C., 1986a.

Baranzangi, M., and L.D., Brown (Eds.), Reflection Seismology: The Continental Crust, Geodynamics Series, 14, 339 pp., American Geophysical Union, Washington, D.C., 1986b.

Barberi, F., G. Ferrara, R. Santacroce, and J. Varet, Structural evolution of the Afar triple junction, in Afar Depression of Ethiopia: Volume 1, edited by A. Pilger, and A. Rosler, pp. 38-54, E. Schweizerbart'sche Verlagsbuchhandlung Stuttgart, 1975.

Barton, P., and R. Wood, Crustal thinning and subsidence in the North Sea: Reply to comment, Nature, 304, 561, 1983.

Barton, P., and R. Wood, Tectonic evolution of the North Sea basin: crustal stretching and subsidence, Geophys. J. Roy. Astron. Soc., 79, 987-1022, 1984.

Beach, A., A deep seismic reflection profile across the northern North Sea, Nature, 323, 53-55, 1986.

Beach, A., T., Bird, and A. Gibbs, Extensional tectonics and crustal structure; deep seismic reflection data from the northern North Sea Viking Graben, in Continental Extensional Tectonics, edited by M.P. Coward, J.F. Dewey, and P.L. Hancock, Geol. Soc. London Spec. Publ. 28, pp. 467-476, 1987.

Best, A., F. Kockel and H. Schöneich, Geological history of the southern Horn Graben, Geol. en Mijnbouw, 62, 25-34, 1983.

Čermák, V., and L. Rybach (Eds.), Terrestrial Heat Flow in Europe, 328 pp., Springer-Verlag, New York, 1979.

Cloetingh, S., H. McQueen and K. Lambeck, On a tectonic mechanism for regional sea level variations, Earth Planet. Sci. Lett., 51, 139-162, 1985.

Drummond, M., and C. Smith (Eds.), Deep Seismic Reflection Profiling of the Continental Lithosphere, Geophys. J. Royl. Astron. Soc., 89, 1-495, 1987.

Füchtbauer, H., Carbonate sedimentation and subsidence in the Zechstein Basin (Northern Germany), in Recent Developments in Carbonate Sedimentology in Central Europe, edited by G. Müller, and G.M. Friedman, Springer-Verlag, New York, pp. 196-204, 1968.

Gibbs, A., Structural evolution basin margins, J. Geol. Soc. London, 141, 609-620, 1984.

Grachevsky, M.M. Paleogeomorphological bases of the oil and gas occurrence, 156 pp., Nauka, Moscow, 1974.

Hamar, G.P., T. Fjaeran, and A. Hesjedal, Jurassic stratigraphy and tectonics of the south-eastern Norwegian offshore, Geol. en Mijnbouw,62, 103-114, 1983.

Hay, J.T.C., Structural development of the northern North Sea, J. Petrol. Geology, 1, 65-77, 1978.

Le Pichon, X., and J.C. Sibuet, Passive margins: a model of formation, J. Geophys Res., 86, 3708-3720, 1981.

Maximov, S.P., (Ed.), Oil and Gas Fields of the Hydrocarbon Province of the Northwest Europe, Nedra, Moscow, 208pp., 1975.

Manghnani, M.H., R., Ramanantoandro, and S.P. Clark, Jr., Compressional and shear wave velocities in granulite facies rocks and eclogites to 10 kbar, J. Geophys. Res., 79, 5427-5446, 1974.

McClay, K.R., and P.G. Ellis, Analogue models of extensional fault geometries, in Continental Extensional Tectonics, edited by M.P. Coward, J.F. Dewey, and P.L. Hancock, Geol. Soc. London Spec. Publ. 28, pp. 109-125, 1987.

Michelsen, O., (Ed.,), Geology of the Danish Central Graben, Geol. Surv. of Denmark, Ser. B., 8, 133 pp., 1982.

Penn., I.E., B.M. Cox and R.W. Gallois, Towards precision in stratigraphy: geophysical log correlation of Upper Jurassic (including Callovian) strata of the Eastern England Shelf, J. Geol. Soc. London, 143, 381-410, 1986.

Proffett, J.M., Jr., Cenozoic geology of the Yerington district, Nevada, and implications for the nature and origin of Basin and Range faulting, Geol. Soc. Amer. Bull., 88, 347-266, 1977.

Sclater, J.C., and P.A. Christie, Continental stretching: an explanation of the post Mid-Cretaceous subsidence of the central North Sea basin, J. Geophys. Res., 85, 3711-3739, 1980.

Taylor, J.C.M., and V.S. Colter, Zechstein of the English Sector of the southern North Sea, in Petroleum and the continental shelf of the North-West Europe, Volume 1; Geology, edited by A.W. Woodland, pp. 249-263, Applied Sci. Publishers Ltd., Inst. Petrol. Great Britain, 1975.

Vail, P.R., R.M. Mitchum, Jr., and S. Thompson III, Global cycles of relative changes of sea level, in Seismic Stratigraphy-Applications to Hydrocarbon Exploration, edited by C.E. Payton, Am. Assoc. Pet. Geol. Mem., 26, pp. 83-97, 1977.

Van Andrichem Boogaert, H.A., and W.F.J., Burgers, The development of the Zechstein in the Netherlands, Geol. en Mijnbouw, 1, 83-92, 1983.

Wood, R., and P. Barton, Crustal thinning and subsidence in the North Sea, Nature, 302, 134-136, 1983.

Ziegler, P.A., North Western Europe: Tectonics and Basin Development Geol. en Mijnbouw, 57, 589-626, 1978.

Ziegler, P.A., Hydrocarbon Provinces of the Northwest European Basin, Can. Soc. Pet. Geol. Mem., 6, 653-706, 1980.

Ziegler, P.A., Evolution of sedimentary basins in North-West Europe, in Petroleum Geology of the Continental Shelf of North-West Europe, edited by L.V. Illing, and G.D., Hobson, pp. 3-39, Institute of Petroleum, London, 1981.

Ziegler, P.A., Geological Atlas of Western and Central Europe, 130 pp., Elsevier, Amsterdam, 1982a.

Ziegler, P.A., Faulting and graben formation in western and central Europe, Phil. Trans. R. Soc. London, 305, 113-143, 1982b.

Ziegler, P.A., Crustal thinning and subsidence in the North Sea, nature, 304, 561, 1983.

Ziegler, P.A., Late Cretaceous and Cenozoic intra-plate compressional deformations in the Alpine Foreland: a geodynamic model, Tectonophysics, 137, 389-420, 1987.

VOLCANISM AND IGNEOUS UNDERPLATING IN SEDIMENTARY BASINS AND AT RIFTED CONTINENTAL MARGINS

Robert S. White

Bullard Laboratories, Madingley Road, Cambridge CB3 0EZ, England

Abstract. When continental lithosphere is thinned by stretching, the underlying asthenosphere wells up to fill the space. As the asthenosphere wells up it decompresses and generates partial melt. The amount of melt that is produced is critically dependent on the temperature: the hotter the asthenosphere the more melt is generated. The partial melt is buoyant, and rises quickly upwards to be added to the overlying crust. Relatively small variations in the asthenosphere temperature of the order of 100°C cause major differences in both the volume of igneous rocks generated and in the subsidence history of the stretched region. A simple model of association of a stretched region with thermal anomalies caused by plumes ('hot-spots') in the underlying mantle can explain the occurrence or absence of volcanically active sedimentary basins and rifted continental margins. This simple model works remarkably well globally.

Introduction

This paper contains a short summary of the mechanisms for generating underplating and volcanism in rifted regions, as presented at the nineteenth IUGG Symposium on the origin and evolution of sedimentary basins. However, because these ideas are developed in a series of papers that are in press at the time of writing, I simply give an outline here rather than repeating work discussed in other papers. The bibliography lists papers in which these ideas are developed: full references to allied work by other authors and to other models will be found in the papers listed in the bibliography, and are not repeated here in this summary.

Melt Generation in Rifted Regions

The model for melt generation hinges around two main results. First is a quantitative model developed by McKenzie and Bickle [1988] for the amount and composition of partial melt produced by decompression when mantle is elevated rapidly. Using McKenzie and Bickle's parameterizations and with knowledge of the temperature structure of the lithosphere and underlying asthenosphere, the amount of melt produced by asthenosphere as it wells up beneath thinning lithosphere can readily be calculated (Figure 1).

The second major result which we need in order to explain the presence or absence of significant quantities of melt in rifted regions is the temperature of the asthenosphere, and the extent and magnitude of lateral variations in this temperature. If the temperature of the asthenosphere at any given depth did not vary globally, then we would expect rather little variation in the amount of igneous activity in different rifted regions, because a given degree of lithosphere thinning

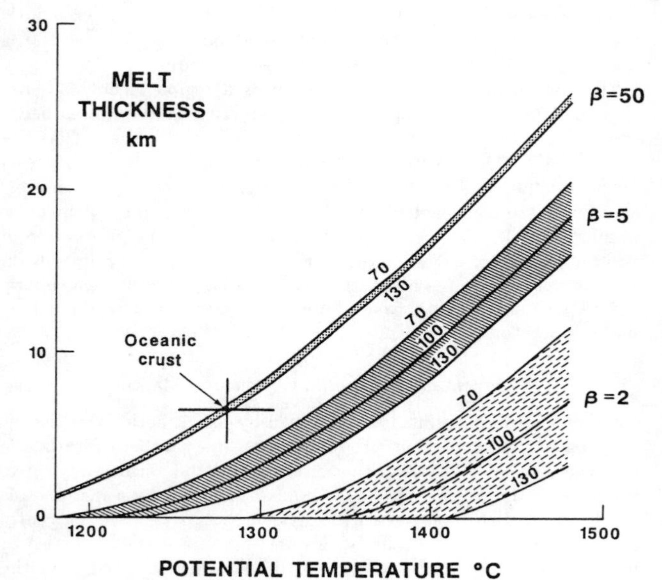

Fig. 1. Thickness of melt generated by adiabatic decompression of asthenospheric mantle over a range of potential temperatures. Curves are shown for initial thicknesses of the mechanical boundary layer of 70, 100 and 130 km, with thinning by factors of 2, 5, and 50. Corresponding lithosphere thicknesses in the thermal plate model are somewhat thicker than the mechanical boundary layer [McKenzie and Bickle, 1988; White, in press]. Annotated mechanical boundary layer thicknesses of 70, 100 and 130 km correspond to equivalent thermal plate thicknesses of 87, 118 and 149 km respectively. Cross shows limits on normal asthenosphere temperatures calculated from the range of measured oceanic igneous crustal thickness from spreading centres.

would always generate approximately the same volume of melt by decompression. If the asthenosphere under some regions was hotter, then we would expect considerably enhanced igneous activity when the lithosphere above those regions was stretched and thinned.

In practice, we find that the normal asthenosphere temperature is surprisingly consistent globally. This is shown by the observation that the oceanic crust generated at spreading centres varies little in average thickness from ocean to ocean [White and McKenzie, in press]. Since the asthenosphere has welled up almost to the surface (i.e. infinite stretching factor) beneath all oceanic spreading centres, the consistency of the crustal thickness and therefore of the volume of melt generated points to a global uniformity of normal asthenosphere temperature: McKenzie and Bickle estimate that the normal asthenosphere potential temperature must be close to 1280°C (the potential temperature is the temperature that it would have if brought to the surface adiabatically without melting).

However, although the asthenosphere has a consistent background temperature there are regions where thermal convection brings plumes of abnormally hot material to the base of the plate. In the oceans these plumes generate the well known 'hot spots' such as Iceland, Hawaii and the Cape Verde. By modelling the bathymetric, geoid and heat flow anomalies above the slow moving Cape Verde Rise, Courtney and White [1986] showed that the rising plume is relatively narrow, typically only 150–200 km in width, but that it is deflected laterally by the overlying plate to form a region of hot mantle typically 1500–2000 km in diameter just beneath the plate. The effect of this hot material is to generate a bathymetric swell 1500–2000 km across above the abnormally hot mantle. The mantle brought up by the plume has a temperature on average about 100°C hotter than the normal asthenosphere temperature.

If now a continental rift occurs across a region of hot asthenosphere, there will be enhanced melt production as the hot asthenosphere wells up passively and decompresses beneath the rift [White et al., 1987]. An increase in asthenosphere potential temperature of 100°C roughly doubles the thickness of melt produced by decompression when the continent is stretched to breaking point: an increase of 200°C trebles the melt production (Figure 1). The production of voluminous igneous rocks in rifting regions thus depends primarily on whether the rift occurs above the broad region of hot asthenospheric mantle surrounding a thermal plume or whether it occurs above normal temperature asthenosphere.

Sedimentary Basins and Continental Margins

Examples of sedimentary basins formed by lithospheric extension at the two extremes of melt production are the southern North Sea and the Aegean regions at one extreme and the Basin and Range province at the other. Although thinned in places by a factor of as much as four, there is very little igneous activity in the currently rifting Aegean basin. Similarly, despite thinning by 60%, there was little volcanism in the southern North Sea when it rifted. On the other hand the extended Basin and Range province shows evidence of considerable igneous underplating. The entire region containing the Basin and Range province is abnormally elevated, providing independent evidence that the underlying asthenosphere is abnormally hot. Thus the Basin and Range, stretched above hot asthenosphere, exhibits considerable igneous activity whilst the Aegean and North Sea regions stretched above normal temperature asthenosphere, are volcanically quiescent. Similar correlations of extensive igneous activity with the presence of hot asthenosphere can be seen on rifted continental margins around the world [White and McKenzie, in press]. In these cases the correlation is particularly striking, because oceanic rifts commonly continue right across the thermal anomaly and beyond it, so it is possible to map the transition from volcanic to non-volcanic margins along strike and to correlate this with the extent of the thermal anomaly.

A classic example is the North Atlantic [White, 1988]. The rifted continental margins on both sides of the northern North Atlantic show extensive volcanism along a 2000 km long stretch. Extrusive basalts were poured out in the Early Tertiary at the time of rifting along the Hatton-Rockall margin off northwest Britain northwards along the Faeroes margin, to the Vøring Plateau and the continental margin off Norway. Similar basaltic outpourings were also produced along the conjugate continental margins of east Greenland and the Jan Mayen continental fragment. The extrusive basalts form easily mappable seawards dipping reflector sequences on seismic reflection profiles: from seismic reflection records together with detailed seismic velocity measurements [White et al., 1987; Spence et al., in press], we can tell that the extrusive basaltic sequences are up to 8 km thick and typically 100 km wide on the rifted margins. They therefore constitute an enormous volume: about 2 million km^3 in the North Atlantic alone, extruded in as little as 1–2 my as the continent rifted. Furthermore, along with the extrusive volcanism there was massive emplacement of igneous material at depth [White et al., 1987; Fowler et al., in press; Morgan et al., in press], increasing the total volume of igneous rocks emplaced by a factor of two or three above the purely extrusive portion.

This massive igneous activity at the time of rifting has an obvious spatial and temporal correlation with the thermal anomaly generated by the mantle plume that is at present centred beneath Iceland but which at the time of continental breakup was beneath east Greenland (Figure 2). The volcanic margins are found extending about 1000 km in either direction away from Iceland: this is just the region over which the hot mantle brought up in the plume spreads laterally beneath the plate. The same region is also marked by abnormally shallow bathymetry in the North Atlantic around Iceland, again indicative of the presence of a large thermal anomaly in the underlying mantle.

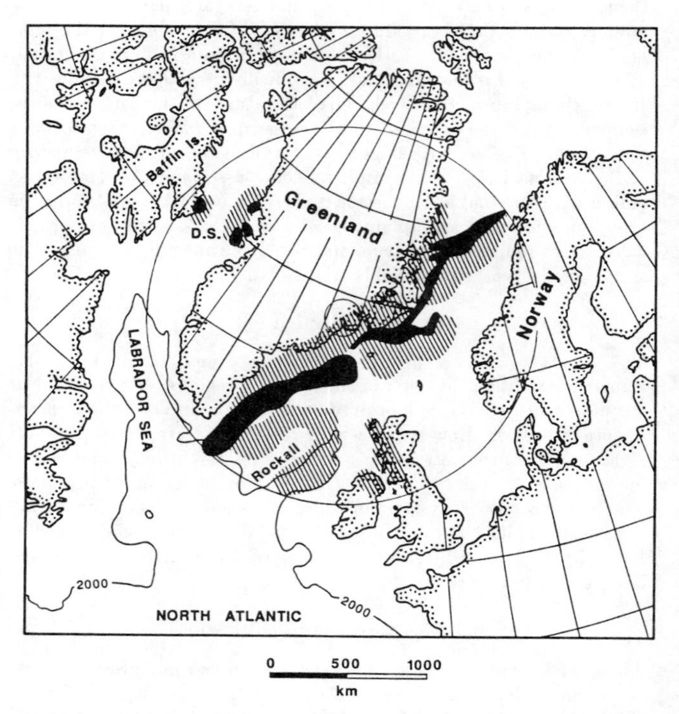

Fig. 2. Reconstruction of the North Atlantic region at magnetic anomaly 23 time, just after the onset of oceanic spreading (from White and McKenzie, in press). Position of extrusive volcanic rocks is shown by solid shading, with hatching to show the extent of early Tertiary igneous activity in the region. The inferred position of the mantle plume beneath east Greenland at the time of rifting and the extent of the mushroom-shaped head of abnormally hot asthenospheric mantle is superimposed. Projection is equal area centred on the mantle plume.

Equally striking is the change to non-volcanic margins as the continental rift cut through lithosphere not underlain by abnormally hot material. Remaining in the North Atlantic, for example, as the rift is followed southwards it becomes non- volcanic off the Biscay margin of France, which was away from the influence of any hot-spot when it rifted. Another good example is the Labrador Sea. This started opening in the Cretaceous before the Iceland hot spot influenced it and produced deep, non- volcanic rifted margins. However, as the rift progressed, it became influenced by the Iceland hot-spot in the Early Tertiary, thereafter generating volcanic margins and with a significant increase in elevation.

Conclusions

This paper contains only an outline of the model relating volcanism in rifted regions to the passive welling up and partial melting of anomalously hot underlying asthenospheric mantle. There are many other tests of the model. We can for example investigate the geochemistry of the melts, which vary systematically with the temperature of the asthenosphere [McKenzie and Bickle, 1988; White, in press]. We can check whether the timing and duration of the igneous activity is coincident with the major lithospheric thinning. By studying the uplift and subsidence of sedimentary basins and rifted margins we can check our models of subsidence patterns derived from calculations of the degree of lithospheric thinning, the asthenosphere temperature and the addition of melt to the crust. From seismic data we can look for evidence of high velocity lower crustal layers indicative of underplating. Not least, we should investigate whether the occurrence of volcanic continental margins is always related to the thermal anomaly created by a known nearby mantle plume. For a comprehensive review the reader is referred to White and McKenzie [in press].

All these tests suggest that volcanically active sedimentary basins and continental margins are indeed caused by partial melting in passively upwelling hot asthenosphere beneath the rifting and thinning lithosphere. The presence of 1500–2000 km broad regions of abnormally hot asthenosphere is caused by upwards convection of hot material in narrow thermal plumes beneath hot spots, and the lateral deflection of the hot material by the overlying plate to generate broad thermal anomalies.

References

Courtney, R. C. and R. S. White, Anomalous heat flow and geoid across the Cape Verde Rise: evidence for dynamic support from a thermal plume in the mantle. *Geophys J.R. astr. Soc.*, *87*, 815–868, plus microfiche GH 87/1, 1986.

Fowler, S. R., R. S. White, G. D. Spence and G. K. Westbrook, The Hatton Bank continental margin—II. Deep structure from two-ship expanding spread seismic profiles. *Geophys. J.*, in press, 1988.

McKenzie, D. P. and M. Bickle, The volume and composition of melt generated by extension of the lithosphere. *J. Petrology*, *29*, 625–679, 1988.

Morgan, J. V., P. J. Barton and R. S. White, The Hatton Bank continental margin—III. Structure from wide-angle OBS and multichannel seismic refraction profiles. *Geophys. J.*, in press.

Spence, G. D., R. S. White, G. K. Westbrook and S. R. Fowler, The Hatton Bank continental margin—I. Shallow structure from two-ship expanding spread seismic profiles. *Geophys. J.*, in press, 1988.

White, R. S., When continents rift. *Nature*, *327*, 191, 1987.

White, R. S., A hot spot model for Early Tertiary volcanism in the N Atlantic in *Early Tertiary volcanism and the opening of the N.E. Atlantic* edited by Morton, A.C. and L.M. Parson, Geol Soc. Lond. Spec. Publn., *39*, 3–13, 1988.

White, R. S. The Earth's crust and lithosphere. *J. Petrology, Special Lithosphere issue*, in press, 1988.

White, R. S. Initiation of the Iceland plume and opening of the North Atlantic in *Extensional tectonics and stratigraphy of the North Atlantic margins* edited by Tankard, A. and H. Balkwill, AAPG Memoir, in press, 1988.

White, R. S. and D. P. McKenzie, Magmatism at rift zones: the generation of volcanic continental margins and flood basalts. *J. Geophys. Res.*, in press, 1988.

White, R. S., G. D. Spence, S. R. Fowler, D. P. McKenzie, G. K. Westbrook and A. N. Bowen, Magmatism at rifted continental margins. *Nature*, *330*, 439–444, 1987.

White, R. S., G. K. Westbrook, A. N. Bowen, S. R. Fowler, G. D. Spence, C. Prescott, P. J. Barton, M. Joppen, J. Morgan and M. H. P. Bott, Hatton Bank (northwest U.K.) continental margin structure. *Geophys. J.R. astr. Soc.*, *89*, 265–271, 1987.

A LOW-TEMPERATURE HYDROTHERMAL MATURATION MECHANISM FOR SEDIMENTARY BASINS ASSOCIATED WITH VOLCANIC ROCKS

Neil S. Summer

Department of Geology, Hebrew University, Jerusalem 91904, Israel

Kenneth L. Verosub

Department of Geology, University of California, Davis, California 95616

Abstract. Data from sediments associated with volcanic rocks around the world demonstrate that a generally unrecognized maturation mechanism is operating in certain geologically active areas. This mechanism is hydrothermal in nature and involves the transport of heat away from intrusive bodies or deep penetrating faults by laterally-flowing aquifers. The mechanism accounts for regional maturation and diagenetic effects which cannot be explained by conductive heat transfer. In many cases economically important hydrocarbon accumulations can be associated with volcanism via a hydrothermal maturation model, wherein thermal fluids play a major role in the maturation of source rocks and assist in migration of the evolved hydrocarbons. Applying the model would not only give new insights into the thermal history of basin sediments but may assist in determining areas of highest exploration potential. Overall, volcanism has played a larger role in the thermal maturation of certain sedimentary basins than has previously been assumed.

Introduction

For many years, conventional wisdom has held that the hydrocarbon potential of sedimentary basins associated with volcanic rocks would be low. Petroleum geologists assumed that local sediments would either be over-cooked or would contain no reservoir rocks, given the clay-rich nature of altered volcanogenic sediments. This view, combined with the high cost of drilling through volcanic cover and the difficulties involved in obtaining good geophysical data from under volcanic strata, has made these basins an unattractive target for hydrocarbon exploration.

However we have recently found in the Pacific Northwest (USA) that sedimentary basins overlain by volcanic rocks may contain significant gas resources and that geothermal fluids associated with volcanic activity were probably responsible for regional and vertical maturation patterns. In this paper we present a generalized version of a hydrothermal maturation mechanism which accounts for the maturation patterns in these and other sedimentary basins whose hydrocarbon reserves appear to be related to volcanic activity.

Background

Vitrinite reflectance data from drillholes in the Pacific Northwest, revealed a systematic pattern of near-constant maturation levels with increasing depth [Summer and Verosub, 1987a,b] (Fig.1). Such steep maturation profiles suggested the presence of temperature fields that had been relatively uniform over thousands of meters of section. Supported by limited Rock-eval data [Summer, 1987] the data indicated that the organic matter in the sediments of these basins were sufficiently mature to represent a significant gas potential.

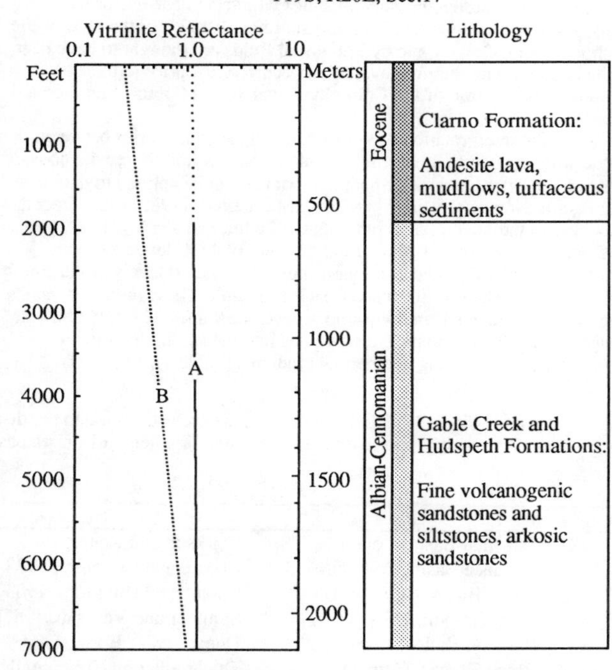

Fig. 1. Data from Summer and Verosub [1987b]. Solid line (A) is a best-fit profile through the data set. B is the "normal" profile of predicted Ro calculated [after Middleton, 1982] from the present day geothermal gradient of 44°C/km. Extrapolation of the profile (A) to a normal value of 0.2%Ro at the surface implies unrealistically that about 8,000m of section has been removed.

Further study of the inorganic diagenesis of sediments from the Ochoco Basin, Oregon revealed a near-constant authigenic mineralogy with depth. This supported the maturation results and led us to conclude

TABLE 1. Interpretation of vitrinite reflectance (Ro) utilising a time/temperature relationship [Middleton, 1982]. Column (a) shows Ro values calculated from the present day geothermal gradient. Column (b) shows the geothermal gradient necessary to attain the measured Ro levels.

Parameter	Units	(a) Predicted Ro		(b) Measured Ro	
Vitrinite Reflectance	%Ro	0.40	0.89	1.11	1.26
Age of strata (T)	Ma	91	113	91	113
Geothermal gradient (dT/dz)	°C/km	44	44	178	58
Depth (z)	meters	608	1962	608	1962
Calculated temperature at depth (z)	°C	37	96	118	124

that the sediments had been strongly affected by the presence of low-temperature (≈200°C) hydrothermal fluids [Summer and Verosub, 1987c]. We believe that these fluids flowed laterally in confined aquifers perched high in the rock column and this interpretation is supported by field evidence and mineralogical data [Summer, 1987; Summer and Verosub, 1987c]. The fluid flow in these aquifers is thought to have been driven by convection associated with heat sources such as cooling intrusive bodies or fed by vertical conduits such as deeply-penetrating faults. The thermal input from such perched thermal aquifers and the steady state geothermal gradient relatively quickly developed near-isothermal conditions in underlying sediments [Ziagos and Blackwell, 1986], resulting in levels of maturation and authigenic mineral suites which are near-uniform throughout the sedimentary section [Summer, 1987]. In the case of the Ochoco Basin, Oregon, the hydrothermal fluids are thought to have been associated with the thermal event that occurred just prior to the voluminous eruption of the Columbia River Basalt Group [Summer and Verosub, 1987b].

Since the interpretation of vitrinite reflectance data varies between those who consider it an absolute geothermometer and those who consider time an important parameter both approaches can be applied to data from the Pacific Northwest (Fig. 1). Using Middleton's [1982] time-dependant model with parameters of depth z=680, 1962m, and age T=91, 113 Ma respectively, the current geothermal gradient of 44°C/km, a surface temperature of 10°C, and continuous burial; predicted levels of vitrinite reflectance are 0.4% (680m) and 0.89% (1962m). These predicted values are far below the measured vitrinite reflectance values of 1.11% and 1.26% respectively. To fit the measured values into the Middleton model requires widely differing geothermal gradients (Table 1), a scenario unlikely to be maintained over geological time, but common under relatively short-lived thermal events. The time independent model [Price, 1983] gives calculated paleotemperatures of approximately 200°C, consistent with those inferred from the authigenic mineralogy. A conciliatory model which suggests that coal ranks require time only to stabilize at the ambient temperature implies a thermal event of duration about 200,000 years (For t= 200°C) [Kisch, 1987,p.298]. Therefore given all the above discussion, the simplest most elegant model to account for the data is one entailing a hydrothermal maturation mechanism.

Hydrothermal Maturation Mechanism

Petroleum geologists, long opposed to any genetic relationship between volcanism and hydrocarbons, generally ignored local volcanic events in prospect or producing areas. Yet in many cases local source rocks became mature contemporaneously with a local volcanic event, and common is the presence of an oil or gas field in an area where the deep sediments of a basin are immature. In fact commercial oil and gas reservoirs are found either in or under every kind of volcanic deposit and this, together with a growing database (Table 2) strongly support a relationship between some hydrocarbon occurrences and local volcanic events.

However coal geologists have known for many years the regional effects of certain volcanic centers. This telemagmatic coalification [Teichmuller and Teichmuller, 1966], related mineralization, coalification and in some cases hydrocarbon accumulation to the thermal effects attributed to cooling volcanic bodies. Even though these effects extended well beyond the aureole that could be expected from purely conductive heating [Horvath et al., 1986], movement of fluids was not invoked to explain telemagmatic occurrences. Observed uniform maturation levels with depth were attributed to differences in thermal conductivities even though sandstone units, presumably more porous and hence aquifers, were found to markedly influence coalification profiles in a positive manner [Teichmuller and Teichmuller, 1968].

On the other hand, hydrogeologists and others acquainted with hydrothermal systems are well aware of the widespread effects of thermal waters. When magma bodies make their way into shallow crustal levels, heated connate and inflowing meteoric waters are flushed away, usually vertically upward. If these thermal waters find stratigraphic paths that allow lateral movement, the heat from the volcanic body can then be

TABLE 2. Oil and gas occurrences attributed to the distillative effects of volcanism on sediments. Critera for inclusion are: supportive maturation data, or where volcanism has been invoked by the sundry authors as causing the hydrocarbon accumulation.

Oil/Gas	Place	References	Geology	Source	Volcanism	Data	Comment
O/G	San Juan Basin, Colorado, USA	Clarkson and Reiter, 1987	V,CB,HT	Cret	Tertiary	Ro	major role of Volcanism
G	Piceance Basin, Colorado, USA	Choate and Ringmire, 1983	V,CB,HT	Cret	Tertiary	A	major role of Volcanism
G	Raton Basin, CO/NM, USA	Choate and Ringmire, 1983	V,CB,HT	Cret	Tertiary	A	major role of Volcanism
G	Mist, Oregon, USA	Summer and Verosub, 1987b	V,FAB,HT	L Cret	Mio-Recent	Ro	local intrusions
O	Railroad Valley, Nevada, USA	Duey, 1983 ; Bortz, 1983	V,B&R,HT	Cret	Oligocene	Ro	Volcanic reservoirs
O/G	Rhine Graben, Germany	Teichmuller and Teichmuller, 1968	I,RB,HT	L. Cret	Cret	Ro	Single large Intrusive body
O/G	West Siberia, USSR	Roberts, 1981; Ammosov, 1979	PB,I,CB	Paleozoic	Mesozoic	A	+100,000 km2 affected
O/G	Taranaki Basin, New Zealand	Pilaar and Wakefield, 1984	RB,I,V	UCret,Tert	Mio-Pleis	Ro	Steep profiles
O/G	Esirito Santo Basin, Brazil	Estrella et al., 1984	CB, PB,I	Aptian,Tert	Eary Eocene	all	"Paleothermal unconformity"
O/G	Amazon Basin, Brazil	Mossman et al., 1984	CB, PB,I	Permian	Jurrassic	Ro	Vertical profiles
O/G	South Mangyshlak Basin ,USSR	Ammosov, 1979	PB,I,CB	Jur/Cret	?	A	Still hot (240°C), intrusive?
O/G	Pripyat Basin, USSR	Ammosov, 1979	PB,I,CB	Devonian	?	A	30x100 km thermal anomaly
G	Godavari Basin	Mukerjee, 1983	PB, RB	Jur/UCret	Tertiary	A	Under Deccan Traps
O/G	Niigata Basin, Japan	Komatsu et al., 1984	V,HT,F	Miocene	Miocene	A	Hydrothermal effects

I = intrusive(s)
F = Faulting
TF = Trust faulting
V = nearby extrusive Volcanic center(s)
HT = hydrothermal area
AV = Active volcanoe(s) (arc)
B&R = Basin and Range
FAB = Fore-arc Basin
PB = Plateau Basalt Province
RB = Rift Basin
BAB = Back-arc Basin
CB = Intra-cratonic Basin
all Ro,TAI, Rock-Eval
Ro Vitrinite Reflectance
A Acknowledged by author

transported far beyond that expected from pure conduction. In fact, the movement of fluids is considered the most efficient form of lateral and vertical heat transport [Smith and Chapman, 1983] and is consistently invoked to account for widespread metamorphism [Hoisch, 1987]. Furthermore, lateral flow is a very common feature of geothermal systems [Healy and Hochstein, 1973], and several contemporary large-scale aquifers have been described [Wood and Low, 1986; Chapman et al., 1984] with widespread thermal perturbations [e.g. Bodner and Sharp, 1988]. In addition, there is evidence for areally-extensive fossil geothermal systems associated with centers of both extrusive and intrusive volcanism [e.g. Criss et al., 1984]. The large scale geothermal manifestations sufficient to account for regional diagenesis have analogs particularly in the flood basalt provinces of the world [e.g. NW India... Saxena and Guptha, 1987] and large scale fossil thermal aquifers have been invoked as the cause of oil and gas fields in Canada [Majorowicz, et al., 1986] as well as lead-zinc deposits in the Mississippi basin [Cathles and Smith, 1983].

The thermal input from the laterally-flowing aquifers can be transferred to the underlying strata either by convection or by conduction. Modelling of the less-efficient conductive heat transfer has shown that the thermal input can be substantial. In fact, the interaction between a steady-state geothermal gradient and the thermal input from an aquifer high in a sedimentary column quickly leads to an inverted or near-constant temperature field [Ziagos and Blackwell, 1986].

Because the effects of hydrothermal circulation on organic matter [e.g. Simoneit, 1983] and on the mineral matrix [e.g. Browne, 1978; Hoisch, 1987] are well-documented, one can predict that the thermal perturbations of the lateral aquifers would manifest themselves vertically as near-constant or inverted hydrocarbon maturation and coal rank profiles, and horizontally over tens of kilometers as an aureole of elevated coal rank or thermally-mature sediments, together with low grade diagenesis of the mineral matrix. Thus, hydrothermal systems associated with volcanism represent a viable mechanism of basin diagenesis, one that can have important economic implications. This mechanism will be termed the hydrothermal maturation (HTM) mechanism.

The HTM model is merely a combination of well-known geological concepts applied to petroleum geology. In the past geologists have compiled data on occurrences of hydrocarbons in metamorphic and igneous rocks [Powers, 1932], and Hedberg [1964] briefly discussed genesis of hydrocarbons from "igneous magma or at least from hydrothermal solutions." These and other papers documented hydrocarbons associated with intrusive activity, hydrothermal solutions and ore deposits. Others [e.g. Roberts, 1981] recognized the association between thermal anomalies, oil pools and groundwater flow but overall conclusions were confined to migration [e.g. Mekhtiev, 1967; Price, 1981] rather than maturation phenomena. Thus the maturation of organic matter to form commercial oil and gas has never really been considered to anything but via a burial scenario. The HTM model suggests otherwise; that sufficiently large volume of sediments can be matured under the influence of thermal aquifers, therefore constituting another valid model for maturation of organic matter to form hydrocarbons.

Hydrocarbon Potential and Volcanic Activity

In simple terms it is generally agreed that oil and gas are created by the distillative effect of heat on the organic matter present in most sedimentary rocks [Tissot and Welte, 1984]. Generally maturation of such source rocks to form oil and gas results from burial to a level where the steady-state geothermal gradient leads to temperatures of between 65-120°C. Further evolution of gas is believed to take place up to temperatures of up to 250°C. The evolved hydrocarbons migrate either as discrete globules or as dissolved organic species to reservoir rocks.

However for many sedimentary basins, it is difficult to geochemically correlate hydrocarbons and local source rocks. In these situations, usually it is simply assumed that the source rocks must be deeper than the deepest drilling in the basin. In some cases, oil and gas fields whose deep sediments are thermally immature are located 5-100 km from a major volcanic center or deep-penetrating fault [Price, 1980]. In others, the maturation of the local source rocks can be shown to be contemporaneous with a local volcanic event [e.g. Rice, 1983]. However, even when this relationship has been noted, geologists have persisted in trying to incorporate this maximum heat flux or "heat flash" into standard burial/depth models with a uniform geothermal gradient despite the fact that the available maturation data showed an regional increase towards the volcanic center (e.g. San Juan Basin, Colorado/New Mexico, USA).

Example: San Juan Basin, Colorado/New Mexico, USA.

A good example of the inappropriateness of classical basin analysis methods applied to an active geological area is the case of the San Juan Basin. This intra-cratonic structural depression is one of the major hydrocarbon producing areas of the Rocky Mountains. The area has been extensively explored, and the geological framework is well established and documented. The basin itself can be divided into a central, inner basin and an outer basin which, in turn, is surrounded by areas of uplift. The central basin is roughly circular but is strongly asymmetric in cross-section (Fig. 2). The basin is very rich in natural resources (Fig. 3). In the near undisturbed central basin, there a large, non-associated gas accumulation [Rice, 1983]. The outer basin has anticlinal folds perpendicular to the dip, and these hold large oil accumulations. Additional oil and gas fields are found in the San Juan Sag under the volcanic cover to the northwest [Gries, 1985] and high rank coal deposits in the Fruitland Formation fringe the basin. Furthermore the 160x60 km Grants Mineral Belt lies in the southwest part of the basin, producing mainly Uranium.

Regional studies of the San Juan Basin and the associated Piceance Basin and Raton Basin have shown that the thermal history of the basins has been strongly influenced by the magmatic and volcanic activity of the San Juan volcanic field [Dolly and Meissner, 1977; Choate and Ringmire, 1982]. The coal rank data for the basin shows a regional gradient towards the volcanic complex in the San Juan Mountains [Rice, 1983](Fig. 4). In fact, heat flow is still currently centered in the volcanic complex (Fig. 3), even though volcanic activity ceased during the Pliocene. The current heat flow pattern and coal maturation within the basin has been attributed to groundwater advection of heat [Clarkson and Reiter, 1987].

In such a geologic setting, it is not surprising that maturation data cannot be incorporated into a conventional burial/maturation model [Bond, 1984]. In fact several studies of the petroleum geology have concluded that the presence of elevated coal rank and maturation levels within the basin is in some way related to the volcanism [Reiter and Clarkson, 1983; Rice, 1983; Bond, 1984].

However, all the available data on the basin have not been synthesized to form a model for hydrocarbon generation and migration. It is clear that thermal fluids have caused a regional maturation pattern in which maturation increases towards the volcanic center. One result is that the highest maturation levels are found in sediments that were not the site of deepest burial. This is supported by isotopically heavier and chemically drier gases, together with higher API gravities of oil towards the volcanic center in the northeast irrespective of stratigraphic position [Rice, 1983]. The matured gases are found in a large unassociated gas field primarily in the low porosity Mesaverde Group. Oil found in Tocito sandstone within the Mancos shale is considered to be Cretaceous in age derived from the Lewis-Mesaverde or Dakota formations [Rice, 1983]. Thus, based on the available data from the San Juan Basin one interpretation is that fluids either connate, meteoric or via shale diagenesis were heated and moved updip both to the SW and NE in any of number of horizons with aquifer potential [e.g. the Morrison Formation]. These fluids aided the distribution of heat from the laccolith across the basin and is still doing so. We suggest that these fluids aided in the maturation and hence distillation of hydrocarbons from the various organic rich rocks in the basin, and aided in the migration of oil to reservoirs nearly 100 km to the southwest.

In addition these thermal waters may have played a role in the concentration of ores in the Grants Mineral Belt, southern San Juan Basin. Several important types of ore deposits have been related to extensive migration of thermal fluids (<200°C), and in particular are

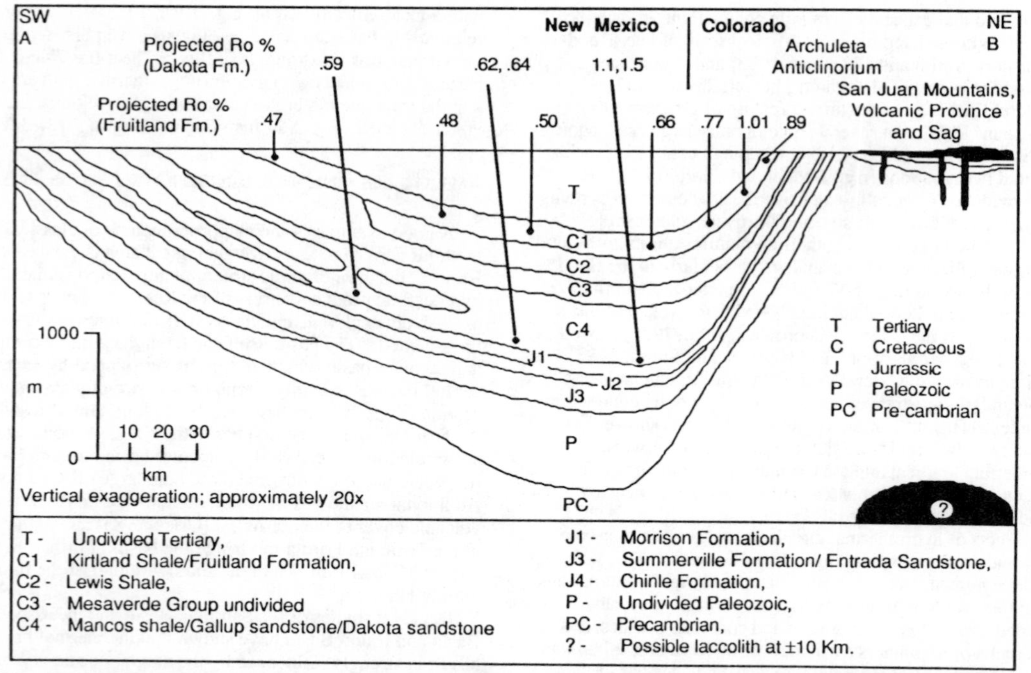

Fig. 2. Generalized cross section (A-B) of the San Juan Basin (Fig. 4) [after Clarkson and Reiter, 1987; Gries, 1985]. Projected vitrinite reflectance for the Fruitland and Dakota Formations are from Rice [1983]. Vitrinite reflectance increases primarily as a function of closeness to the San Juan volcanic center and secondarily as a function of depth.

unconformity related Uranium deposits and the previously mentioned Mississippi valley Pb-Zn deposits [Sheppard, 1984]. One model of ore formation in the Grants Mineral Belt involves updip flow of metallic-rich fluids from the basin, mixing with down flowing huminate-rich meteoric water [e.g. Saxby, 1976] resulting in the precipitation of the organic compounds in the Morrison Formation with the adsorbed metallic species. Thus the uranium deposit in the Southern San Juan Basin may also be related to hydrothermal waters related to the volcanism.

Although the concept for the HTM mechanism arose from our studies of thermal maturation in the Pacific Northwest [Summer and Verosub, 1987a,b], that area is very large, sparsely drilled and primarily mantled by volcanic deposits. On the other hand the San Juan mountains are a localized volcanic center surrounded by sedimentary basins. These basins have been extensively drilled and have been subjected to intense geological study. In addition, some of the sedimentary basins are exposed while others are buried under the volcanic cover, and they all contain commercial ore deposits, coalfields and oil and gas fields. Therefore the San Juan, Piceance and Raton basins may represent the definitive area for study of the HTM mechanism.

Other Examples

Association between hydrocarbon production and volcanic activity is not uncommon and the HTM mechanism may have been active in a variety of geologic settings. In the Basin and Range Province of the western United States, the connection between volcanism, hydrothermal systems and hydrocarbon occurrences is especially strong. This geologically complex area has been the site of tectonic and volcanic activity since Pre- Cambrian times. During the Miocene the area was severely disrupted by the extrusion of massive ignimbrites and by large-scale east-west extension. The area developed into a series of pull-apart basins with parallel north-south trending ranges (horsts) and valleys (grabens). The valleys accumulated up to 5,000 m of sediment

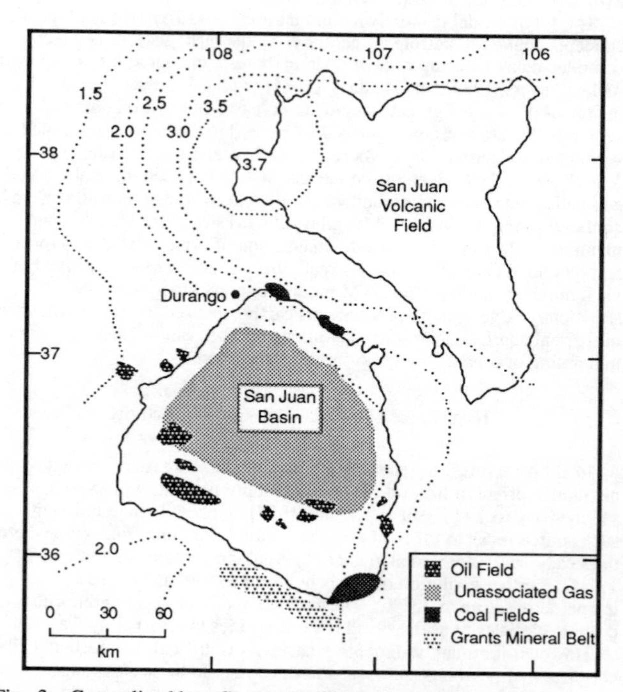

Fig. 3. Generalized heat flow (in HFU) map of San Juan Basin and volcanic field showing oil, gas and coal fields, and the Grants mineral belt. Heat flow lines are after Choate and Ringmire [1982]. The uniform heat flow across the basin is believed to be due to the advection of groundwater [Clarkson and Reiter, 1987].

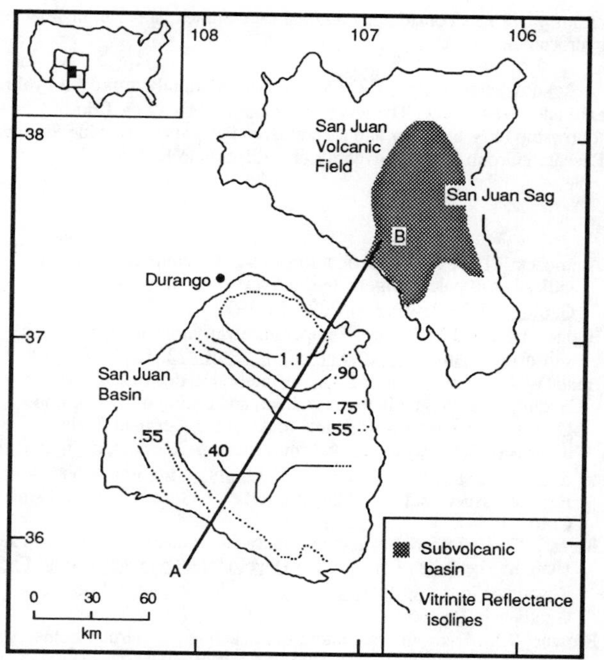

Fig. 4. Map of San Juan Basin and volcanic field. Vitrinite reflectance contours are for Fruitland Formation [Rice, 1983]. Section A-B is shown in Fig. 2.

accompanied by sporadic magmatism. Volcanic activity has continued until the present time, and the area is rich in geothermal resources as well as ore deposits and oil. Most hydrocarbon reservoirs in Nevada (USA) are found in volcanic rocks, primarily ignimbrites and several commercial oil fields have been attributed to localized heating by intrusive rocks [Duey, 1983; Bortz, 1983]. Maturation data from these areas is sparse but available data do show constant maturation profiles with depth [Poole et al., 1983]. Consistent with the HTM mechanism is the fact that large-scale hydrothermal systems are common [Edmiston and Benoit, 1984] and although these systems are usually confined to individual grabens, some have effects that extend over adjacent ranges and basins. Another important fact is that in some basins which are presently affected by active geothermal systems the sediments are under near isothermal conditions [Ross and Moore, 1985], indicating that the HTM mechanism is still active. Therefore the HTM model has important ramifications for oil exploration in the Basin and Range Province as structures associated with active and fossil geothermal systems may hold reservoired hydrocarbons distilled by local hydrothermal activity.

Solitary laccoliths are known to have regional maturation effects and in Germany, some oil and gas fields have been directly attributed to the distillative effects of heating from the Bramshe Massif, a large subsurface intrusive body [Teichmuller and Teichmuller, 1968]. Sandstones in the sedimentary section show elevated vitrinite reflectance levels and this was once interpreted to be the result of differing thermal conductivities. In fact it was noted that the metallic mineralization occurred near faults at about the same time as the coal maturation [Teichmuller and Teichmuller, 1968]. Therefore a more plausible suggestion is that thermal fluids found conduits within the more porous sandstones, resulting in the elevated coal ranks. The effects of the Bramshe massif are more widespread (130 x 50 km) than can be explained by purely conductive heat transfer. Therefore thermal fluids must have played a major role in the distribution of heat. Other examples of regional coal rank aureoles resulting from volcanic activity possibly via the HTM model are listed in Table 3.

At mid-oceanic ridges the effect of hydrothermal fluids on the organic matter of recent sediments has been documented [e.g. Simoneit, 1983] and lateral flow of thermal fluids has been postulated to account for the maturation and thermal anomalies there [Einsele et al., 1980]. These examples are important a they show that hydrocarbons can be distilled and transported in solution by thermal fluids under very shallow burial conditions and within geologically instantaneous events [Simoneit, 1983].

With regard to the numerous examples around the world of oil and gas fields associated with volcanic rocks, at the present time maturation or coalification data for most of these is not available, and thus the evidence is circumstantial, being mainly of close proximity. The remainder either have evidence of abnormally high maturation data or the relationship to volcanism has been invoked by the various authors to account for the local occurrence of hydrocarbons (Table 2).

Implications of the HTM Mechanism

Most of the world's easily-found oil reserves have been developed, and current exploration is focussed on areas of the world that are geologically active. Many of these areas contain sedimentary basins near volcanic centers. Better understanding of the HTM mechanism would not only give insights into the thermal history of these basins but would also help in the prediction of pathways for the migration of hydrocarbons. The conventional trap and reservoir rock concepts developed by petroleum

TABLE 3. Maturation or coalification anomalies due to the effects of volcanism.

Place	References	Area effected	Geology
Bramshe massif, Germany	Teichmuller and Teichmuller, 1968	50 x 130 km	I,RB
Erklenz massif, Germany	Damberger, 1974	30 x 40 km	I,RB
Central Italy	Teichmuller and Teichmuller, 1986	over 150 km	I,TF
Greenland	Surlyk et al., 1983	50 x 20 km	I,F
Eastern US basins, USA	Damberger, 1974	±10,000 km^2	I,CB,
Canning Basin, Australia	Reekman and Mebberson, 1984	100x100 km	I,CB,
Natal Coalfields, S.Africa	Mackowsky, 1968	over 200 km	PB,I,
San Juan Area, CO/NM, USA	Choate and Ringmire, 1983	280km radius	V,CB,HT
Pacific Northwest, USA	Walsh and Phillips, 1983	65x300 km	AV,I,F
South Wales, UK	Gill et al., 1979	over 65 km	TF
Ireland	Wright, 1975	over 60 km	PB

I = intrusive(s)
F = Faulting
TF = Trust faulting
V = nearby Volcanic center(s)
HT = hydrothermal area
AV = Active volcanoe(s) (arc)
PB = Plateau Basalt Province
CB = Intra-cratonic Basin
B&R = Basin and Range
FAB = Fore-arc Basin
BAB = Back-arc Basin
RB = Rift Basin

geologists would still apply as traps may have formed prior to extrusion of the volcanics, but additional traps might be found within the volcanic cover [e.g. Komatsu et al., 1984].

As noted above, in many cases the sediments of an immediate reservoir area are immature, and lateral migration over distances as great as 300 km have been invoked to account for the source of hydrocarbons. However, the volume of sediments matured by local volcanic rocks may be sufficient to account for the nearby accumulations of hydrocarbons. A case in point is the Mist Gas field, Oregon, which had no known source until a connection was made between the gas field and locally-intruded sediments [Summer and Verosub, 1987b].

Furthermore since the concept of transport of hydrocarbons by thermal fluids [Price, 1981] is being more readily accepted, migration of evolved hydrocarbons assisted by the thermal fluids to traps tens to hundreds of kilometers away is more commonly invoked [e.g. Zielinski and Bruchausen, 1983]. Thus, study of paleoflow directions based on the regional maturation patterns may be used to determine migration vectors and thus assist in delineating areas with the highest exploration potential.

Another important consideration is the volume of mature source rocks resulting from the HTM mechanism. In basins where maturation results from heating by the geothermal gradient, a relatively thin layer of mature sediments is found at depth. However, as the HTM mechanism involves thermal input from above, it has the potential of producing uniform maturity throughout the sedimentary column. Because of the much larger volume of sediments involved, even a relatively poor source rock may generate enough hydrocarbons to produce commercial accumulations of oil and gas.

A re-assessment of what is considered non-prospective basement is required because in some instances sedimentary sequences have been found to lie under volcanic strata with entrapped hydrocarbons [e.g. Gries, 1985]. We suggest that basement be considered massive plutonic rocks or moderate to high grade metamorphic rocks and accumulations of volcanics be drilled to be sure that the drillhole has indeed reached "economic basement".

Most examples of hydrocarbons related to volcanism are found in aborted-rift or block fault basins. However, the HTM mechanism implies that other types of magmatic activity should also be viewed as having the potential to produce hydrocarbons. One such possibility is hot-spot volcanism, especially in North Africa were the trace of a hot spot coincides with oil - rich basins. This hot spot is currently centered at Mt. Etna, Sicily ,and may have played a role in the formation of the oil and gas reserves in and around the island. However the mainly unexplored flood basalt provinces of the world, hold the most promise and are definite areas in which the HTM mechanism has been active.

Conclusions

There appears to be a consistent relationship between hydrocarbon production and volcanic activity in certain sedimentary basins. This relationship can be interpreted as evidence for a low-temperature hydrothermal maturation mechanism which operates in magmatically active areas. The existence of this mechanism can explain the role of local intrusions of magma in producing commercial gas and oil deposits in sediments which are otherwise immature. It can also explain the uniformity of maturation profiles on a regional scale in areas affected by large hydrothermal circulation systems, and it suggests a method for the long distance transport of hydrocarbons via thermal fluids. Furthermore it accounts for observed telemagmatic effects around intrusive bodies that previously were ascribed to purely conductive heating.

The hydrothermal maturation mechanism provides a new basis for hydrocarbon exploration in sedimentary basins associated with volcanic rocks. Analysis of the paleo-flow regime of hydrothermal systems may provide insights into the location of accumulations of hydrocarbons. Because the hydrothermal maturation mechanism involves a larger volume of sediments, a relatively poor source rock may generate enough hydrocarbons to produce commercial accumulations of oil and gas. Therefore a better understanding of the hydrothermal maturation mechanism has important economic implications for certain sedimentary basins previously considered marginal for commercial accumulations of hydrocarbons.

Acknowledgements. This paper was substantially improved by three anomalous reviewers. The work was supported by a Lev Tzion fellowship to N. Summer and a grant from the University-wide Energy Research Group of the University of California to K. Verosub.

References

Ammosov, I.I., Petrographic features of solid organic materials as indicators of paleotemperatures and oil potential, International Geological Review, 23, 4, 238-248, 1979,.

Bodner, D. and J.M. Sharp, Temperature variations in south Texas subsurface, Am. Assoc. Pet. Geol. Bull., 72, 1, 21-32, 1988.

Bond, W.A., Application of Lopatins method to determine burial history, Evolution of the geothermal gradient, and timing of hydrocarbon generation in Cretaceous source rocks in the San Juan Basin, northwestern New Mexico and southwestern Colorado, in Hydrocarbon Source Rocks in the Greater Rocky Mountain Region, Woodward, J., F.F. Meissner, and J.L. Clayton, (eds.), 433-447, RMAG, Denver, Colo., 1984.

Bortz, L.C., Hydrocarbons in the northern Basin and Range, Nevada and Utah, in The Role of Heat in the Development of Energy and Mineral Resources in the northern Basin and Range Province, Spec. Rpt.13, Geotherm. Resour. Counc., Davis, Calif., 179-197, 1983.

Browne, R.L., Hydrothermal alteration in active geothermal fields, Ann. Rev. Earth Planet. Sci., 6, 229-250, 1978.

Cathles, L.M., and A.T. Smith, Thermal constraints on the formation of Mississippi Valley-type lead-zinc deposits and their implications for episodic basin dewatering and deposit genesis, Econ. Geol., 78, 983-1002, 1983.

Chapman, D.S., T.H. Keho, M.S. Bauer, and M.D. Picard, Heat flow in the Unita basin determined from bottom hole temperature (BHT) data, Geophysics, 49, 453-466, 1984,.

Choate, R. and C.T. Ringmire, Influence of the San Juan Mountain geothermal anomaly and other Tertiary igneous events on the coalbed methane potential in the Piceance, San Juan and Raton Basins, Colorado and New Mexico, in Unconventional Gas Recovery Symposium. SPE/DOE10805., Soc. Petrol. Eng., 151-164, 1982.

Clarkson, G. and M. Reiter, The thermal regime of the San Juan Basin since late Cretaceous times and its relationship to San Juan Mountains thermal sources, J. Volc. Geotherm. Res., 31, 217-237, 1987.

Criss, R.E., E.B. Ekren, and R.F. Hardyman, Castro Ring Zone, A 4500 Km^2 fossil hydrothermal field, central Idaho, Geology, 12, 331-334, 1984.

Damberger, H.H., Coalification patterns of Pennsylvanian coal basins of the eastern United States, in Carbonaceous Materials as Indicators of Metamorphism, R.R. Dutcher, A. Hacquebard, J.M. Schopf, and J.A. Simon, (eds.), Geol. Soc. Am., Spec. Paper 153, 53-74, 1974.

Dolly, E.D., and F.F. Meissner, Geology and gas exploration potential upper Cretaceous and lower Tertiary strata, northern Raton Basin, Colorado, in Exploration Frontiers of the Central and Southern Rockies, Veal, H.K., ed., 247-260. RMAG, Denver, Colo., 1977.

Duey, H.D., Oil generation and entrapment in Railroad Valley, Nye County, Nevada, in The Role of Heat in the Development of Energy and Mineral Resources in the northern Basin and Range Province, Spec. Rpt.13, Geotherm. Resour. Counc., Davis, Calif., 199-205, 1983.

Edmiston, R.C., and W.R. Benoit, Characteristics of Basin and Range geothermal systems with fluid temperatures of 150°C to 200°C, Trans. Geotherm. Resour. Counc., 5, 417-424, 1984.

Einsele, G., J.M. Gieskes, J. Curray, D.M. Moore, E. Aguayo, M. Aubry, D. Fornari, J. Guerrero, M. Kastner, K. Kelts, M. Lyle, Y. Matoba, A. Molina-Cruz, J. Niemitz, J. Rueda, A. Saunders, H.

Schrader, B. Simoneit, and B. Vacquier., Intrusion of basaltic sills into highly porous sediments, and resulting hydrothermal activity, Nature, 283, 441-445, 1980.

Estrella, G., M. Rocha Mello, P.C. Gaglianone, R.L.M. Azevedo, K. Tsubone, E. Rosetti, J. Concha, and I.M.R.A. Bruning, The Espirito Santo Basin (Brazil), source rock characterization and petroleum habitat, in Petroleum Geochemistry and Basin Evolution, Demaison G. and R.J. Murris, (eds.), Am. Assoc. Pet. Geol., Memoir 35, 253-269, Tulsa, Oaklahoma, 1984.

Gill, W.D., F.I. Khlaf, and M.S. Massoud, Organic matter as indicator of the degree of metamorphism of the Carboniferous rocks in the South Wales Coalfields, J. Petrol. Geol., 1, 4, 39-62, 1979.

Gries, R.R., San Juan sag, Cretaceous rocks in volcanic covered basin, south central Colorado, Oil and Gas Journal, Nov 25, 83, 109-115, 1985.

Healy, J., and M. Hochstein, Horizontal flows in geothermal systems, J. Hydrol. (N.Z.), 12, 71-82, 1973.

Hedberg, H.D., Geological aspects of the origin of petroleum, Am. Assoc. Pet. Geol. Bull., 48, 11, 1755-1803, 1964.

Hoisch, T.D., Heat transport by fluids during Late Cretaceous regional metamorphism in the Big Maria mountains, southeastern California, Geol. Soc. Am. Bull., 98, 5, 549-553, 1987.

Horvath, F., P. Dovenyi, and I. Laczo, Geothermal effect of magmatism and its contribution to the maturation of organic matter in sedimentary basins, in Paleogeothermics, Buntebarth, G., and L. Stegena, (eds.), 173-183. Springer-Verlag, Berlin, 1986.

Kisch, H.J., Correlation between indicators of very low-grade metamorphism, in Low Temperature Metamorphism, edited by M. Frey, 227-299, Blackie, London, 1987.

Komatsu, N, Y. Fujita, and O. Sato, Cenozoic volcanic rocks as potential hydrocarbon reservoirs, in Proceedings, Eleventh World Petroleum Congress, 2, 411-420, 1984.

Mackowsky, M. Th., European Carboniferous coalfields and Permian Gondwana coalfields, in Coal and Coal-bearing Strata, Murchison, D. and T.S. Westoll, (eds.), 325-345, Elsevier, New York. 1968.

Majorowicz, J.A., F.W. Jones and W.M. Jessop, Geothermics of the Williston Basin in Canada in relation to hydrodynamics and hydrocarbon occurrence, Geophysics, 51, 767-779, 1986.

Mekhtiev, SH. F., Oil and gas pools formation in Cenozoic sediments of Azerbaijan in the light of the latest investigations in Proceedings, Seventh World Petroleum Congress, 2, 543-551, 1967.

Middleton, M.F., Tectonic history from vitrinite reflectance, Geophysical Journal of the Royal Astronomical Society, 68,121-132, 1982.

Mossman, R., F.U.H. Falkenhein, A. Goncalves, and F.N. Filho, Oil and gas potential of the Amazon Paleozoic basins, in Future Petroleum Provinces, edited by M. Halbouty, Am. Assoc. Pet. Geol., Memoir 38, 207-241, Tulsa, Oaklahoma, 1984.

Mukerjee, M.K., Petroleum prospects of cretaceous sediments of the Cambay Basin, Gujarat, India, J. Petrol. Geol., 5, 3, 275-286, 1983.

Pilaar, W.F.H., and L.L. Wakefield, Hydrocarbon generation in the Taranaki Basin, New Zealand, in Petroleum Geochemistry and Basin Evolution, Demaison G. and R.J. Murris, (eds.), Am. Assoc. Pet. Geol., Memoir 35, 405-423, Tulsa, Oaklahoma, 1984.

Poole, F.G., G.E. Claypool, and T.D. Fouch, Major Episodes of petroleum generation in part of the northern great basin, in The Role of Heat in the Development of Energy and Mineral Resources in the northern Basin and Range Province, Spec. Rpt.13, Geotherm. Resour. Counc., Davis, Calif., 207-213, 1983.

Powers, S. Symposium on occurrence of petroleum in igneous and metamorphic rocks, Am. Assoc. Pet. Geol. Bull., 16, 741-768, 1932.

Price, L.C., Utilisation and documentation of vertical oil migration in deep basins, J. Petrol. Geol., 4, 2, 353-387, 1980.

Price, L.C., Primary petroleum migration by molecular solution, consideration of new data, J. Petrol. Geol., 4, 1, 89-101, 1981.

Price, L.C., Geologic time as a parameter in organic metamorphism and vitrinite reflectance as an absolute geothermometer, J. Petrol. Geol., 6, 5-38, 1983.

Reeckmann, S.A., and A.J. Mebberson, Igneous intrusions in the north-west Canning Basin and their impact on oil exploration, in Canning Basin, W.A. - Petroleum Geology and Mineral Potential, edited by G. Purcell, Proceedings of the PESA/GSA Canning Basin symposium. Perth. 389-399, 1984.

Reiter, M. and G. Clarkson, Relationships between heat flow, paleotemperatures, coalification, and petroleum maturation in the San Juan Basin, northwest New Mexico and southwest Colorado, Geothermics, 12, 4, 323-339, 1983.

Rice, D.D., Relation of natural gas composition to thermal maturity and source rocks type in San Juan Basin, northwestern New Mexico and southwestern Colorado, Am. Assoc. Pet. Geol. Bull., 67, 8, 1199-1218, 1983.

Roberts,W.H., Some uses of temperature data in petroleum exploration, in Unconventional Methods in Exploration for Petroleum and Natural Gas, Symposium II, edited by B. Gottlieb, Southern Methodist University Press, Dallas. 8-49, 1981.

Ross, H., J. and Moore, Geophysical investigations of the Cove Fort - Sulphurdale geothermal system, Utah, Geophysics, 50, 11, 1732-1745, 1985.

Saxby, J.D., The significance of organic matter in ore genesis, in Handbook of Stratibound and Stratiform ore deposits, edited by K.H. Wolff, 111-133, Elsevier, Amsterdam, 1976.

Saxena, K. and M.L. Gupta, Evaluation of reservoir temperatures and local utilization of geothermal waters of the Konkan coast, India, J. Volc. Geotherm. Res., 33, 4, 337-342, 1987.

Sheppard, S.M.F., Stable isotope studies of formation waters and associated Pb-Zn hydrothermal ore deposits, in Thermal Phenomena in Sedimentary Basins, edited by B. Durand, 301-317, Editions Technip, Paris, 1984.

Simoneit, B.R.T., Organic matter maturation and petroleum genesis - geothermal versus hydrothermal, in The role of heat in the development of energy and mineral resources in the northern Basin and Range Province, Spec. Rpt.13, Geotherm. Resour. Counc., Davis, Calif., 215-241, 1983.

Smith, L., and D.S. Chapman, On the thermal effects of groundwater flow, 1 - Regional scale systems, J. Geophys. Res., 88, B1, 539-608, 1983.

Summer, S., 1987, Maturation, Diagenesis and Diagenetic Processes in Sediments Underlying Thick Volcanic Strata, Oregon, M.S. Thesis, University of California, Davis, 87pp.

Summer, S., and K.E. Verosub, Maturation Anomalies in sediments underlying thick volcanic strata, Oregon, Evidence for a thermal event, Geology, 15, 1, 1987a.

Summer, S., and K.E. Verosub, Extraordinary maturation profiles of the Pacific Northwest, Oregon Geology, 49. 11, 135-140, 1987b.

Summer, S., and K.E. Verosub, Diagenesis in sediments under thick volcanic strata, Ochoco Basin, Oregon (abstract), Am. Assoc. Pet. Geol. Bull., 71, 5, 1987c.

Surlyk, F., J.M. Hurst, S. Piasecki, F. Rolle, A., Scholle, L. Stemmerik, and E. Thomsen, The Permian of the western margin of the Greenland Sea - A future exploration target, in Future Petroleum Provinces, edited by M. Halbouty, 239-276, Am. Assoc. Pet. Geol., Memoir 38, Tulsa, Oaklahoma, 1986.

Teichmuller, M. and R. Teichmuller, Geological causes of coalification, in Coal Science, edited by P. Given, Advances in Chemistry Ser. 55, American Chemical Society, 133-155, 1966.

Teichmuller, M. and R. Teichmuller, Geological aspects of coal metamorphism, in Coal and Coal-bearing Strata, Murchison, D. and T.S. Westoll, (eds.), Elsevier, New York, 233-267, 1968.

Teichmuller, R. and M. Teichmuller, Relationships between coalification and paleogeothermics in Variscan and Alpidic foredeeps of western Europe, in Paleogeothermics, Buntebarth, G., and L. Stegena, (eds.), 53-78. Springer-Verlag, Berlin, 1986,

Tissot, B., and D.H. Welte, Petroleum Formation and Occurrence, Springer-Verlag, Berlin. 699, 1984.

Walsh, T.J. and W. M. Phillips, Rank of Eocene coals in western and central Washington State - A reflection of Cascade plutonism?, Wash. Dept. Nat. Res., Open-File Rpt. 83-16., 21p., 1983.

Wood, W.W., and W.H. Low, Aqueous geochemistry and diagenesis in the eastern Snake River Plain aquifer system, Idaho: Geol. Soc. Am. Bull.,97, 1456-1466, 1986.

Wright, J.E., The geology of the Irish Sea, Can. Soc. Petrol. Geol., Memoir, 4, 295-312, 1975.

Ziagos, J., and D.D. Blackwell, A model for the transient temperature effects of horizontal fluid flow in geothermal systems, J. Volc. Geotherm. Res., 27, 3/4, 371-397, 1986.

Zielinski, G.W., and M. Bruchausen, Shallow temperatures and thermal regime in the hydrocarbon province of Tierra del Fuego, Am. Assoc. Pet. Geol. Bull., 67, 1, 166-177, 1983.

HYDROCARBON MATURATION IN THRUST BELTS: THERMAL CONSIDERATIONS

Kevin P. Furlong

Department of Geosciences, Pennsylvania State University, University Park, Pennsylvania 16802

Janell D. Edman

Exlog/Brown & Ruth Laboratories, Inc., 8100 South Akron, Suite 310, Englewood, Colorado 80112

Abstract. Sedimentary strata within thrust belts experience transient thermal histories which perturb the maturation paths of organic material contained within the rocks. Calculation of the thermal history, including perturbations which occur with overthrusting, for a particular sequence of tectonic events, allows us to evaluate the timing of maturation reactions and the remaining generative potential in the source rock during the evolution of the geologic terrain. In addition, thermal-maturation indicators can be used to constrain tectonic models for a region and eliminate nonviable geologic interpretations. We have utilized a numerical model to evaluate the thermal response to burial, erosion and thrusting. This model allows us to specify reasonably complex (and geologically reasonable) tectonic histories, including time varying erosion and sedimentation, syn-thrusting erosion, and multiple thrusting events. Such complexities are not easily incorporated in analytic thermal models of thrust belt evolution. In case studies of the western overthrust belt of Wyoming, thermal modeling of geologic histories provides insight into maturation processes, timing and geometries of thrust sheets, and pre-thrusting tectonism. In particular, the timing of thrust sheet motion in the Wyoming portion of the thrust belt appears to be younger than normally thought. Much of the thin skinned thrusting evaluated here appears (from a thermal perspective) to be of similar age to Laramide thrusting in the region.

Introduction

The transient nature of the temperature field in regions of overthrust tectonism has been recognized for quite some time. Models of this thermal evolution have been applied in a variety of settings including basement thrusting [Oxburgh and Turcotte, 1974; Brewer, 1981] and thin skinned overthrusts such as the Western Overthrust Belt of the United States [Angevine and Turcotte, 1983; Furlong and Edman, 1984; Edman and Surdam, 1984]. In these areas of thin skinned thrusting, the thermal perturbations are more subtle, but nonetheless play an important role in the evolution of organic materials within the sedimentary strata. The models referenced above provided a framework from which to evaluate the thermal evolution during and after thrusting. For thin skinned thrust regimes for which rates and timing of erosion and fault motion play a crucial role in the thermal evolution, these analytical models have not provided the needed resolution. It has become clear to us in our work that the analytic models used previously can provide a first order estimate of the temperature history of units of interest, but the application of relatively simple numerical models allow a better constrained evaluation of the maturation history of the thrust belt. Such models allow the direct incorporation of realistic erosion histories for the thrust sheets, multiple thrust events, and thermal evolution during thrusting.

Utilization of thermal evolution models in thrust belts can provide several different types of information. The use to which the models are put depends strongly on the type and quality of available data and can be used to evaluate tectonic histories, maturation histories and/or reservoir evolution. From an exploration perspective, thermal modeling provides the needed 'look backwards' to the time of oil generation, allowing inferences on migration paths and availability of hydrocarbon traps. A second very important (though often overlooked) role for thermal modeling is the determination of tectonic histories for thrust belts using organic markers such as vitrinite reflectance to constrain tectonic parameters including timing of thrust movement, erosion history, or regional heat flow. This latter use provides tectonic data not easily obtained via traditional geologic studies.

Thermal-Tectonic Overview

The principal feature of all overthrust thermal models is the development and relaxation of a non-equilibrium thermal field produced by the advection of heat with movement of the thrust sheet. A dramatic feature of many of these models is the initial development of a "sawtooth" geotherm with a temperature inversion from the base of the upper plate into the upper part of the lower plate. This feature, although potentially important in regions of thick (> 10 km) overthrusts, is geologically ephemeral in thin skinned thrusting. For typical thrust sheet thicknesses of 5 km or less, a temperature inversion would last less than 200,000 years. As shown by Furlong and Edman [1984], when the non-dimensional time (τ) which depends on thrust plate thickness and thermal diffusivity is ≤ 0.2 (which corresponds for a 5 km thick thrust sheet to approximately 160,000 years), the temperature inversion has disappeared, and temperatures monotonically increase downward from the surface. Since this is undoubtably a shorter time span than represented by the thrusting event itself, such a temperature inversion has little if any importance in thin skinned thrust environments. Rather, what is more important in these terrains, is the rapid cooling of the upper plate that occurs with thrusting, heating of the upper part of the lower plate, reduction in temperature gradient in both the upper and lower plates, and the long thermal recovery period (order 50 Ma). A crucial tectonic parameter during this recovery period is the erosion history of the thrust belt. Relatively small differences in this erosion behavior can have important effects on the maturation of hydrocarbons within the thrust belt basins.

After thrusting, a typical erosion history coupled with thermal relaxation results in rapid cooling of the upper plate which never recovers pre-thrusting temperatures because of the unroofing associated with erosion. For studies of hydrocarbon maturation, this effect causes the cessation of maturation reactions in the upper plate with thrusting, while reaction rates in the lower plate are typically enhanced by the thrusting event. Upper plate units normally record the effects of the pre-thrusting thermal regime, while lower plate units may record pre-, syn- and post- thrusting thermal effects. This dichotomy in thermal record (coupled with the typically non-linear time-temperature behavior of organic reactions) allows significant tectonic detail to be recovered even from data obtained at single well sites.

The modeling presented here represents a middle ground between the analytic models which have been used in the past and an ultimate goal of fully three-dimensional dynamic models. We have utilized numerical one-dimensional models which allow us to incorporate realistic rates of erosion and sedimentary burial. These 1-D models fall short in not including horizontal (conductive) heat transfer and any effects from fluid flow along the fault plane itself or within the thrust sheets, however such heat transfer represents only a very small part of the total heat budget in most thin skinned thrust belts. Data available to us do not warrant the computational complexity and expense of 2-D models. At present we typically have data from single well sites; without better spatial coverage, it is not possible to constrain the 2-D models to provide better resolution than our 1-D modeling approach.

Modeling Algorithm

The modeling algorithm we have used is an implementation of a simple finite difference method solution to the transient heat equation. We have used an implicit formulation (for solution stability) and have found that with a 200 meter space step, we can utilize a 0.15 Ma time step, allowing relatively high spatial resolution with geologically appropriate temporal resolution. The model spans 30 km in depth (essentially one crustal thickness) with a heat flux bottom boundary condition and specified temperature upper boundary condition. This model (30 km thick) is adequate for the typical thin skinned thrusts we have evaluated (thickness < 5 km), providing essentially 'half-space' results in the upper 10 km of the model. It would need to be increased for thicker thrust sheets. Erosion, sedimentary burial, and thrusting are simulated by shifting of temperature information along our spatial discretization as appropriate for the process occurring. The model allows burial and erosion before and after thrusting, syn-thrust erosion, surface and sub-surface thrust planes, and multiple thrust events. The numerical model has been tested against the analytic model for simple thrust histories and produces equivalent results.

A site specific geologic history is used as input to the model, and the temperature histories of specific stratigraphic units are monitored. These temperature histories can than be used to calculate other temperature dependent parameters such as vitrinite reflectance [Lopatin, 1971; Waples, 1980], biomarkers [Beaumont et al., 1985; Mackenzie et al., 1985], and production kinetics [Ungerer and Pelet, 1987; Tissot and Espitalié, 1975]. Vitrinite reflectance and biomarkers can be used, as indicators of the total effect of the thermal history since they provide present day data. Production kinetics allow the production history of a source unit (for a computed thermal history) to be evaluated in light of tectonic and geologic histories. In this way the relative timing of maturation and reservoir availability can be tested. As described later, we have found this to be a useful tool for discriminating among a suite of tectonic histories which produce similar present day results but have quite different production histories.

Time-Temperature Indicators

In our analyses, we have focused on two types of time and temperature dependent indicators of the thermal history of thrust belt units. We have used Vitrinite Reflectance (R_0) profiles in wells to calibrate thermal-tectonic histories, and Hydrocarbon Production Kinetics to evaluate the timing of source rock maturation (relative to the geologic history). The R_0 observations have been compared to TTI (time-temperature index) calculations [Lopatin, 1971] via the empirical correlation function described by Waples [1980]. In the TTI calculations, rates of reaction are assumed to double with every 10°C increase in temperature leading to an index which increases at a rate given by:

$$TTI = \Sigma \, \Delta t \cdot 2^n \tag{1}$$

where Δt is a time interval in Ma (equal to the model time step) and n is the temperature index given by n = (Temperature - 100)/10. With this definition, a temperature of 100°C serves as the base for TTI accumulation. TTI has been related to R_0 by Waples [1980] based on analysis of many wellsites in relatively simply subsiding basins. This correlation has some shortcomings particularly in that it may underestimate the temperature dependence and overestimate the time dependence for vitrinite reflectance development [Price, 1983] and the correlative capabilities between R_0 and stages of oil development [Quigley and Mackenzie, 1988]. We have tried to overcome this potential weakness by combining our R_0 calculations with 'kinetic' models of liquid hydrocarbon production to both test integrated temperature histories and the actual timing of production of liquid hydrocarbon products.

Although not extremely well calibrated, the correlation between R_0 and TTI has been used to constrain final maturation levels in our case studies. Biomarker reactions such as aromatization and isomerization [Beaumont et al., 1985; Mackenzie et al., 1985] could be similarly used, however such biomarker maturity parameters were not generally applicable at the sites we studied because of the high maturity levels of organic material in the units of interest. The use of the R_0 - TTI calibration appears to work reasonably well in our modeling as shown by the strong correlation of observed and predicted R_0 values for vertical profiles in the wells studied. Additional constraints provided by other thermal indicators would clearly be welcome.

Our analyses of hydrocarbon production and source rock potential follow the 'kinetic' approach of Tissot and Espitalié [1975] using kinetic parameters recently published by Ungerer and Pelet [1987]. This methodology allows us to evaluate probable timing, volumes and rates of production of liquid hydrocarbons from source rock kerogens. Since it monitors only a subset of the organic reactions which occur, it does not allow the separation of the various steps in oil and gas generation, but does allow us to monitor the primary conversion of kerogen to oil. This approach assumes an Arrhenius like behavior for maturation reactions:

$$\frac{d\chi_i}{dt} = -\chi_i A_i \exp[-E_i/RT]$$

where χ_i is the mass fraction of the i kerogen present (dependent on kerogen type), E_i is effective activation energy, A_i is a scaling factor, R is the universal gas constant, t is time (in units complementary to the units of A_i), and T is absolute temperature (K). Evaluation of this equation leads to:

$$\chi_i(t) = \chi_{i0} \exp\{-\int_0^t A_i \exp[-E_i/RT]\, dt\} \qquad (2)$$

and

$$Q = \sum_{i=1}^{n} (\chi_{i0} - \chi_i) \qquad (3)$$

where the χ_{i0} are initial mass fractions, and Q is the total mass fraction kerogen converted at time t. The number of kerogen components (n) vary with kerogen type ranging from seven components for Type I (lacustrine), nine components for Type II (marine) to thirteen components for Type III (continental) kerogens. In our modeling we have assumed Type II kerogen behavior, as the source units monitored are primarily marine shales or shale/sandstone mix. For Type II kerogens, total initial mass fractions of convertible organic material ($\Sigma \chi_{i0}$) is 630 mg/g organic carbon [Ungerer and Pelet, 1987].

Case Studies

This modeling approach has been applied to a series of well sites in the Western Overthrust Belt of Wyoming-Idaho-Utah, USA. Results from two sites are discussed here, and show similar behavior to other well sites in the Western Overthrust Belt. Geologic cross sections for these sites are shown in Figure 1. The Whitney Canyon site (Fig. 1b.) shows the effect of thrust movement on a sub-surface thrust plane, minimizing the thermal perturbation from thrusting, but still sufficiently altering the hydrocarbon maturation history. At the Tip Top (Mobil 22-19G) site (Fig. 1c.), thrusting was (at the site) along essentially a surface thrust plane. Lower plate thermal perturbations were significant, resulting in alteration of the maturation path.

Whitney Canyon

The Whitney Canyon well samples upper plate units of early Paleozoic through Cretaceous age and Cretaceous lower plate units. The R_0 profile from the well (triangles, Figure 2) shows a 'sawtooth' pattern common for many thrust regions with repeated stratigraphy. Lower plate Cretaceous units show higher R_0 values than equivalent upper plate units. The existence of the sawtooth pattern after the thrusting event indicates that much of the maturation occurred prior to thrusting. However, higher levels of maturation in lower plate units indicates maturation reactions continued after thrusting. Our modeling of the thermal evolution allows us to quantify this observation.

R_0 observations from late Paleozoic units immediately above the thrust plane and Cretaceous units below the thrust plane show levels of maturation indicating prior passage through the oil window for these units of quite different ages. Of interest to petroleum exploration studies is the relative timing of the maturation of these units, and the source potential of various organic rich units. We have modeled the burial-thrusting-erosion history of these two units to monitor the maturation progress. The assumed burial history is shown in Figure 3a. Thrusting is assumed to have occurred at approximately 71 Ma. The thrusting placed a thrust sheet approximately 4.2 km thick over the lower plate resulting in only 600 m net burial of the lower plate because of the sub-surface location of the thrust plane (depth of 3.6 km). The Paleozoic units immediately above the thrust originated at a depth of approximately 9.6 km, but syn-thrusting erosion removed approximately 5.4 km of overburden. This burial history is consistent with data presented by Warner and Royse [1987].

The resulting thermal history is shown in Figure 3b. Upper plate rocks reached maximum temperatures of approximately 150°C just prior to thrusting. The combined thermal effects of syn-thrust erosion and thrust related cooling rapidly lower temperatures in the upper plate by approximately 70°C with a subsequent rewarming of only 20°C since that time. The thermal history of the lower plate shows essentially monotonically increasing temperatures with a slight (10 - 15°C) step increase at the time of thrusting. This is approximately 25% of the syn-thrusting temperature increase predicted from the analytic thrust models and reflects the effects of erosion during thrusting and the effects of upper plate cooling prior to emplacement at the wellsite. Temperatures in the units of interest in the lower plate never reach the maximum levels of the (pre-thrusting) upper plate, although present day R_0 values are similar.

Maturation histories for these units, based on the calculated thermal histories are shown in Figure 4. TTI (and equivalent R_0) histories (Fig. 4a.) are calculated using equation (1), while production histories (Fig. 4b.) are

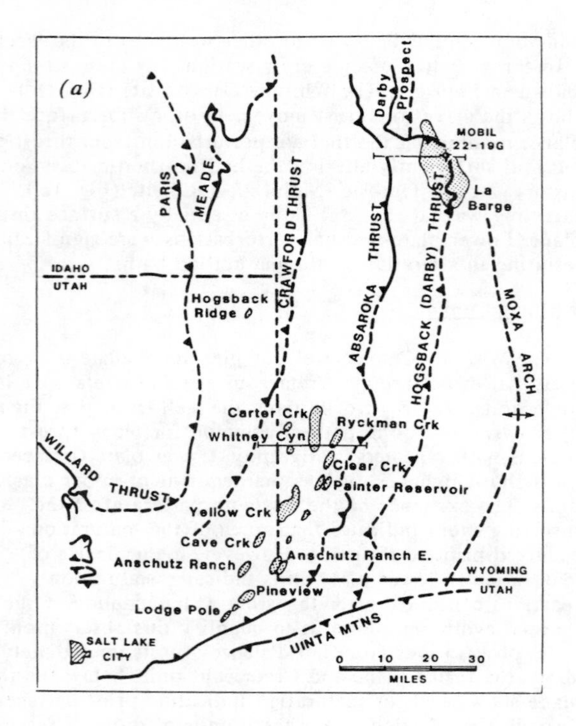

calculated from equations (2) and (3) using parameters for Type II (marine) kerogens from Ungerer and Pelet [1987]. Although present R_0 values are post oil generation for these two units, upper plate rocks reached present maturity prior to thrusting with little change in TTI after the thrust event. Production modeling indicates that not only was hydrocarbon production halted by cooling associated with thrusting, but also that virtually all potential source material would be exhausted by the time of thrusting. Thus although thrusting dramatically reduced temperatures in the upper plate, its effect on the overall maturation history (as far as primary conversion of kerogen to liquid hydrocarbon is involved) was minimal.

The maturation history for the lower plate units differs from the upper plate in that the thermal events during and soon after thrusting resulted in a period of high productivity and maturation. Additionally, these lower plate units could serve as potential source rocks for 20 - 30 Ma after thrusting; permitting oil produced to benefit from permeability increases generated by thrust tectonics, allowing for easy migration to structural traps produced during thrusting. The migration and trapping history of oils produced within the Paleozoic units of the upper plate is independent of the thrusting event and requires knowledge of the pre-thrust tectonic environment for evaluation.

The burial history used here produces a match to observed R_0 values through the entire borehole, reproducing

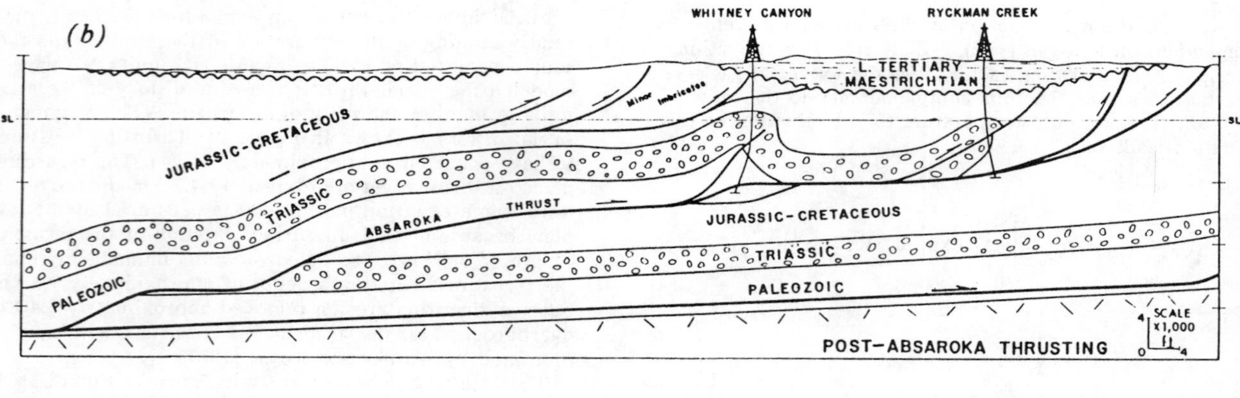

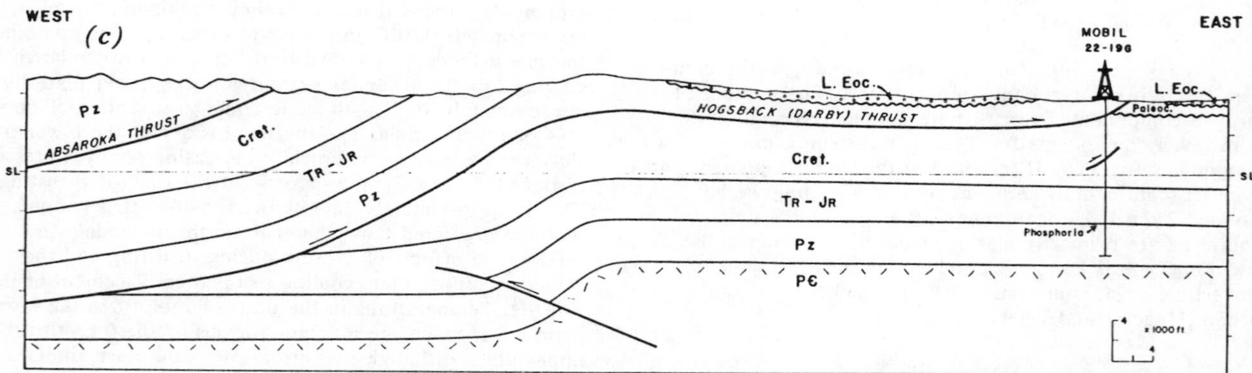

Fig. 1. Index map and West to East geologic cross sections for thrust belt sites evaluated here. (a) Index map of Wyoming-Idaho-Utah thrust belt. (b) Geologic cross section at Whitney Canyon site near Wyoming - Utah border. (c) Geologic cross section at Tip Top (Mobil 22-19G) site. (adapted from Warner and Royse [1987], reprinted by permission of American Association of Petroleum Geologists).

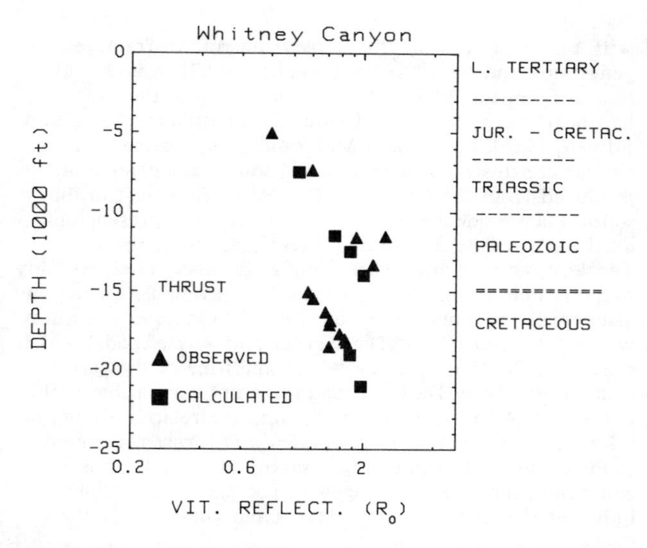

Fig. 2. R_0 profile for Whitney Canyon wellsite. Triangles are observed R_0 values [Warner and Royse, 1987], squares are calculated R_0 values. In depth range of 15,000 to 20,000 feet calculated R_0 values are omitted as they overlie observed values. Thermal modeling described here produces the observed 'sawtooth' pattern in R_0 values. Generalized stratigraphy is shown on the right of the figure.

the "sawtooth" pattern seen in many thrust belts (squares, Figure 2). It is important to recognize that this pattern in R_0 is unrelated to the "sawtooth" temperature regime predicted by analytic thermal models of thrusting, but rather is a consequence of superimposing a pre-thrust thermal-maturation history for the upper plate with a pre- and post-thrust history for the lower plate.

Tip Top

Thermal modeling of the thermal evolution of units sampled at the Tip Top (Mobil 22-19G) wellsite coupled with geologic observations allow one to distinguish between alternative tectonic models of thrust history. Recently Warner and Royse [1987] have argued that in much of the western overthrust belt, syn-thrust erosion effectively reduces thrust sheets to present day thicknesses, resulting in minimum thermal perturbations during thrusting. Present levels of organic maturation in lower plate units can be used in conjunction with geochemical analysis of produced liquids and petrographic analysis of core samples available from the Tip Top well to determine timing of maturation relative to thrusting, acceptable pre- and syn-thrust tectonic scenarios, and strategies for exploration for oil from source rocks in the lower plate units.

The present setting of the Tip Top site (Fig. 1c.) shows that virtually all of the thrust plate has eroded (present thickness < 400 meters), with most of the erosion accomplished prior to the Lower Eocene, or within about 10 Ma of the thrusting. Of importance to the maturation of the Permian Phosphoria (a likely source rock for regional reservoirs [Edman, 1982]) is whether the thrust sheet erosion extended until the Lower Eocene, or was complete by the end of thrust motion (as suggested by Warner and Royse [1987]). If the thrust sheet was essentially at present day thickness at the time of emplacement at Tip Top, significant pre-thrusting burial and erosion events are needed to produce observed R_0 values. We have modeled both the burial and erosion history proposed by Warner and Royse [1987] (Figure 5a.) which incorporates this pre-thrust burial/erosion event and also a burial/erosion history proposed by Edman and Surdam [1984] which involves some pre-thrust erosion but includes a 2.8 km thick thrust which erodes during the 7 Ma following thrusting (Figure 5a.).

Both burial histories produce footwall thermal histories (Figure 5b.) consistent with observed R_0 values for the Phosphoria. Maturation histories differ dramatically in the two cases (Figure 6) with all oil production occurring 10-12

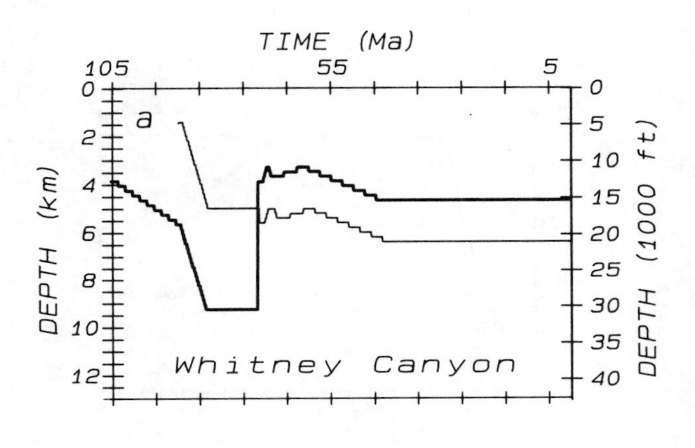

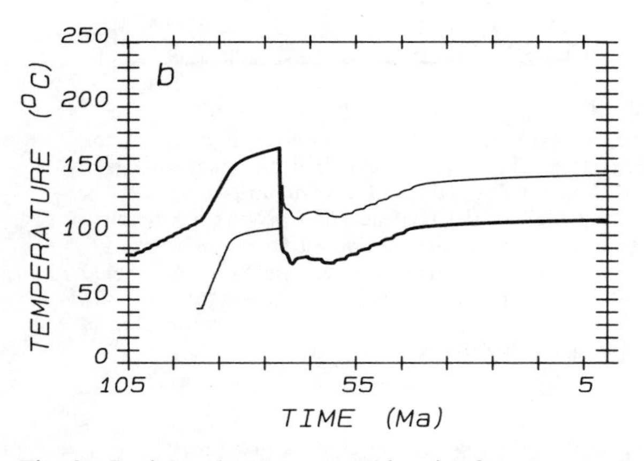

Fig. 3. Burial and temperature histories for upper and lower plate units at Whitney Canyon site. (a) Depth vs Time plot shows assumed burial-thrust-erosion history for Whitney Canyon site. Heavy lines are history of upper plate units, Lighter lines are history for lower plate units. ('Stairsteps' are result of discrete (200 meter) vertical motions in numerical models.) Thrusting occurred at approximately 71 Ma. Units monitored are late Paleozoic units above the thrust in the upper plate, and Cretaceous Aspen-Bear River Formations in lower plate. (b) Calculated thermal history for burial history shown in part (a) of figure.

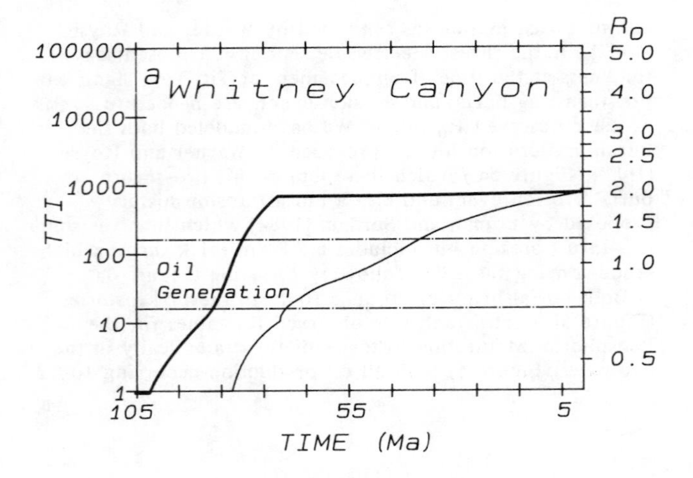

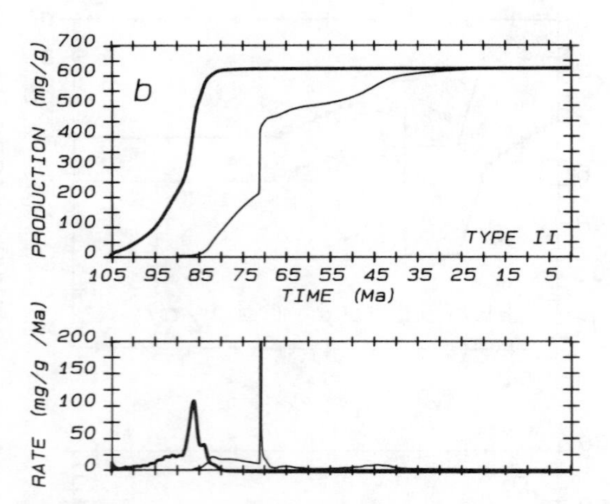

Fig. 4. TTI and Hydrocarbon Production history for Whitney Canyon site . (a) TTI and equivalent R_0 history. Approximate TTI range of principal oil generation is labeled. Present day values of R_0 for units monitored is approximately 2.0. (b) Hydrocarbon production history for well site. Rate curves are computed from derivative of production curves. Production and rate axes are labeled in units of mg of production per gram of organic carbon. For Type II kerogen, maximum production is assumed to be 630 mg/g [Ungerer and Pelet, 1987].

Ma prior to thrusting for the case of pre-thrust burial/erosion (WR model), while significant oil production occurs during and immediately after thrusting in the Edman and Surdam model. The maturation histories indicate that at the time of thrust emplacement, it is unlikely that there would be any liquid oil within the sub-thrust Phosphoria under the Warner and Royse tectonic assumptions.

A test of these two models is available in the form of analysis of liquid hydrocarbons produced from the overlying Nugget sandstone and analysis of 'dead oil' which fills fractures in breccias in the lower plate core samples. These fractures are believed to have formed during deformation associated with thrusting, although overpressuring associated with oil generation may have played a role in fracture generation and development [Spencer, 1987]. Analysis of liquids from the Nugget at Tip Top indicates the hydrocarbons were sourced from the Phosphoria [Cook and Edman, 1988]. Seifert and Moldowan [1981] came to a similar conclusion of a Phosphoria source for condensate produced from the Nugget at Dry Piney field just to the south. Such evidence in conjunction with the presence of dead oil that fills fractures in brecciated Phosphoria, Tensleep, and Madison core samples [Edman, 1982] strongly suggests Phosphoria oil must have been available during and after thrusting when the fractures and producing structures were probably formed. The Warner and Royse model which shows the Phosphoria to be within the zone of dry gas generation during Darby thrusting, is not compatible with these data. As a result, understanding the relative timing of oil generation fracture development and thrust movement become critical to evaluating reservoir potential. Thus combining thermal modeling with geologic observations indicates that the present thickness and geometry of the

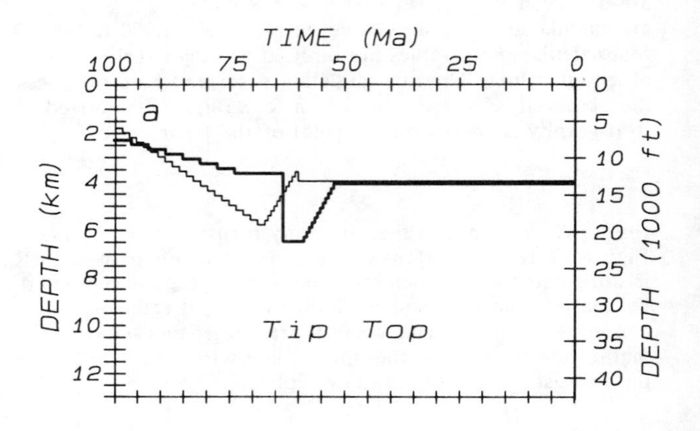

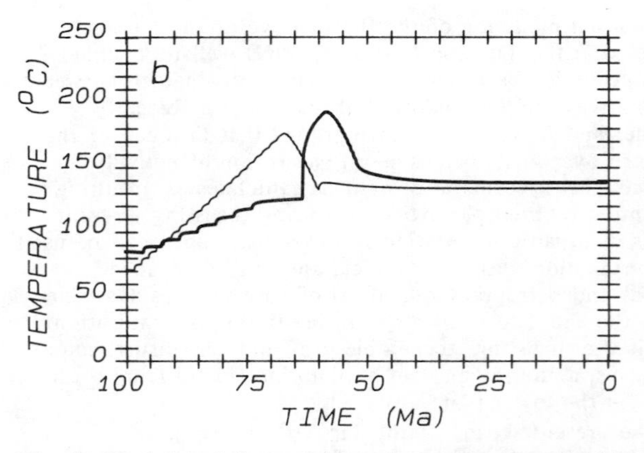

Fig. 5. Burial and thermal histories at Tip Top (Mobil 22-19G) site. (a) Burial-thrust-erosion history. Heavy line is history proposed by Edman and Surdam [1984], lighter line is history proposed by Warner and Royse [1987]. Thrusting in ES model is at 63 Ma, thrusting in WR model is at 60 Ma. Unit monitored is the Permian Phosphoria Formation. (b) Thermal history for models shown in part (a) of figure.

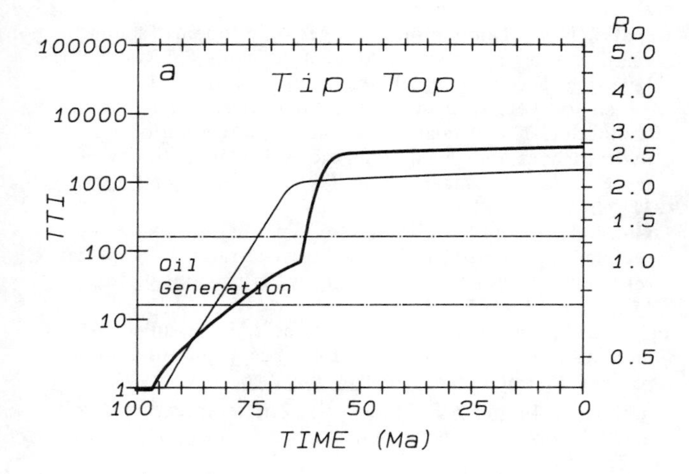

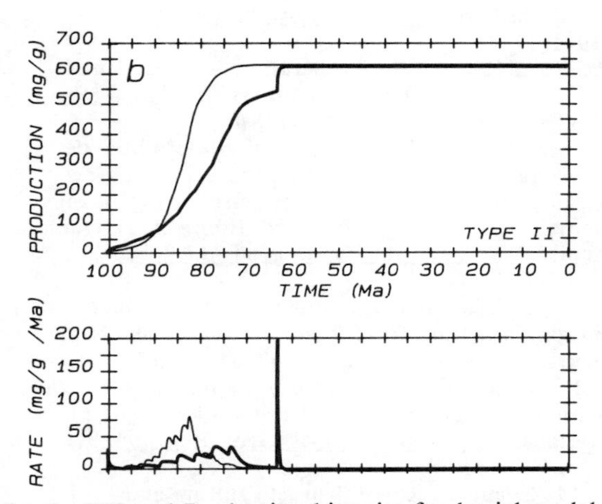

Fig. 6. TTI and Production histories for burial models given in Figure 5. See Figure 4 for explanation of axes. Observed values of R_0 for units monitored is approximately 2.6 (measured on bitumen).

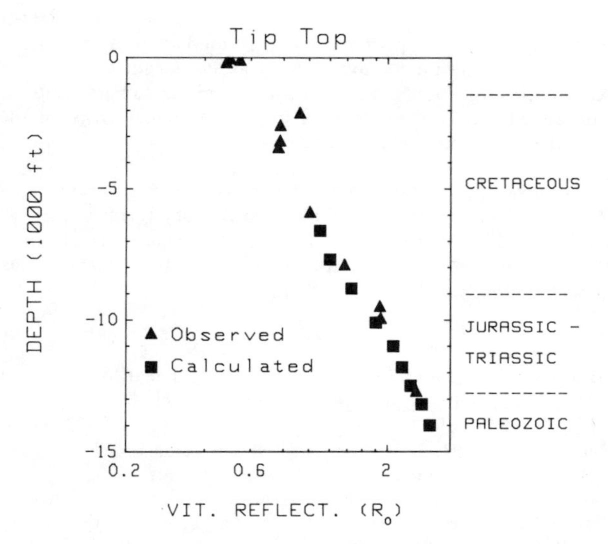

Fig. 7. R_0 profile in Tip Top well. Triangles are observed values of R_0 [Warner and Royse, 1987], squares are calculated values from Edman and Surdam [1987] burial model and calculations described here.

Darby thrust sheet in the vicinity of Tip Top likely formed by erosion during the 10 Ma period following thrusting and that at the time of thrust burial, thrust sheet thickness of at least 2.5 - 3 km is needed to provide the needed burial and thermal effects to produce present day observed R_0 values throughout the well (Figure 7).

Discussion

The modeling described here serves to indicate some of the potential use of thermal modeling of thrust belts in evaluating tectonic and hydrocarbon maturation processes. The data sets available in many such regions are often limited and unfortunately are often restricted in access. The addition of other information pertinent to thermal processes can serve to better constrain the tectonic histories. Temperature information from fluid inclusions, fission track apparent ages, and isotopic and geochemical data can provide 'point readings' of temperature and also information on additional integrated temperature effects.

An important application of this modeling, which has been only touched on here, is the constraining of thrust timing and rates of motion. Our modeling indicates that in most cases in the Western Overthrust Belt thrust emplacement occurred rapidly (order cm/yr rates), although more complete spatial coverage would allow us to better constrain the rates. Additionally, timing of much of the thrust movement in the Wyoming thrust belt appears (from thermal arguments) to be younger than often reported and thus nearly contemporaneous with Laramide (basement involved) thrusting. This result is consistent with flexural modeling of Green River Basin evolution [Hagen et al., 1985] which required loading effects from both the western thrust belt and the Wind River Mountains (simultaneously) in order to match basin geometry and stratigraphy.

A fundamental problem with thrust belt thermal modeling is poor constraint on paleo-heat flow in thrust regions. Any complex variation of regional (basal) heat flux through time adds further complications. In the modeling described here we have taken a conservative approach of using relatively standard values for basal heat flow (50 - 70 mW/m^2) which is assumed to remain constant throughout the period modeled. Heat flow within the thrust belt varies through time in response to the effects of thrusting but the basal heat flow (lower boundary condition) is held constant. At present we are unable to isolate the heat flow history from the tectonic history, but with three-dimensional data sets (good horizontal coverage and well profiles) we may begin to be able to evaluate these fine details.

We believe that thermal modeling of thrust belt evolution when coupled with geologic and geochemical data is a powerful tool for evaluating the tectonic and maturation history of a region. Knowledge of the timing and rates of oil maturation relative to tectonism is crucial for evaluating source rock potential, migration histories, and thus reservoir sites. In geologically complex areas such as the thrust belt of

Wyoming, such an approach has provided insight into the evolution of the petroleum sources and reservoirs. Application of such models to other regions can provide similar information and may also provide constraints on the viability of various tectonic scenarios.

Acknowledgements. The Donors of the Petroleum Research Fund, American Chemical Society (Grant 16560-AC2 to KPF) are acknowledged for support of this research. A careful and helpful review by W.J. Perry, Jr. is greatly appreciated.

References

Angevine, C.L., and D.L. Turcotte, Oil generation in overthrust belts, Amer. Assoc. Petrol. Geol. Bull., 67, 235-241, 1983.

Beaumont, C., R. Boutilier, A.S. Mackenzie, and J. Rullkötter, Isomerization and aromatization of hydrocarbons and the paleothermometry and burial history of Alberta foreland basin, Amer. Assoc. Petrol. Geol. Bull., 69, 546-566, 1985.

Brewer, J., Thermal effects of thrust faulting, Earth Planet. Sci. Lett., 56, 233-244, 1981.

Cook, L., and J.D. Edman, Tip Top field, Atlas of Treatise of Petroleum Geology, Amer. Assoc. Petrol. Geol., Tulsa, OK, (in press), 1988.

Edman, J.D., Diagenetic history of the Phosphoria, Tensleep, and Madison Formations, Tip Top field, Wyoming, Ph.D. Dissertation, 229 pp., University of Wyoming, Laramie, WY, 1982.

Edman, J.D., and R.C. Surdam, Influence of overthrusting on maturation of hydrocarbons in Phosphoria Formation, Wyoming-Idaho-Utah overthrust belt, Amer. Assoc. Petrol. Geol. Bull., 68, 1803-1817, 1984.

Furlong, K.P., and J.D. Edman, Graphic determination of hydrocarbon maturation in overthrust terrains, Amer. Assoc. Petrol. Geol. Bull., 68, 1818-1824, 1984.

Hagen, E.S., M.W. Shuster, and K.P. Furlong, Tectonic loading and subsidence of intermontane basins: Wyoming foreland province, Geology, 13, 585-588, 1985.

Lopatin, N.V., Temperature and geologic time as factors in coalification (in Russian), Akademiya Nauk SSSR Izvestiya, Seriya Geolgicheskaya, 3, 95-106. 1971.

Mackenzie, A.S., C. Beaumont, R. Boutilier, and J. Rullkötter, The aromatization and isomerization of hydrocarbons and the thermal and subsidence history of the Nova Scotia margin, Phil. Trans. R. Soc. Lond., Ser. A, 315, 203-232, 1985.

Oxburgh, E.R., and D.L. Turcotte, Thermal gradients and regional metamorphism in overthrust terrains with special reference to the eastern Alps, Schweiz. Mineral. Petrog. Mitt., 54, 641-662, 1974.

Price, L.C., Geologic time as a parameter in organic metamorphism and vitrinite reflectance as an absolute paleogeothermometer, J. Petrol. Geol., 6, 5-38, 1983.

Quigley, T.M., and A.S. Mackenzie, The temperature of oil and gas formation in the sub-surface, Nature, 333, 549-552, 1988.

Seifert, W.K., and J.M. Moldowan, Paleo-reconstruction by biological markers, Geochem. Cosmochem. Acta, 45, 783-794, 1981.

Spencer, C.W., Hydrocarbon generation as a mechanism for overpressuring in Rocky Mountain region, Amer. Assoc. Petrol. Geol. Bull., 71, 368-388, 1987.

Tissot, B.P., and J. Espitalié, L'evolution thermique de la matiere organique des sediments; applications d'ure simulation mathematique potential petrolier des bassins sedimentaires et reconstitution de l'histoire thermique des sediments, Revue de l'Institut Francais du Petrole, 30, 742-777, 1975.

Ungerer, P. and R. Pelet, Extrapolation of the kinetics of oil and gas formation from laboratory experiments to sedimentary basins, Nature, 327, 52-54, 1987.

Waples, D., Time and temperature in petroleum formation: application of Lopatin's method to petroleum exploration, Amer. Assoc. Petrol. Geol. Bull., 64, 916-926, 1980.

Warner, M.A., and F. Royse, Thrust faulting and hydrocarbon generation: discussion, Amer. Assoc. Petrol. Geol. Bull., 71, 882-889, 1987.

TECTONICALLY INDUCED TRANSIENT GROUNDWATER FLOW IN FORELAND BASIN

Shemin Ge and Grant Garven

Department of Earth and Planetary Sciences, The Johns Hopkins University, Baltimore, Maryland 21218

Abstract. Deep groundwater flow in sedimentary basins can be driven by several mechanisms. In the case of evolving foreland basins, large scale compression and thrusting could develop high pore pressures in the foreland sag and initiate transient fluid flow. It has been argued also that fluid flow associated with the compression of foreland basins could play a role in the formation of sediment-hosted ore deposits and in petroleum migration. One goal of our research is to quantitatively evaluate the role of compressional tectonics in driving regional fluid flow in foreland basins. A numerical model has been constructed to study the coupled processes of mechanical deformation and pore-pressure dissipation. Part of the modeling study examines the mechanical behavior of a foreland basin as it is subjected to specified tectonic forces of compression and thrusting. The finite element method is used to analyze the mechanical deformation as a plane strain problem, from which the computed displacement field is used to define transient boundary conditions for regional groundwater flow in the sedimentary basin. The second part of our modeling study examines the influence of the tectonic forces on inducing transient fluid flow. Preliminary computations suggest that flow rates on the order of $10^{-3} \sim 10^{-2}$ $m\ yr^{-1}$ are possible soon after compression of the foreland, and that the flow field dissipates in about 10^3 to 10^4 years, but longer diffusion times can exist in very low permeability strata.

Introduction

During the evolution of a foreland basin, groundwater systems in the basin undergo a continuous transition in response to changes in basin geometry, sediment compaction, continental uplift, and external tectonic loadings caused by thrust sheets advancing onto the craton platform (Fig. 1). Understanding the transient history of subsurface fluid flow in a deforming foreland basin could help elucidate the role of deep groundwater flow in processes such as overthrusting, diagenesis, stratabound ore genesis, and petroleum migration (Fig. 2). A number of hydrologic studies of sedimentary basins have documented certain stages of the basin evolution problem. Garven and Freeze [1984a,b] and Garven [1985, 1986] examined the theoretical aspects of topography-driven flow as it relates to ore formation. Sharp [1978], Cathles and Smith [1983], Bethke [1985,1986] and Shi and Wang [1986] have calculated the effects of compaction due to rapid sedimentation. The effects of erosional unloading on groundwater flow have also been explored by Neuzil and Pollock [1983], Tóth and Millar [1983], and Tóth and Corbett [1986]. The close association between regional fluid migration, ore mineralization, and the deformation of orogenic belts is also outlined by Oliver [1986, 1987].

Our general understanding of the role of tectonics in the hydrologic evolution of foreland basin has improved greatly as

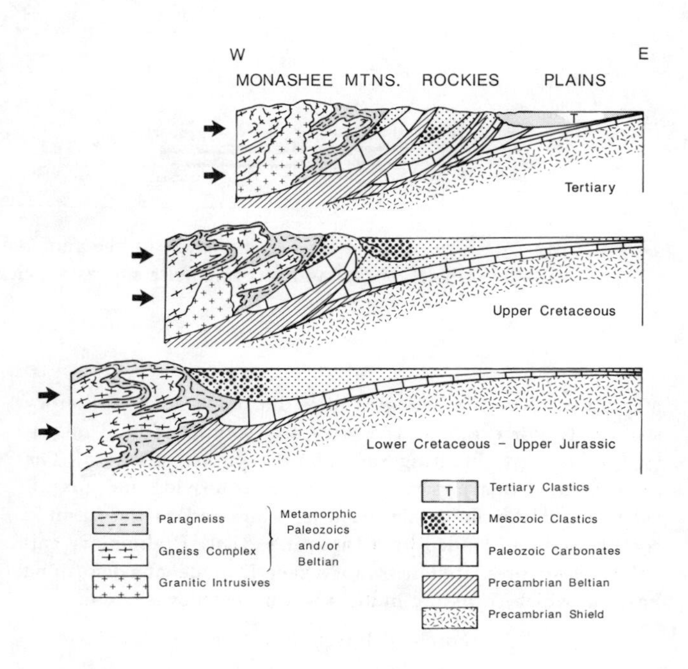

Fig. 1. The evolution of a foreland basin as illustrated by the tectonic history of the fold belts of the southern Canadian Rockies [after Bally, Gordy, and Stewart, 1966].

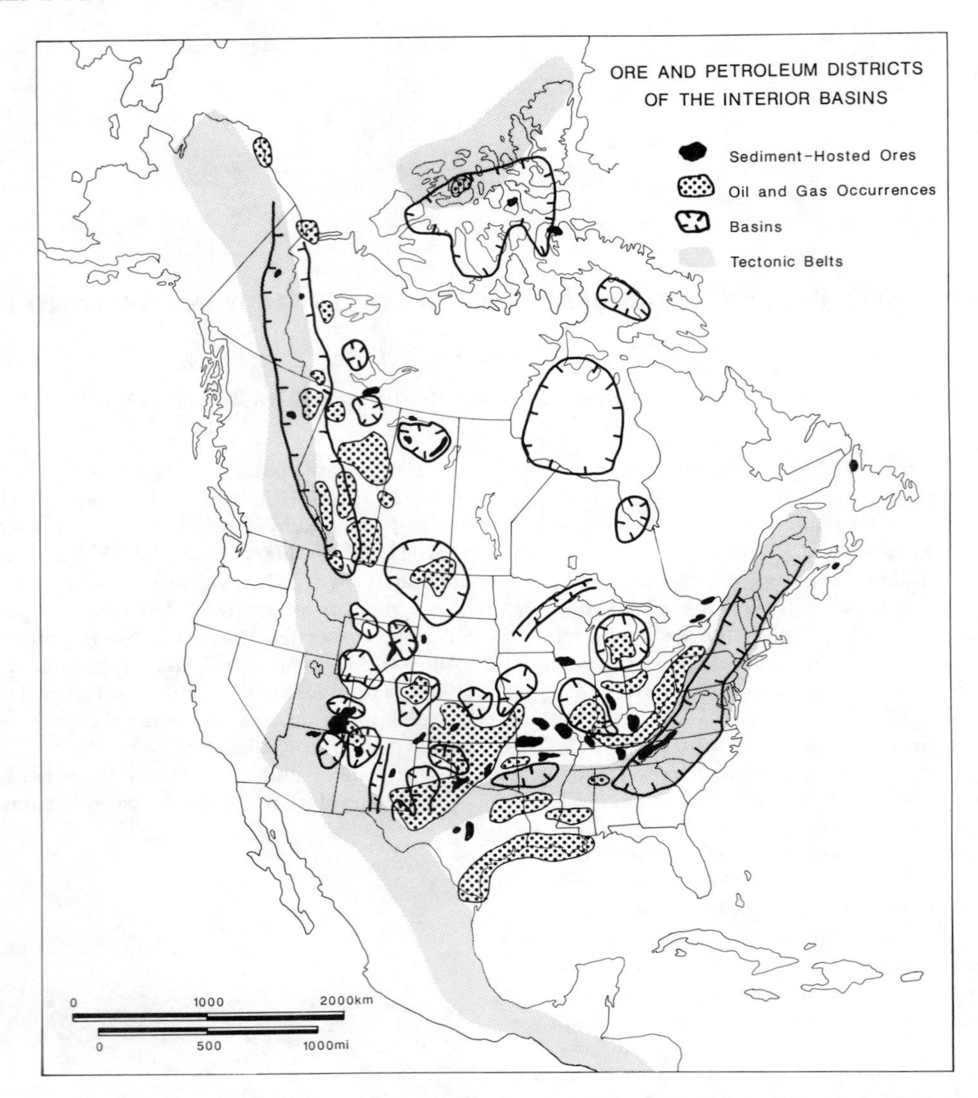

Fig. 2. Ore, oil and, gas occurrences in the midcontinent region of the United States. The geographical association with basins and tectonic belts suggests a clear genetic control on fluid migration [after Oliver, 1986, 1987].

a result of these theoretical studies, but gaps in the full story still exist. For example, the rates and volume of fluid migration induced by thrusting and compression are unknown. The main goal of this paper, therefore, is to provide one quantitative model for fluid migration in a foreland as the basin is compressed and loaded by a thrusting event. Preliminary calculations are presented below for a generic model of a deforming basin in which thrusting induces a transient flow system.

Foreland Basin Evolution

Over the past decade, several geodynamic models have been proposed to depict the evolution of foreland basins. Most models assume a thin elastic [Jordan, 1981] or viscoelastic lithosphere [Beaumont, 1978, 1981] overlying an inviscid fluid asthenosphere in which the plates are loaded and deformed in a vertical direction. At the early stage of foreland basin development, compressional forces and the weight of adjacent orogenic loads cause basement strata to downwarp. The subsidence of the basement allows for more sediment deposition. Depending on the rates of subsidence and lithologies, over-pressured zones may develop in the foreland wedge during sediment compaction [Bethke et al., 1988]. Further tectonic compression and high pore pressures in certain regions may trigger new thrust faulting. As a thrust sheet advances towards the basin, it would again cause compaction of sediments and fluid expulsion. These processes could cycle (Fig. 3) as a basin is repeatedly com-

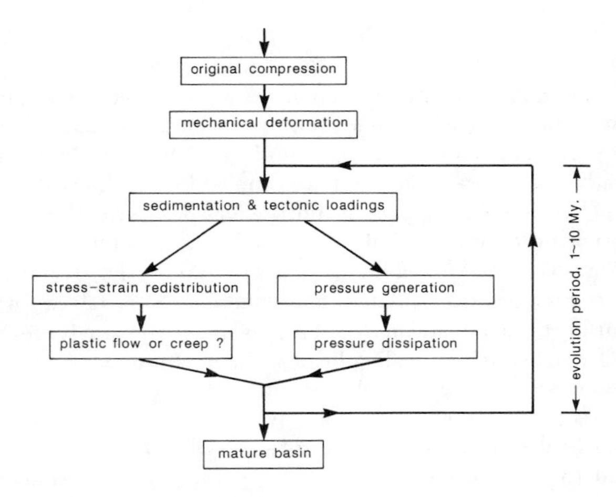

Fig. 3. The tectonic cycle of foreland basin evolution.

pressed in subsequent events, over the time periods on the order of a few million years [Suppe, 1985]. Erosional unloading of the fold and thrust belt will ultimately lead to uplift of the foreland platform and the development of topography-driven flow systems [Garven, 1982]. Topography-driven flow dominates the fluid-flow regime in mature cratonic basins [Garven and Freeze, 1984a,b].

There are two coupled aspects of the hydrodynamic problem involved in the basin evolution process which have not been addressed in earlier hydrologic studies. The first is the mechanical aspect concerning the stress-strain redistribution in the foreland basin under external loading. The second is the hydrological aspect concerning the transient fluid flow response induced in the foreland platform by changes in boundary conditions and deformation of the sediments. The following sections discuss these two aspects in sequence.

Stress-Strain Modeling

In order to study the mechanical and associated hydrodynamic responses of a sedimentary basin subjected to tectonic forces, we have developed a numerical model to perform stress-strain analysis coupled to fluid flow. The governing equation of a plane strain model for a deforming continuum in two dimensions is the force equilibrium equation. The integral form of the equation can be written as [Malvern, 1969 and Zienkiewicz, 1977] :

$$\int_{\Omega} [B]^T \{\sigma_T\} d\Omega + \int_{\Omega} [N]^T \{b\} d\Omega + \{f\} = 0 \tag{1}$$

where Ω is the study domain, $[B]^T$ is the strain matrix, $\{\sigma_T\}$ the total stress vector, $[N]^T$ an interpolation function, $\{b\}$ the body forces vector, and $\{f\}$ the external force vector. Equation (1) is based on virtual work principles and it is valid for a wide range of material behaviors (see Appendix I). A simple elastic model is used in our modeling exercises. The physical interpretation of (1) is more easily explained by multiplying each term by a displacement vector $\{u\}$ in which case the first term on the left side of the equation is the internal work done by stresses, the second term is the work done by distributed body forces, and the third term is the work done by external forces.

In the analysis of deformation in porous media, it is important to note that the total stress is supported in part by the skeleton of the solid particles and in part by the pore fluid. Hence, the principle of effective stress must be employed as introduced by Terzaghi [1936]: $\{\sigma\} = \{\sigma_T\} - \{P\}$, where $\{\sigma\}$ is the effective stress and $\{P\}$ is the pore pressure. The final form of the governing equation for mechanical modeling can be derived from equation (1), the effective stress equation, and constitutive laws with strain-displacement relations [see Appendix I; Reddy, 1984; Zienkiewicz and Humpheson, 1977]. The governing equation can now be expressed in terms of displacement $\{u\}$:

$$\int_{\Omega} [B]^T [C][B]\{u\} d\Omega + \int_{\Omega} [B]^T \{P\} d\Omega + \int_{\Omega} [N]^T \{b\} d\Omega + \{f\} = 0 \tag{2}$$

where $[C]$ is the material matrix containing mechanical parameters such as Young's modulus E and Poisson's ratio ν. Equation (2) is solved using the finite element method. Input data for numerical solutions include the basin geometry, boundary conditions (i.e. constraints on the solid displacement field), material properties, pore pressure, and the loading history. Output from a numerical solution includes the displacement and stress fields as functions of time.

Transient Fluid Flow Analysis

The variations in pore fluid pressure caused by externally imposed stress changes reflect the interaction between fluid pressure and stress. The classic theory of consolidation provides the basic relations between pore pressure, stress and strain in an elastic porous medium [Biot 1941, 1955; Rice and Cleary; 1976; Das, 1983]. The mass conservation equation for pore fluid can be written as the following [Bear, 1972; Domenico and Palciauskas, 1979; Walder, 1984] :

$$(\alpha + \phi\beta)\frac{\partial P}{\partial t} = \nabla \cdot \left[\frac{\bar{\mathbf{k}}}{\mu}(\nabla P - \rho\vec{g})\right] + \alpha\frac{\partial \sigma_t}{\partial T} \tag{3}$$

where: α the bulk compressibility, $[M^{-1}Lt^2]$; ϕ is the porosity, [dimensionless]; β the isothermal compressibility of water, $[M^{-1}Lt^2]$; $\bar{\mathbf{k}}$ the intrinsic permeability, $[L^2]$; μ the fluid viscosity, $[ML^{-1}t^{-1}]$; P the pore pressure, $[ML^{-1}t^{-2}]$; ρ the fluid density, $[ML^{-3}]$; σ_t the mean total stress, $[ML^{-1}t^{-2}]$; and t the time, $[t]$. The term on the left side of the equation (3) represents the time rate of change of fluid mass per unit volume of medium. On the right side, the first term represents the net mass flux due to the driving forces, and the second term takes into account the effects of total stress changes. The finite element method is used in this study to solve equation (3) numerically [Huyakorn and Pinder, 1983] for the primary unknown, pore pressure P. With the pore-pressure field computed, the fluid flux components are simply calculated using Darcy's law:

$$\vec{q} = -\frac{\bar{\mathbf{k}}}{\mu}(\nabla P - \rho\vec{g}) = -\bar{\mathbf{K}}\nabla h \tag{4}$$

where $\vec{q}$ is the Darcy velocity, $[Lt^{-1}]$.

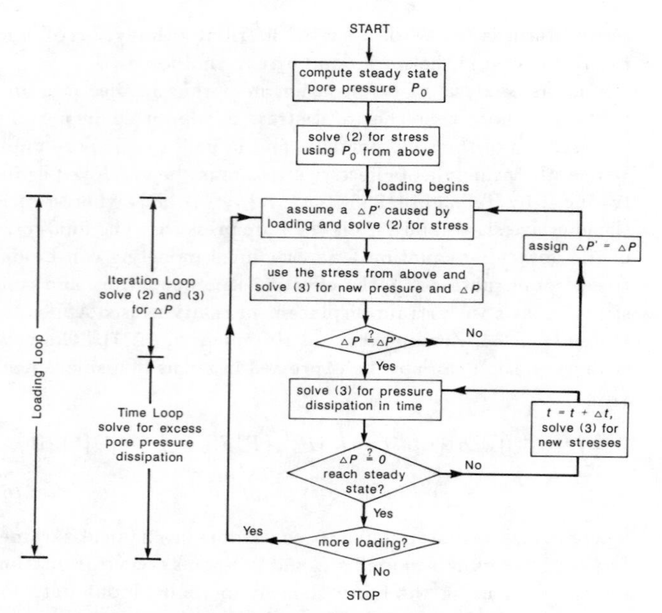

Fig. 4. Flow chart for the iterative algorithm used to solve the coupled stress-strain and fluid flow system.

Alternatively, equation (3) can be written with hydraulic head ($h = P/\rho g + Z$) as the dependent variable, hydraulic conductivity $\bar{\mathbf{K}} = \bar{\mathbf{k}}\rho g/\mu$ as the permeability parameter, and specific storage $S_s = \rho g(\alpha + \phi\beta)$ as the compressibility property. Assuming incompressible solid grains, specific storage S_s can be related to the mechanical properties by the following expression [Van der Kamp and Gale, 1983]:

$$S_s = \rho g \left(\frac{1}{K} + \frac{\phi}{K_f} \right) \tag{5}$$

Where K is the bulk modulus for the sediment, $[ML^{-1}t^{-2}]$, given by Young's modulus E and Poisson's ratio ν:

$$K = \frac{E}{3(1-2\nu)}$$

K_f is the bulk modulus for the fluid, $[ML^{-1}t^{-2}]$.

Solution Technique

Fluid flow and stress-strain fields in a sedimentary basin are interrelated. Mechanical loading causes fluid flow, and drainage of pore fluid enables compaction of the sediments. Mathematically, equations (2) and (3) are coupled because pore pressure and stress terms appear in both equations. An iterative solution procedure is used here to solve the coupled equations (Fig. 4). At the beginning of a simulation, the steady-state pore pressure is calculated, and equation (2) is solved for the initial stress and strain distributions under the gravity force of self weight. When loading begins, the pressure increase $\Delta P'$ is predicted and used to solve equation (2) for stress. This new stress field is incorporated in solving (3) for the new pressure and hence the computed pressure change ΔP. Equations (2) and (3) are solved repetitively until the predicted change in pressures agrees within a specified tolerance. The change in pore pressure ΔP induced by a loading event is allowed to dissipate between loading events. For subsequent loading steps, the computation procedure is repeated.

Numerical Results

Figure 5 shows a conceptual model of a foreland basin in a late period of thrusting and uplift. A newly developed thrust sheet compresses the platform sediments from the left and a vertical load can result from the mass of the thrust sheet. A gentle topographic slope exists across the basin platform due to flexural rebound following an earlier tectonic event. Figure 6a is the numerical representation of the foreland basin platform. Tectonic compression and loading are represented by distributed horizontal and vertical forces along the contact of the thrust sheet and the sedimentary basin. In this model, we assume that emplacement of the loads is instantaneous. The bottom thin layer is a regional aquifer that extends over the 500 *km* length of the basin. Less permeable sediments overly the aquifer with a thickness ranging from 500 m to 2500 m. Figure 6b is the finite element mesh. The boundary conditions are shown in Figure 6a. At the bottom AB, a contact between sediments and crystalline basement rock is assumed, where dis-

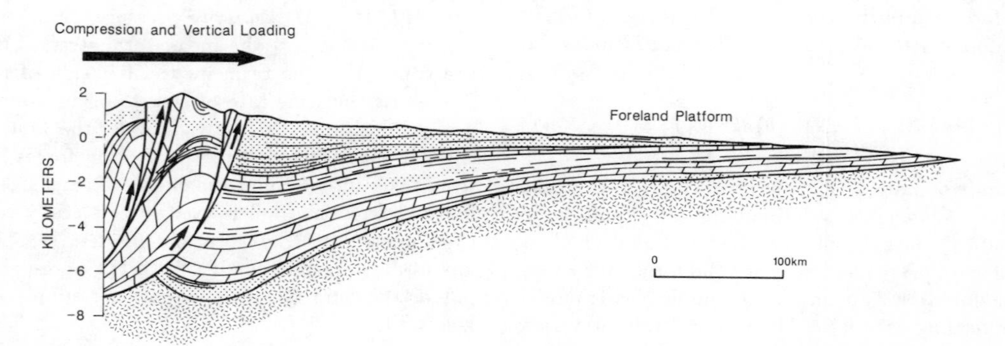

Fig. 5. Conceptualized foreland basin and adjacent fold belt. The newly developed thrust sheet compresses the basin from the left and loads the sedimentary pile.

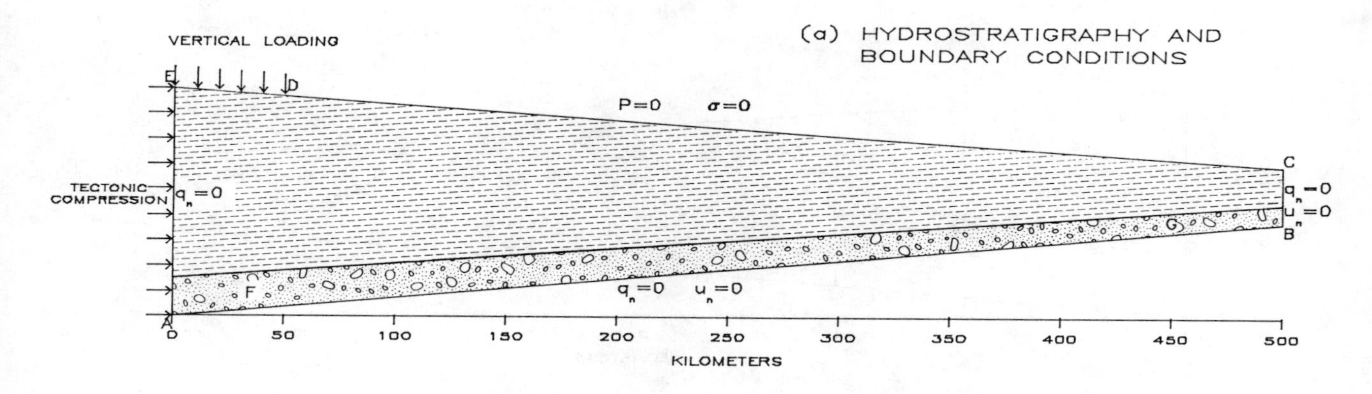

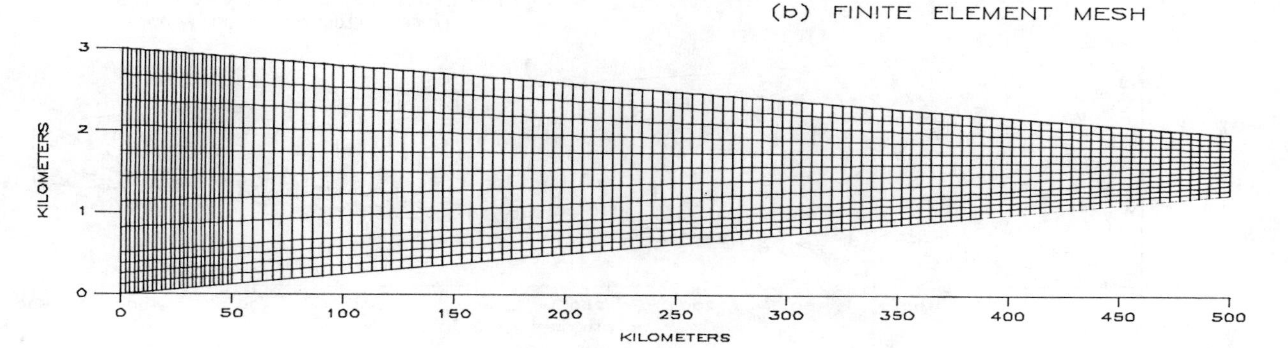

Fig. 6. (a) Numerical representation of a sedimentary basin with the hydrostratigraphic units and boundary conditions, (b) Finite element mesh.

placements are possible only in the direction along the boundary. No horizontal displacements are allowed at right end of the basin BC. The contact between thrust load and the basin, AE and ED, are specified stress boundaries. The top boundary DC is a free-stress ground surface. For hydrologic calculations, the two lateral boundaries and the lower boundary are considered to be impermeable (i.e. flow flux $q_n = 0$). The upper boundary is assumed to be coincident with the top of the saturated zone where pore pressure $P = 0$. The slope of the water table is 1:500.

A suite of simulations were performed to investigate the roles of permeability and loading magnitude. The material properties used in one sample simulation (Fig. 7 and 8) are listed in Table 1. Specific storage in Table 1 is defined as: $S_s = \rho g(\alpha + \phi\beta)$, where ρ, g, α, ϕ, and β were defined previously in equation (3) and assumed to be constants in this study: $\rho = 1000\ k_g\ m^{-3}$, $g = 9.8\ m\ s^{-2}$, $\alpha = 3 \times 10^{-11}\ m^2\ N^{-1}$ (aquifer), $\phi = 0.3$, and $\beta = 4.4 \times 10^{-10}\ m^2\ N^{-1}$. In practice, porosity ϕ decreases with depth and compressibility α also varies with the stress state [Neuzil, 1986; Bethke and Corbet,

TABLE 1. Rock Properties

	Hydraulic parameter			Mechanical properties	
	Horizontal conductivity $K_x(m\ s^{-1})$	Vertical conductivity $K_z(m\ s^{-1})$	Specific storage $S_s(m^{-1})$	Young's modulus $E(Pa)$	Poisson's ratio ν
Upper unit (aquitard)	3.2×10^{-7}	3.2×10^{-8}	1.45×10^{-6}	2×10^{10}	0.3
Lower unit (aquifer)	3.2×10^{-6}	3.2×10^{-7}	1.15×10^{-6}	6×10^{10}	0.2

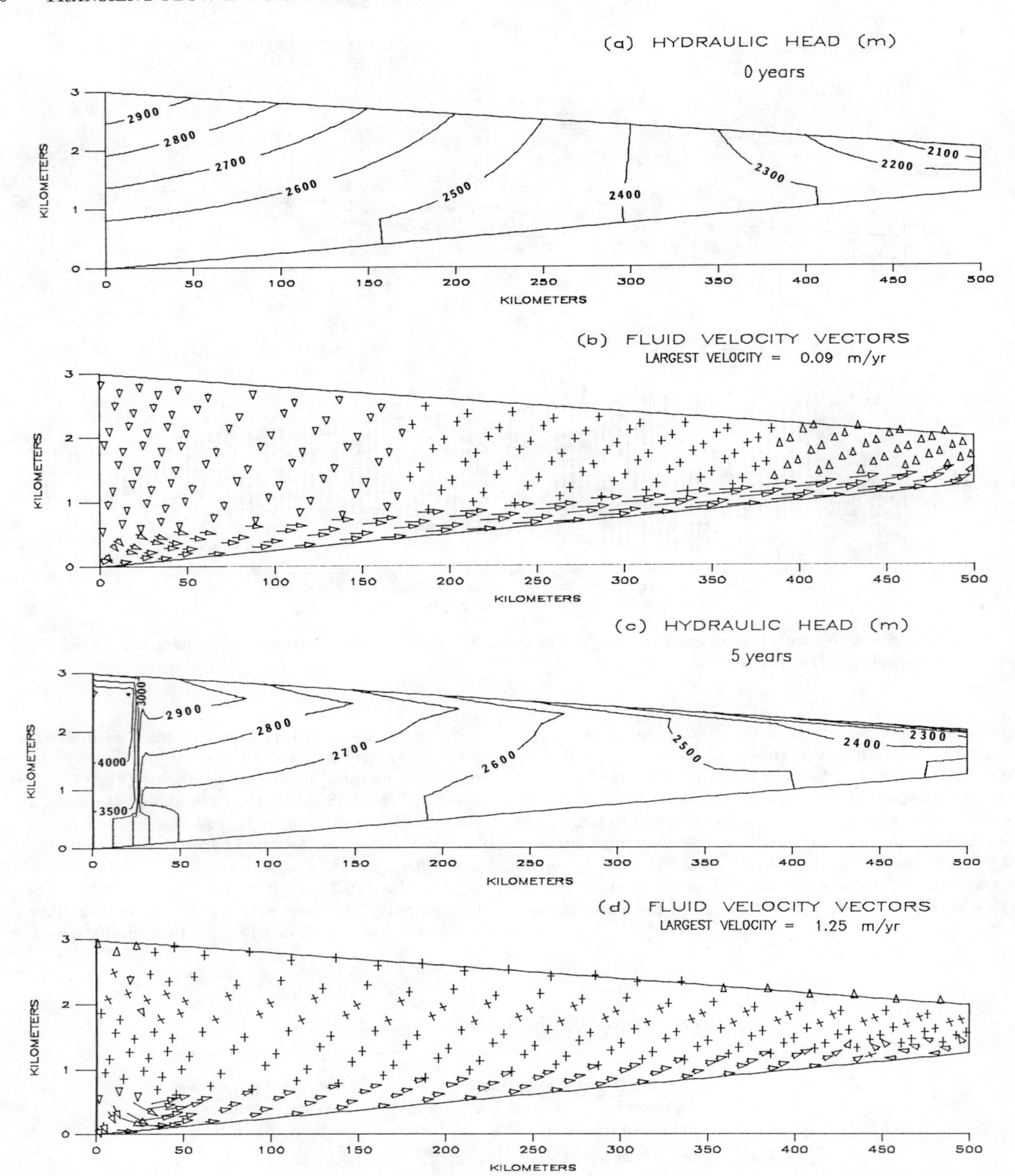

Fig. 7. Hydraulic head contours and velocity fields at several time stages. (a) and (b) Initial steady state, flow is driven by gravity with a 1:500 water-table slope. (c) and (d) At 5 years, hydraulic head contours are highly distorted and large flow velocities are concentrated near the loading area. (e) and (f) At 332 years, flow field is in an intermediate transient state. (g) and (h) 7973 years, most of the excess pressures have dissipated, and the flow field is gradually reaching a new steady state.

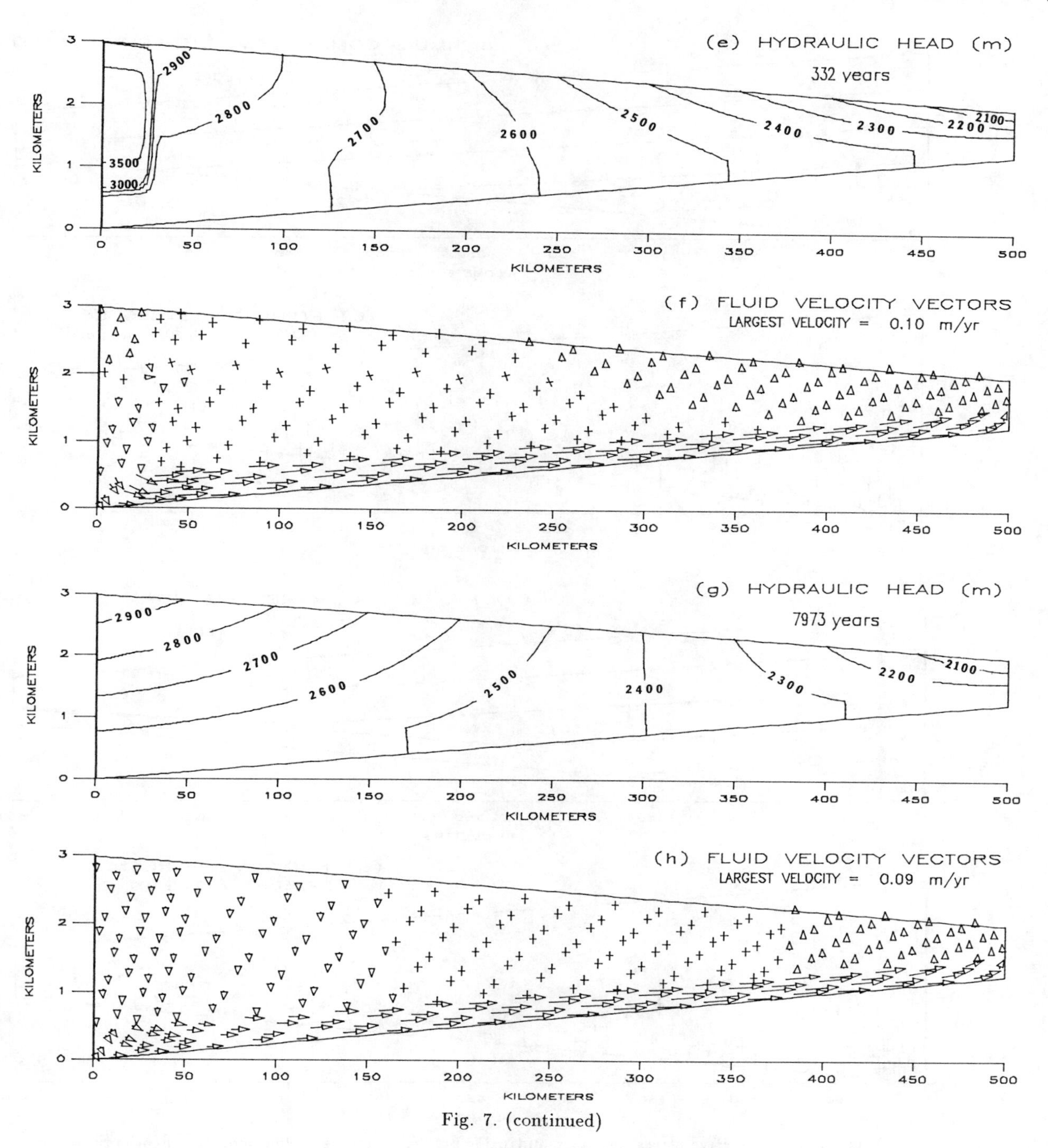

Fig. 7. (continued)

1988]. The upper unit represents a shale or mudstone aquitard and the lower unit represents a highly permeable sandstone or carbonate aquifer. An anisotropy ratio of 10:1 was chosen to represent bed-scale heterogeneity and fracturing [Freeze and Cherry, 1979]. The mechanical properties, Young's modulus and Poisson's ratio are chosen as possible values for sedimentary rocks [Jaeger and Cook, 1979].

Figure 7a and 7b display the numerical solution of the initial condition prior to loading. The steady state flow is driven by gravity. No fluid temperature or salinity gradients exist in the model. In Figure 7a, the hydraulic heads are plotted with a 100 m contour interval. At the water-table boundary, the hydraulic head is equal to the elevation. Therefore, the contours have the values of the elevation of the point at which they inter-

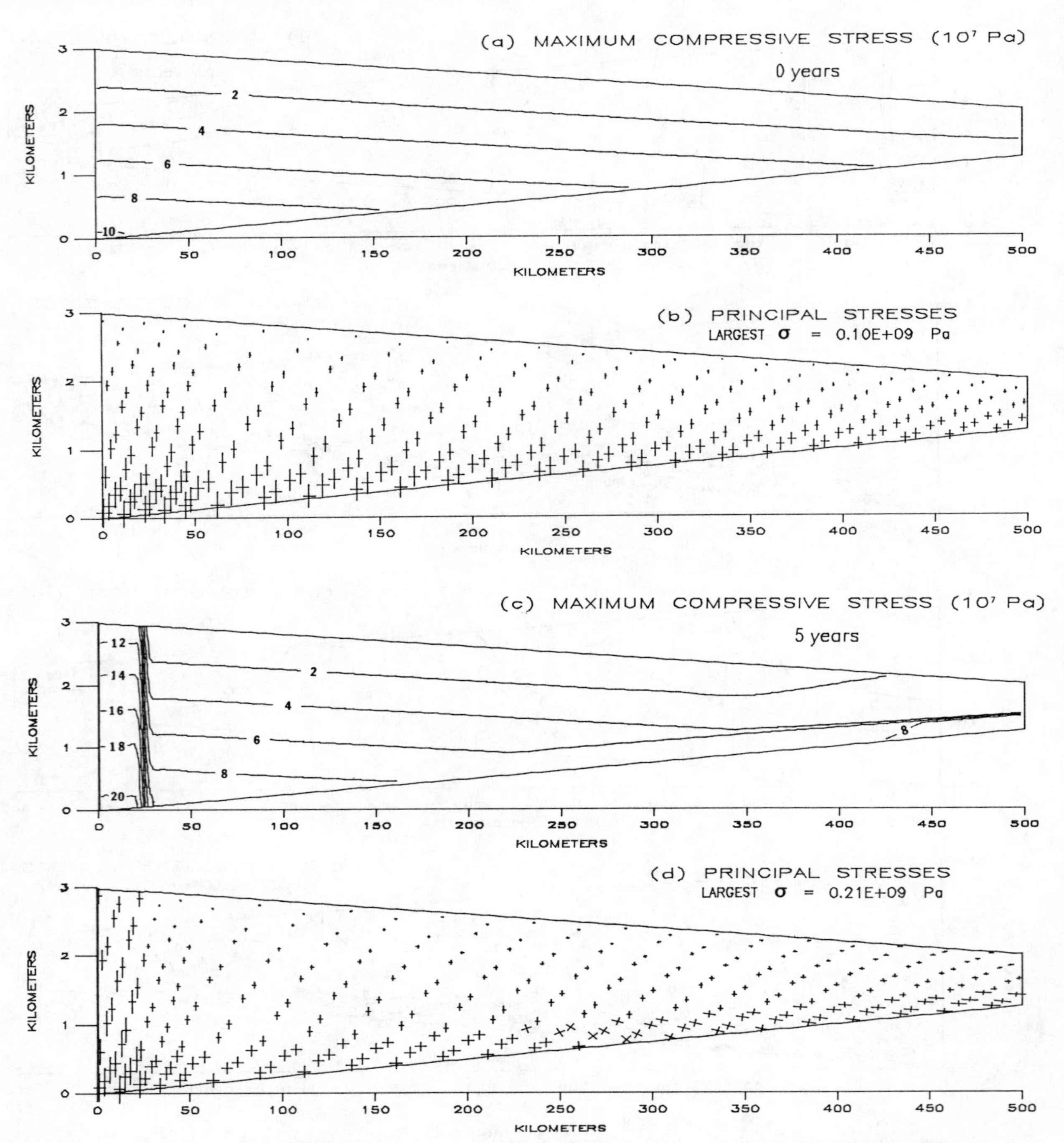

Fig. 8. Maximum compressive stress contours and principal stress crosses. The length and orientation of the crosses indicate the maximum and minimum stress direction and magnitude. (a) and (b) Pre-loading, the compressive stress resulted from gravitational forces increases with depth. (c) and (d) After 5 years of thrusting, significant stress increases occur near the loading area at left. The horizontal stresses increase near the right boundary is primarily due to the no-horizontal displacement imposed on the boundary condition.

sect the water-table surface. The presence of the basal aquifer creates a conduit for flow that is being recharged from above in the elevated region and discharges through vertical flow near the platform margin. The velocity field shown in Figure 7b is calculated using equation (4) and the computed hydraulic head distribution. The crosses in the velocity field plots denote velocities which are 10^{-3} times smaller than the maximum velocity.

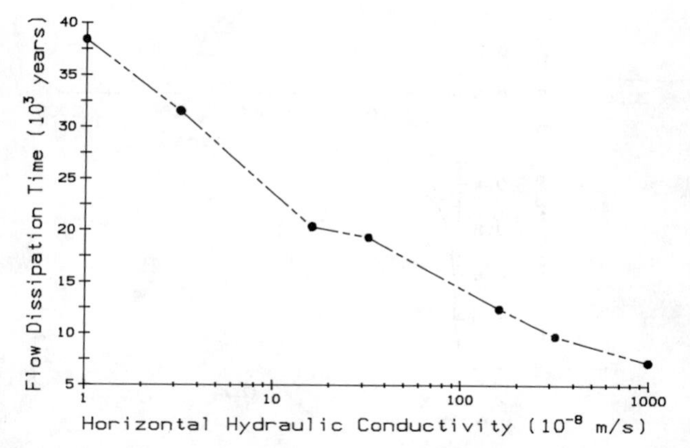

Fig. 9. Flow dissipation time versus horizontal hydraulic conductivity K_x. In all simulations, $K_x/K_z = 10$. The dissipation time refers to the time when the groundwater velocities at observation point F (see Fig. 6a) reach a steady value.

A uniform load with the magnitude of 1×10^8 *Pa* is then applied along ED and 1×10^7 *Pa* along AE (Fig. 6). The vertical loading applied is equivalent to a $3 \sim 4$ *km* thick rock mass thrust onto the sediments. The horizontal stress used is a reasonable value for tectonic compression in continental settings [Turcotte, 1983]. Figure 7c through 7h show the successive transient states of the fluid flow fields. Shortly after loading, at time = 5 years (Fig. 7c), the hydraulic head contours are highly distorted in the left portion of the basin near the loading area. Large flow velocities are concentrated in the aquifer at lower left corner. Some pore water flows towards the ground surface, but most is focused into the basal aquifer (Fig. 7d). After 332 years (Fig. 7e), the hydraulic head contours are much smoother although the flow field is still in a transient state. Figure 7g shows the contour field after about 8000 years, when most of the excess fluid pressure has dissipated, and the flow field is gradually reaching a new steady state. The maximum flow velocity increase in the recharge area at reference point F (see Fig. 6a) is about 6.5×10^{-1} m yr^{-1}, while an increase of 1.8×10^{-2} m yr^{-1} occurs near the right margin of the basin aquifer at x=450 *km*.

The stress field changes are shown in Figure 8. The static compressive stress resulted from the gravitational force increases with depth (Fig. 8a, 8b). Maximum and minimum stresses direction and magnitude are indicated by the orientation and length of the crosses. Significant stress increases are observed near the loading area and right boundary (Fig. 8c, 8d). The maximum compression is in the vertical direction at the left

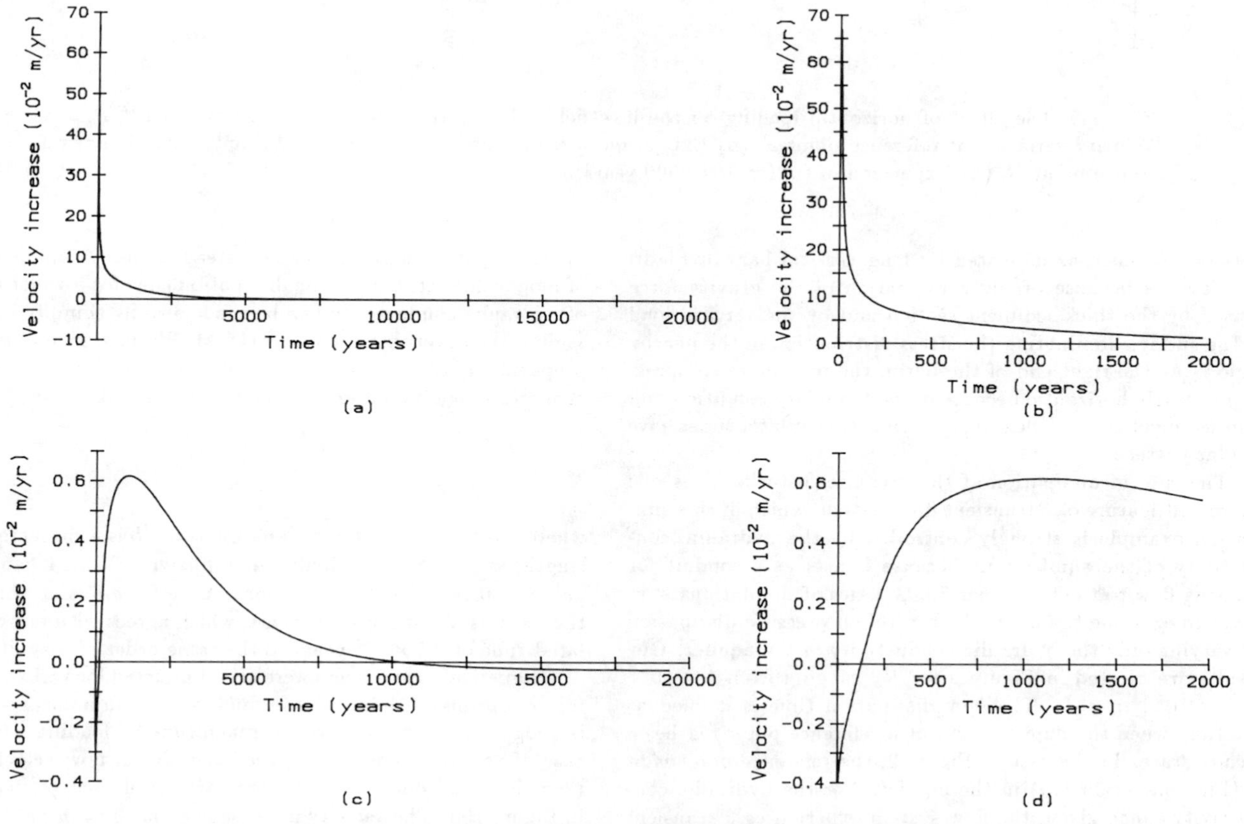

Fig. 10. The effect of vertical loading on the flow field. The vertical load applied is 1×10^8 P_a. (a) Velocity variation at at reference point F (x=35 *km*). (b) Expansion of (a) for first 2000 years. (c) Velocity variation at reference point G. (d) Expansion of (c) for first 2000 years.

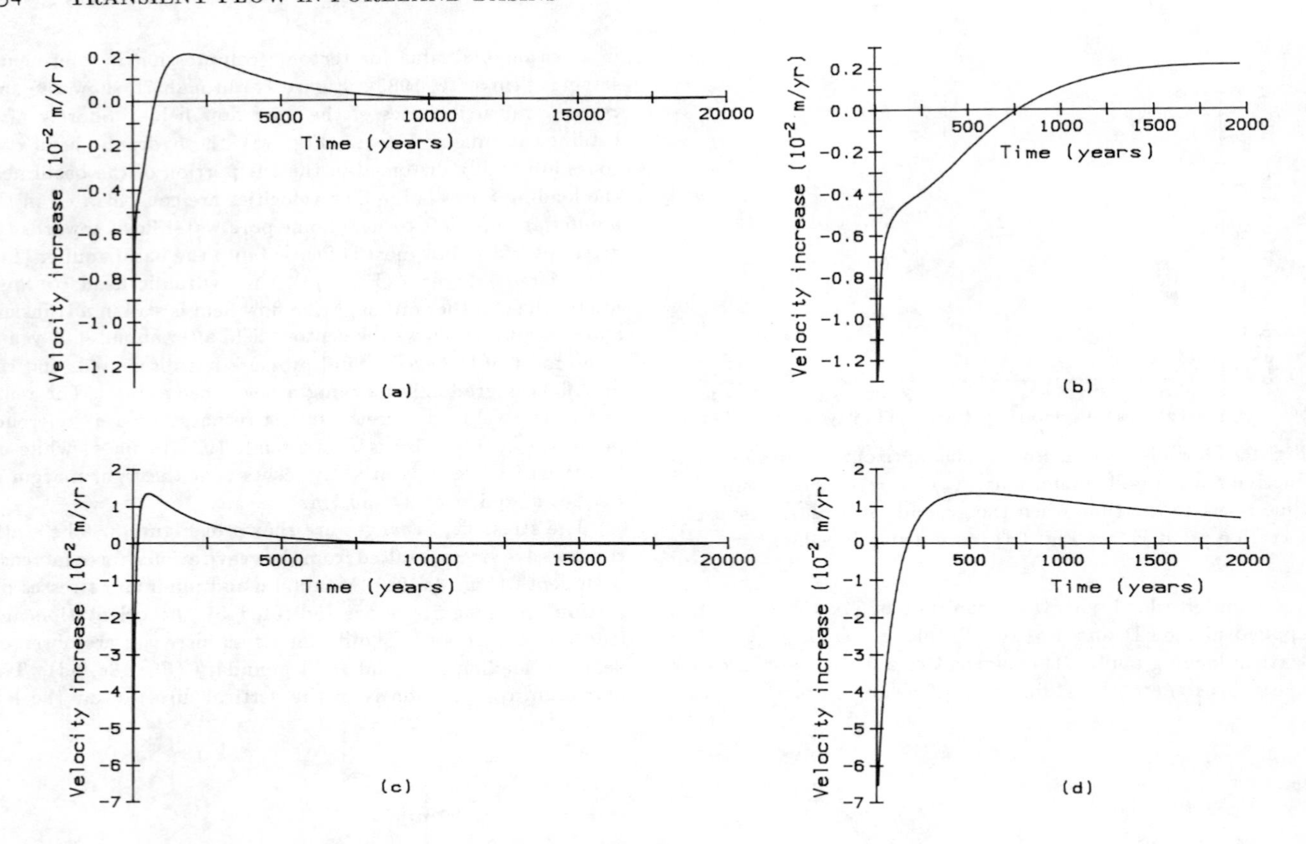

Fig. 11. The effect of horizontal loading on the flow field. The horizontal load applied is 1×10^7 P_a. (a) Velocity variation at reference point F. (b) Expansion of (a) for first 2000 years, (c) Velocity variation at reference point G. (d) Expansion of (c) for first 2000 years.

part of the basin as indicated by long vertical bars in Figure 8d. This is because of the combination of the gravity force caused by the thick sediment section and by the vertical load at the surface dominating the stress distribution in the nearby region. At the right end of the basin, the maximum compression is nearly horizontal because of the boundary condition and thin sediment cover. The stress distributions at later times have similar patterns.

The rate of dissipation of the pressure disturbance is one important feature of a transient flow system, which in this simulation example is strongly controlled by the hydraulic conductivity of the aquifer unit because it acts as a conduit for focusing flow to the basin margin. A series of simulations were made to examine the hydraulic control on pressure dissipation by varying only the hydraulic conductivity of the aquifer. Our results are plotted in Figure 9 for K_x ranging from 1×10^{-8} to 1×10^{-5} $m\ s^{-1}$. The flow dissipation time is defined as the time when the fluid velocity at a reference point reaches a steady state. In the case of Figure 9, the reference point is at F (Fig. 6a, x=35 km) in the aquifer. For the hydraulic conductivity range given, the flow system experiences a transient period on the order of $10^3 \sim 10^4$ years. After a few thousand years, the changes in flow velocities become much smaller (Fig. 8) as the system approaches a new steady state. A simple order of magnitude estimate of the dissipation time for a wider range of hydraulic conductivity can be made also by simplifying and scaling the governing equations (O. M. Phillips, manuscript in preparation, 1988). For the case of incompressible media, the characteristic dissipation time can be expressed as :

$$\tau = \frac{L^2}{2\kappa} \tag{6}$$

where τ is the dissipation or response time, L is a characteristic length, $\kappa = K/S_s$ is the hydraulic diffusivity. With L=450 km and $\kappa = 0.14$ m^2 s^{-1}, the response time for diffusion through the basin is about 2.3×10^4 years, which agrees with our calculated time of 2.0×10^4 years to the same order of magnitude.

Numerical experiments were also conducted for various loading conditions. The increase in fluid velocity in a basal aquifer is roughly proportional to the magnitude of loading. In the case of vertical loading only, the variation in flow velocity at point F (x=35 km) in the recharge area is plotted versus time in Figure 10a. The velocity increase is about 0.65 $m\ yr^{-1}$ (Fig. 10b) under a load equivalent to a 4 km thick rock mass. In Figure 10c, the velocity at point G (x=450 km) in the discharge area is displayed: the velocity increase is only about 6×10^{-3} $m\ yr^{-1}$. In Figure 10d there is a velocity drop before the curve climbs up to its peak value. This occurs because flow is mostly concentrated near the left end of the aquifer at the beginning of loading (Fig. 10a) where the stresses change most significantly. Flow elsewhere, including at point G, is relatively reduced. The high velocity flow propagates and migrates through the basin with time. Figure 11 shows the results with only a horizontal compressive load of 1×10^7 Pa on the left boundary. At point F, the velocity increase is about 2×10^{-3} $m\ yr^{-1}$ (Fig. 11a) and at point G the increase is 1.3×10^{-2} $m\ yr^{-1}$ (Fig. 11c) in the discharge area. Horizontal compression tends to have larger impact on the far field hydrology, with the flow velocity distribution showing an increased flow towards the ground surface.

Conclusion

Our numerical calculations have shown that tectonic compression can create significant transient flows in foreland basins, with excess flow rates is on the order $10^{-3} \sim 10^{-2}$ $m\ yr^{-1}$ for thrust-sheet loads 1 to 10 km thick. Most of the excess pressure generated by compression appears to dissipate in about 10^4 years, and a new steady state can be reached in about 10^5 years. Fluid velocities increase more significantly in the vicinity of the loading area relative to far-field discharge areas. However, the propagation of thrusting across the foreland wedge could ultimately result in the displacement of deep basinal fluids over long distances, albeit the volume of fluid migration is relatively small compared to the mechanism of gravity-driven flow in which flow rates of 1 to 10 $m\ yr^{-1}$ could be sustained over millions of years [Garven and Freeze, 1984a, 1984b]. Our preliminary study suggests, however, that cycles of transient fluid flow could exist repeatedly during the history of foreland basin thrusting. Fluid velocity surges could periodically invade a foreland basin several times during tectonic compression, which may also induce short-term thermal and chemical anomalies on the regional system. Thermal effects have not been included in our study at the present stage, but future experiments will include temperature variations, along with models for large deformation in the foreland fold and thrust belt.

Appendix I. Derivation of Equilibrium Equation

The equilibrium of an element body is given when the sum of work done by external forces and internal stresses is zero [Zienkiewicz, 1977, p.26]:

$$W_{ex} + W_{in} = 0 \qquad (I1)$$

Where W_{ex} is the work done by external forces, defined as the sum of the products of the individual force components and corresponding displacement:

$$W_{ex} = \{u\}^T\{f\} \qquad (I2)$$

In this equation $\{u\}$ is the nodal displacement vector and $\{f\}$ is the force vector. Supscript T denotes the transpose matrix. W_{in} is the work done by internal stresses and distributed forces (body forces) integrated over the volume of the body:

$$W_{in} = \int_\Omega \left(\{\epsilon\}^T\{\sigma_T\} + \{u'\}^T\{b\}\right) d\Omega \qquad (I3)$$

Where $\{\epsilon\}$ is the strain, $\{\sigma_T\}$ is the total stress, $\{b\}$ is the body force, and $\{u'\}$ is the internal displacement.

Strain-displacement relations The strain at any point can be determined from a displacement field. The relationship can be written as:

$$\{\epsilon\} = [B]\{u\} \qquad (I4)$$

Under the assumption of small deformation, the operator [B] is written in a two dimensional system as:

$$[B] = \begin{bmatrix} \frac{\partial}{\partial x} & 0 \\ 0 & \frac{\partial}{\partial y} \\ \frac{\partial}{\partial y} & \frac{\partial}{\partial x} \end{bmatrix}$$

If we define the internal displacement in terms of nodal displacement as $\{u'\} = [N]\{u\}$, where [N] is the interpolation function, then

$$W_{in} = \int_\Omega \left(\{u\}^T[B]^T\{\sigma_T\} + \{u\}^T[N]^T\{b\}\right) d\Omega$$

$$= \int_\Omega \left[\{u\}^T([B]^T\{\sigma_T\} + [N]^T\{b\})\right] d\Omega \qquad (I5)$$

Substitution of ($I5$) and ($I2$) into ($I1$) produces:

$$\{u\}^T\{f\} + \{u\}^T \int_\Omega ([B]^T\{\sigma_T\} + [N]^T\{b\})d\Omega = 0 \qquad (I6)$$

Equilibrium equation is obtained as:

$$\int [B]^T\{\sigma_T\}d\Omega + \int [N]^T\{b\}d\Omega + \{f\} = 0 \qquad (I7)$$

A derivation based on virtual work principles is valid for any material behaviour [Zienkiewicz, 1977, p.458].

Constitutive relation can be introduced to relate stresses and strains:

$$\{\sigma\} = [C]\{\epsilon\} \qquad (I8)$$

Where $\{\sigma\} = \{\sigma_T\} - \{P\}$ is the effective stress, $\{P\}$ is the pore pressure, $[C]$ is the material matrix. For an isotropic elastic body, the matrix is given in terms of Young's modulus E and the Poisson's ratio ν. In a plane strain case:

$$[C] = \begin{bmatrix} \frac{E(1-\nu)}{(1+\nu)(1-2\nu)} & \frac{E\nu}{(1+\nu)(1-2\nu)} & 0 \\ \frac{E\nu}{(1+\nu)(1-2\nu)} & \frac{E(1-\nu)}{(1+\nu)(1-2\nu)} & 0 \\ 0 & 0 & \frac{E}{2(1+\nu)} \end{bmatrix}$$

By substituting ($I4$) into ($I8$) and ($I8$) into ($I7$), the final form of governing equation in terms of displacement becomes :

$$\int_\Omega [B]^T [C][B]\{u\}d\Omega + \int_\Omega [B]^T \{P\}d\Omega + \int_\Omega [N]^T \{b\}d\Omega$$

$$+\{f\} = 0 \qquad (I9)$$

Acknowledgements. We would like to thank G. Quinlan and J. Tóth for their helpful reviews and O. M. Phillips for his valuable comments.

The work reported in this paper was supported by grants from the National Science Foundation (EAR-84096 09 and EAR-8553019).

References

Bally, A. W., P. L. Gordy and G. A. Stewart, Structure, seismic data, and orogenic evolution of southern Canadian Rocky Mountains, Bull. Can. Petrol. Geol., 14, no. 3, 337-381, 1966.

Bear, J., Dynamics of Fluids in Porous Media, 764pp., Elsevier, New York, 1972.

Beaumont, C., The evolution of a sedimentary basin on a viscoelastic lithosphere: theory and examples, Royal Astron. Soc. Geophys. Jour., 55, 471-497, 1978.

Beaumont, C., Foreland basin, Royal Astron. Soc. Geophys. Jour., 65, 291-329, 1981.

Bethke, C. M., A numerical model of compaction-driven groundwater flow and heat transfer and its application to paleohydrology of interacratonic sedimentary basins, Jour. Geophys. Res., 90, no. B8, 6817-6828, 1985

Bethke, C. M., Hydrologic constraints on the genesis of the Upper Mississippi Valley mineral district from Illinois basin brines, Economic Geology, 81, no. 2, 233-249, 1986.

Bethke, C. M. and T. F. Corbet, Linear and nonlinear solution for one-dimensional compaction flow in sedimentary basins, Water Resour. Res., 24, no. 3, 461-467, 1988.

Bethke, C. M., W. J. Harrison, C. Upson, and S. P. Altaner, Supercompter analysis of sedimentary basins, Science, 239, 261-267, January 1988.

Biot, M. A., General theory of three-dimensional consolidation, Jour. Applied Physics, 12, 155-164, 1941.

Biot, M. A., Theory of elasticity and consolidation for a porous anisotropic solid, Jour. Applied Physics, 26, 182-185, 1955.

Cathles, L. M. and A. T. Smith, Thermal constrains on the formation of Mississippi Valley-type lead-zinc deposits and their implications for episodic basin dewatering and deposit genesis, Economic Geology, 78, no. 5, 983-1002, 1983.

Das, B. M., Advanced Soil Mechanics, 511pp., McGraw-Hill, New York, 1983.

Domenico, P. A. and Palciauskas, V. V., Thermal expansion of fluids and fracture initiation in compacting sediments, Geol. Soc. Am. Bull., Part II, 90, 953-979, 1979.

Freeze, R. A. and J. A. Cherry, Groundwater, 604pp., Prentice Hall, New Jersey, 1979.

Garven, G., The Role of Groundwater Flow in the Genesis of Stratabound Ore Deposits : A Quantitative Analysis, Ph.D. Dissert., 304pp., University of British Columbia, Vancouver, Canada, October 1982.

Garven, G., The role of regional fluid flow in the genesis of the Pine Point deposit, Economic Geology, 80, no. 2, 159-169, 1985.

Garven. G., A hydrogeological model for the genesis of the giant oil sands deposits of the Western Canada sedimentary basin (abstract), American Geophysical Union Transactions, 67, no. 16, 273, 1986.

Garven, G. and R. A. Freeze, a, Theoretical analysis of the role of groundwater flow in the genesis of strata- bound ore deposits : 1. Mathematical and numerical model, Am. Jour. Sci., 284, 1085-1124, 1984.

Garven, G. and R. A. Freeze, b, Theoretical analysis of the role of groundwater flow in the genesis of strata- bound ore deposits : 2. Quantitative results, Am. Jour. Sci., 284, 1125-1174, 1984.

Huyakorn, P. S. and G. F. Pinder, Computation Methods in Subsurface Flow, 473pp., Academic Press, New York, 1983.

Jaeger, J. C. and N. G. W. Cook, Fundamentals of Rock Mechanics, 593pp., Methuen, London, 1979.

Jordan, T. E., Thrust loads and foreland basin evolution, Cretaceous, Western United States, Am. Assoc. Petrol. Geol. Bull., 65, 2506-2520, 1981.

Malvern, L. E., Introduction to the Mechanics of a Continuous Medium, 713pp., Prentice-Hall, Inc., Englewood Cliffs, New Jersey, 1969.

Neuzil, C. E., Groundwater flow in low-permeability environments, Water Resour. Res., 22, no. 8, 1163-1195, 1986.

Neuzil, C. E. and D. W. Pollock, Erosional unloading and fluid pressure in hydraulically tight rock, Jour. Geol, 14, no. 12, 179-193, 1983.

Oliver, J., Fluids expelled tectonically from orogenic belts; their role in hydrocarbon migration and other geological phenomena, Geology , 14, no. 12, 99-102, 1986.

Oliver, J., COCORP and Fluids in the Crust, Studies in Geophysics, National Research Committee Series, 1987.

Reddy, J. N., An Introduction to the Finite Element Method, 495pp., McGraw-Hill, New York, 1984.

Rice, J. R. and M. P. Cleary, Some basic stress diffusion solutions for fluid-saturated elastic porous media with compressible constituents, Rev. Geophys. Space. Phys., 14, no. 2, 227-241, 1976.

Sharp. J. M., Energy and momentum transport model of the Ouachita basin and its possible impact on formation of economic mineral deposits, Economic Geology, 73, 1057-1068, 1978.

Shi, Y. and C. Y. Wang, Pore pressure generation in sedimentary basins: overloading versus aquathermal, Jour. Geophys. Res., 91, B2, 2135-2162, 1986

Suppe, J., Principles of Structure Geology, 537pp., Pren- tice-Hall, Inc., Englewood Cliffs, New Jersey, 1985.

Terzaghi, K, The shearing resistance of saturated soil and the angle between the planes of shear, Proc. 1st Conf. Soil Mech. and Foundn. Eng. 1, 54-56, Harvard, Mass., 1936.

Tóth, J. and T. Corbett, Post-paleocene evolution of regional groundwater flow-system and their relation to petroleum accumulations, Taber area, southern Canada, Bull. Can. Petrol. Geol., 34, no. 3, 339-363, 1986.

Tóth, J. and R. F. Millar, Possible effects of erosional changes of the topographic relief on pore pressure at depth, Water Resour. Res., 19(b), 1585-1597, 1983.

Turcotte, D. L. and G. Schubert, Geodynamics, 450pp., John Wiley & Sons, New York, 1982.

Van der Kamp, G. and J. E. Gale, Theory of earth tide and barometric effects in porous formations with compressible grains, Water Resour. Res., 19, no. 3, 538-544, 1983.

Walder, J. S., Coupling between Fluid Flow and Deformation in Porous Crustal Rocks, Ph.D. Dissert., 252pp., Stanford University, 1984.

Zienkiewicz, O. C., The Finite Element Method, 787pp., McGraw-Hill, Inc., London, 1977.

Zienkiewicz, O. C. and C. Humpheson, Viscoplasticity: a generalized model for description of soil behavior, in Numerical Methods in Geotechnical Engineering, Edited by C. S. Desai and J. T. Christian, 783pp., McGraw-Hill, New York, 1977.

EARLY PRECAMBRIAN CRUSTAL EVOLUTION AND MINERAL DEPOSITS, PILBARA CRATON AND ADJACENT ASHBURTON TROUGH

J. G. Blockley, A. F. Trendall, and A. M. Thorne

Department of Mines, Geological Survey of Western Australia, Perth, Australia

Abstract. The early Precambrian rocks making up the Pilbara Craton (comprising the Pilbara Block and the Hamersley Basin) and contiguous Ashburton Trough in northwestern Australia resulted from more-or-less continuous operation of various geological processes from 3.5 Ga to 1.7 Ga. This suggestion contrasts with the long-held view that there was a major time gap, of the order of 400 to 500 Ma, between the stabilization of the 'Archean' Pilbara Block and the deposition of the overlying 'Proterozoic' rocks of the Hamersley Basin. The continuum of geological events preserved in the rocks allows an assessment of the contained mineral deposits over their 1.8 Ga history. Gold, nickel and chromium are restricted mainly or entirely to units older than 3.0 Ga, reflecting the common association of these metals with ultramafic rocks; tin-tantalum deposits, normally rare in rocks of this age, are related to post-tectonic granites intruded during the last stage of the cratonization of the Pilbara Block at about 2.8-3.0 Ga; iron and manganese occur mainly in rocks dated at about 2.5 Ga, corresponding to the maximum development of chemical sediments; while copper, lead and zinc appear at intervals throughout the entire column without any apparent overriding age control.

Introduction

The original intent of this paper, when offered to the Symposium, was to describe the evolution and metallogeny of the early Precambrian Hamersley Basin. However, recent field and isotopic evidence strongly suggests that the Hamersley Basin represents but one stage in a more-or-less continuous succession of geological events which began with the extrusion of the greenstones of the underlying Pilbara Block at 3.5 Ga, and culminated in the folding of the sediments of the Ashburton Trough at about 1.7 Ga. For this reason the content of the paper was modified for presentation at the Symposium and is further expanded here to include descriptive material that was not possible to present in a 20 minute talk.

The area described is shown in Figure 1 It comprises an early Precambrian nucleus, the Pilbara Craton, flanked on its southern side by the middle Precambrian Ashburton Group [Gee, 1979]. The Pilbara Craton consists of two distinct parts: the older Pilbara Block made up largely of a granite-greenstone assemblage with associated clastic sediments, and the younger Hamersley Basin, developed in the southern part of the craton and filled with thick sequences of mafic volcanic rocks, chemical sediments and, in its upper part, clastic sediments.

Until recently, it was common to regard the Pilbara Block as an Archean basement which stablized well before the development of the Proterozoic Hamersley Basin. However, it is here argued that this concept needs revision.

The Ashburton Trough was formerly considered to be an integral part of the Hamersley Basin [MacLeod, 1966] but was subsequently shown to be separated from the basin sequence by a major unconformity [Trendall, 1979; Gee, 1979].

Contained within the rocks of the Pilbara Craton and Ashburton Trough is a wide variety of mineral deposits, of which the iron ores of the Hamersley Basin are of world importance. The region therefore affords an opportunity to examine metallogenetic processes over a considerable part of Precambrian time, from the early initiation of granite-greenstone terrain, through the period of widespread iron-formation deposition, to the development of younger fold belts marginal to the consolidated craton.

The Pilbara Block

The Pilbara Block, consisting of classical granite-greenstone terrain, is illustrated in Figure 2. Anastomosing curvilinear greenstone belts encircle and define broad granitoid domes, or batholiths. The greenstone belts, although often complexly sliced, are generally synclinorial, and have a regionally rational stratigraphy which has been well mapped and defined by Hickman [1983] as the Pilbara Supergroup. The Pilbara Supergroup has a total aggregate thickness of 25 km, although Hickman has emphasized that there is no single section showing a succession as thick as this. There are two main constituent groups (Figure 3), the lower Warrawoona Group, whose main constituents are mafic volcanics, with subordinate ultramafic and acid volcanics and shales; and the overlying Gorge Creek Group, a thick sequence of mainly coarse clastic sediments (sandstones). The Whim Creek Group, and some locally overlying volcanic rocks, complete the succession.

The intervening broad granitoid domes contain a complex mixture of migmatite, gneiss, and granitoid ranging in

Published in 1989 by
International Union of Geodesy and Geophysics
and American Geophysical Union.

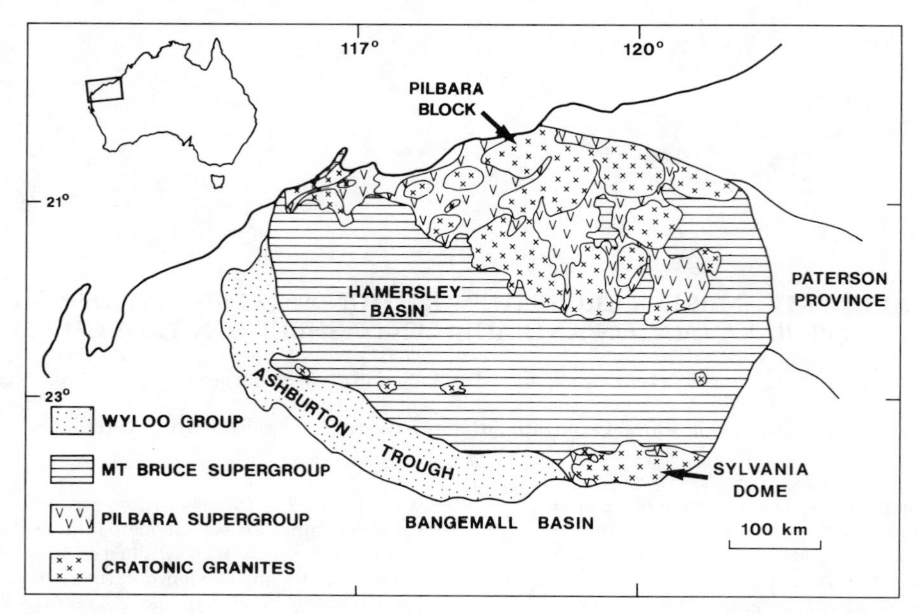

Fig. 1. Maps showing position of the area under discussion and its major tectonic subdivisions. The Pilbara Craton includes the Pilbara Block, the Hamersley Basin, and the Sylvania Dome.

composition from tonalite to adamellite, the latter locally forming unfoliated post-tectonic bodies. Wherever contacts are visible the granitoids are always discordant to, and intrude, the layering of the adjacent greenstone belts.

The Hamersley Basin

The sedimentary and volcanic rocks of the Hamersley Basin - the Mount Bruce Supergroup [MacLeod, 1966] - now crop out over an area some 500 km long (east-west) and 200 km wide (Figure 4) and were probably laid down in a barred basin, open to the west, which may not have extended far beyond the present outcrop area [Trendall, 1983]. It had an area of at least 100 000 square kilometres.

The stratigraphic succession within the basin is divided into three groups, from the base upwards named the Fortescue, Hamersley, and Turee Creek Groups (Figure 5). The basal

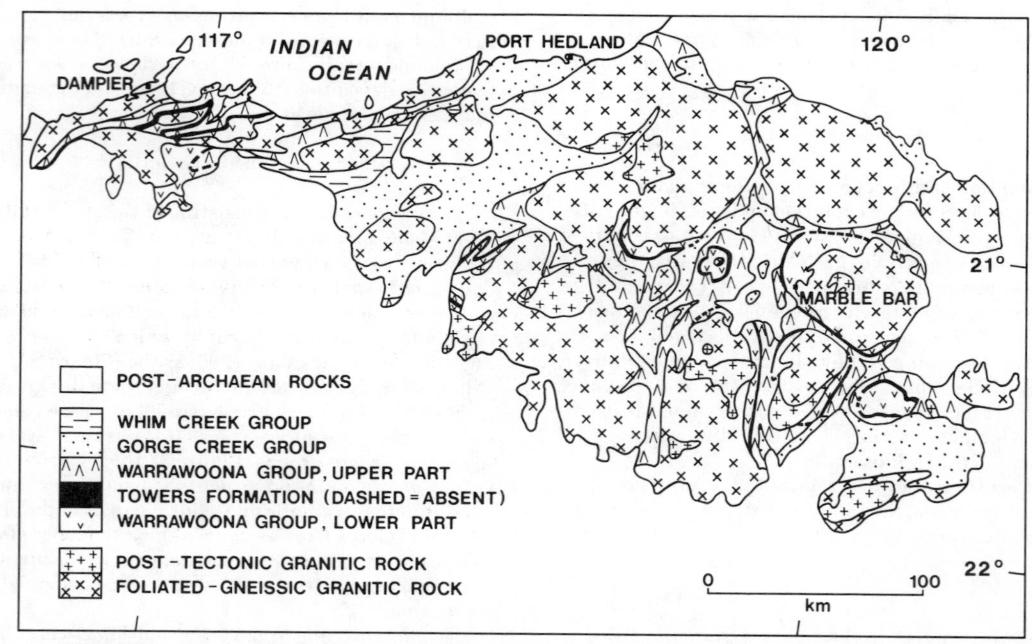

Fig. 2. Simplified geological map of the Pilbara Block, showing major stratigraphic subdivision.

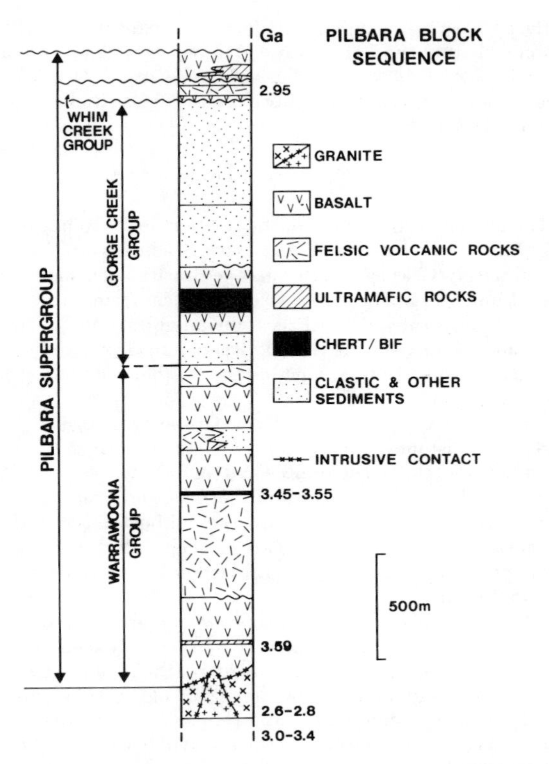

Fig. 3. Major stratigraphic subdivision of the Pilbara Block, showing generalized lithology.

Fortescue Group reaches a maximum thickness of 4.3 km in the central part of the outcrop area. It normally has a thick basal sandstone of granitic debris, although locally the sandstone is underlain by basalt infilling paleovalleys. The sandstone is succeeded upward by alternating thick units of mafic lavas and sediments largely composed of pyroclastic material of similar composition, but including also shales and stromatolitic carbonates. The uppermost unit is a thick regionally uniform shale.

The Hamersley Group overlies the Fortescue Group with perfect conformity. It has a total thickness of 2.5 km. Of this thickness well over 1 km consists of banded iron-formation (BIF), arranged in six separate units each with distinctive characteristics [Trendall and Blockley, 1970]. One of these units, the Dales Gorge Member of the Brockman Iron Formation, has been described in great detail. It displays a remarkable succession of cyclicities of stratification, on scales ranging downwards from tens of metres to millimetres. Like all typical cherty BIF the material consists of layers, or bands of relatively iron-poor chert and dark bands rich in iron oxides, carbonates, and to a lesser extent, silicates Many of the chert bands contain fine regular internal lamination on a scale of 0.1-2 mm. A remarkable feature of the Dales Gorge Member is that not only individual chert bands, but also many laminae within them can be shown to be continuous through the outcrop area of the Hamersley Group. These fine laminae have been interpreted to be varves, or annual layers, and this interpretation has withstood some 20 years of scrutiny since it was first proposed. The BIFs are by far the most noteworthy units of the Hamersley Group. The remaining material consists of shale (peculiar and iron-rich), carbonate, dolerite sills, and an interesting thick stratabound felsic unit, termed

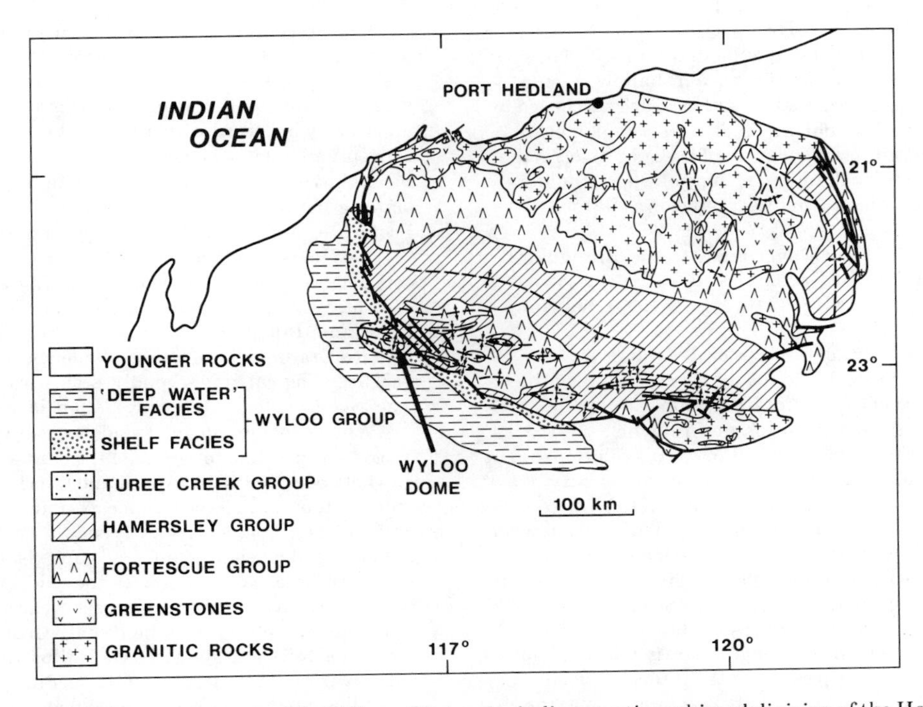

Fig. 4. Generalized geological map of the Pilbara Craton, including stratigraphic subdivision of the Hamersley Basin and facies subdivision of the Ashburton Trough.

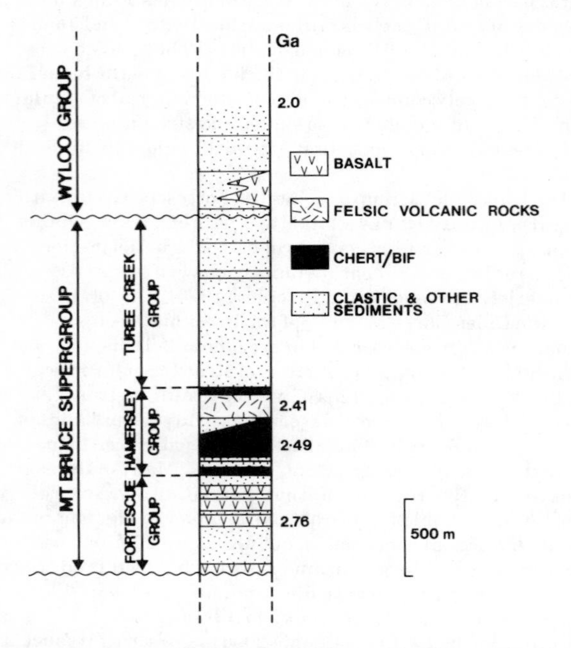

Fig. 5. Major stratigraphic subdivision of the Hamersley Basin and Ashburton Trough, showing generalized lithology.

the Woongarra Volcanics, now thought to be a sill [Trendall, 1976a].

The Hamersley Group is succeeded with perfect conformity by the 4 km thick Turee Creek Group which occurs in scattered outcrops in the southern part of the Basin (Figure 4). The lower 3 km of this succession consists of deep and shallow marine mudstones, siltstones and sandstones. These are succeeded by 1 km of deltaic and shallow marine sandstones, conglomerates and dolomites. An interesting feature of the succession is the presence of a 300 m thick diamictite, some of whose boulders are striated. The unit has been interpreted as a tillite [Trendall, 1976b].

The Turee Creek Group is overlain with sharp, angular unconformity by the Wyloo Group, the thick sequence of clastic, carbonate and volcanic rocks which occupies the Ashburton Trough (Figure 4).

Structurally the rocks of the Hamersley Basin are remarkably undeformed. Along the northern margin of the present outcrop area of the Mount Bruce Supergroup the basal sandstones and basalts of the Fortescue Group lie unconformably on the older Pilbara Block - this relationship is discussed in more detail below. Above the unconformity, the rocks of the Fortescue Group dip very gently (5 to 10 degrees) southwards in an almost undisturbed state. They are succeeded southwards in orderly stratigraphic succession by the Hamersley Group, whose iron-formations give the spectacular scenery of the Hamersley ranges. Only in the central part of the outcrop area, south of the Hamersley Range, does the Mount Bruce Supergroup become significantly folded (about axes trending roughly east-west), and the intensity of this folding increases steadily to the south, reaching its maximum close to the curving southern limit of its outcrop.

Recent geochronological data [Compston and others, 1981] show clearly that the Dales Gorge Member has a depositional age of about 2.5 Ga, while the base of the Fortescue Group is less precisely dated, but probably close to 2.8 Ga [Richards and Blockley, 1984].

The Ashburton Trough

The principal component of the Ashburton Trough is the Wyloo Group which attains a maximum thickness of approximately 12 km and outcrops over an area of some 17 000 square kilometres. The Wyloo Group is unconformably overlain by several younger Precambrian units: the Capricorn Formation, Mount Minnie Group, Bresnahan Group and Bangemall Group, the last forming the southern boundary of the trough.

The lowermost 4 km of the Wyloo Group consists of quartzite sandstones, BIF-derived sandstones and conglomerates, shales, dolomites and mafic volcanics laid down in terrestrial, shallow marine and shelf-edge environments. The remainder of the Wyloo Group, the Ashburton Formation, is comprised of approximately 8 km of 'deep water' shale and sandstone, with minor conglomerate, chemical deposits and felsic-mafic volcanics.

The Wyloo Group has undergone two major phases of deformation, the effects of which are most clearly seen away from the southern and western margins of the Hamersley Basin. In the south and southwestern part of the Ashburton Trough a phase of regional metamorphism and granitoid plutonism reached its climax in the interval between these compressional events.

Relationship of the Hamersley Basin and Pilbara Block

Until recently, the interpretation of the Hamersley Basin as a younger ("Proterozic") intra-cratonic sedimentary basin resting conformably on an older basement of ("Archean") granite-greenstone terrain was a comfortable concept apparently in accord with all available evidence. It was linked, conceptually, with the idea of an Archean, granite greenstone block that had long been stabilized, uplifted, and eroded, before sinking once more to receive the deposits of the younger, Proterozic, and tectonically unrelated sedimentary basin.

Two lines of evidence have led to a radical revision of this concept, and replaced it by one in which there was complete continuity of tectonic and depositional evolution of the Pilbara Block and Hamersley Basin, which are referred to jointly as the Pilbara Craton. These lines of evidence are firstly, geochronological, and secondly, structural and stratigraphic. The geochronological evidence is considered first.

When geochronological evidence was fairly sparse, a decade ago, it was possible to reconcile the data with a significant time gap between the Pilbara Block and the Hamersley Basin. As more isotopic data have become available [see reviews by Blake and McNaughton, 1984; and Trendall, 1983] it has become clear not only that no significant time gap exists, but that there may even be a time overlap between the oldest rocks of the Hamersley Basin and the youngest rocks of the Pilbara Block. The sequence of events in the Pilbara Craton as a whole during the period 3.5 Ga to about 2.0 Ga is illustrated in Figure 6 in which various stratigraphic units (sedimentary and volcanic) of the Pilbara Craton, large-scale acid magma generation, and the longevity of these events, are illustrated. In view of this

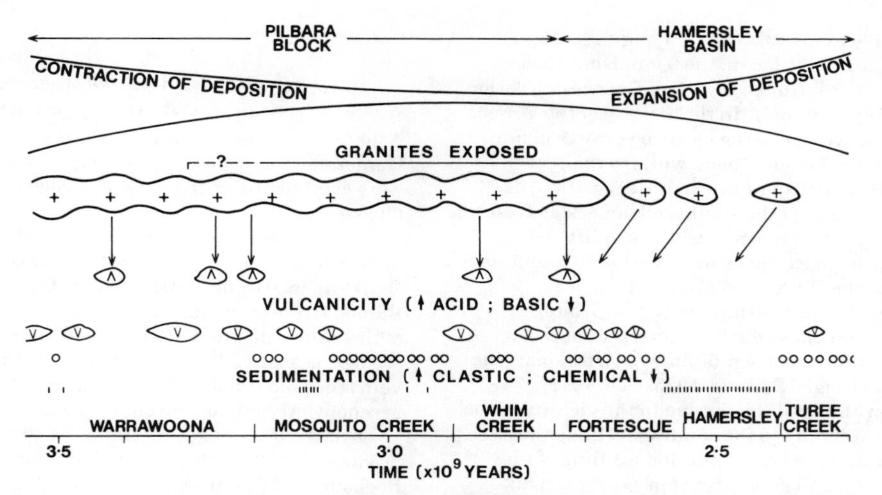

Fig. 6. Diagrammatic summary of the geological evolution of the Pilbara Craton. Of the longer arrows on the diagram, those pointing vertically downwards indicate extrusion of acid lavas from deep plutonic magma chambers; those pointing obliquely left indicate upward migration of similar acid magma which was emplaced as shallow intrusive sheets. Other symbols are explained on the diagram.

apparent continuity, what is the significance of the conspicuous basal angular unconformity of the Fortescue Group?

It turns out that this great regional unconformity is most clearly expressed where the Fortescue Group directly overlies granite. When the unconformity is carefully examined where it overlies greenstone belts of the Pilbara Block the diagrammatic relationship shown in Figure 7 is observed, with a complex series of minor local unconformities indicating that, in the cores of the synclinorial greenstone belts, deposition was effectively continuous.

The depositional history of the Pilbara Craton thus began with a regionally continuous outpouring of thick volcanics, represented by the Warrawoona Group. Below this pile at randomly spaced centres granitic plutons began to form and rise. When these breached the succession their debris accumulated in the Gorge Creek Group, whose deposition was restricted to the greenstone belts. As the diapiric rise of the granites ceased, deposition spread over them again to give the basal Fortescue Group unconformity.

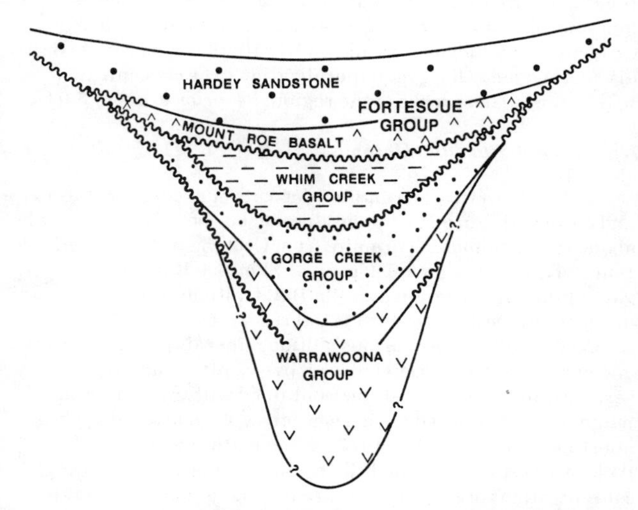

Fig. 7. Diagrammatic representation of the typical transitional stratigraphic relationships between the uppermost "Archean" greenstone succession (Gorge Creek or Whim Creek Group) and the lowermost "Proterozoic" Fortescue Group units in synclinal greenstone belts.

The early units of the Fortescue Group, comprising basalt, sandstone and conglomerate, were laid down in an essentially terrestrial environment, with both flows and sediments tending to follow valleys on the unconformity surface. Later units, from the Kylena Basalt upwards, were laid down under increasingly marine conditions, and individual units can be traced over large distances. The uppermost formation of the Fortescue Group comprises a sequence of mainly fine-grained clastic sediments, dolomite and chert with isolated flows of pillow basalt. It marks the transition to the predominantly chemical Hamersley Group.

The transition from the Hamersley Group to the Turee Creek Group indicates that the final infilling and shallowing of the southern Hamersley Basin was associated with a gradual increase in terrigenous clastic supply. In the southwestern part of the basin this culminated in the development of a thick deltaic body at the top of the Turee Creek Group. The presence of BIF fragments in these deposits indicates that the Hamersley Group was at least partly uplifted and subaerially exposed by the time this unit was laid down. This uplift probably marked the intital stages of pre-Wyloo Group deformation which resulted in the folding, uplift and erosion of Fortescue, Hamersley and Turee Creek Groups in the southwestern Hamersley Basin.

Relationship Between the Hamersley Basin and the Ashburton Trough

Though marked by a sharp angular unconformity in the southwest of the Hamersley Basin, the contact between the Turee Creek Group and the overlying Wyloo Group is apparently conformable to the east of 117° Longitude (Figure 4). In the southwestern part of the basin, the tectonism which gave rise to this unconformity continued so as to influence deposition of much of the lower Wyloo Group. In the vicinity of the Wyloo Dome, pulses of uplift gave rise to the localized growth of fan-delta lobes [Thorne and Seymour, 1986], while to the east, tidally influenced deltaic and shallow marine deposition extended over the remainder of the southern Hamersley Basin. Following a period of basalt extrusion, tectonic uplift and associated fan-delta growth extended over most of this southern margin. Deposition of the Duck Creek Dolomite, a one kilometre thick carbonate unit within the Wyloo Group, coincided with a tensional phase during which the southern edge of the Pilbara Craton was down-faulted to form a marginal basin, the Ashburton Trough [Daniels, 1975; Gee, 1979]. Regional compression and uplift beginning in the vicinity of the Sylvania Dome, and extending to the southwest Hamersley Basin, resulted in rapid sediment supply and infilling of the Ashburton Trough by several submarine fan systems. This phase of compression reached its climax with the deformation and cratonization of the Wyloo Group, an event which took place by approximately 1.7 Ga.

Mineralization

Figure 8 shows the geological column developed within the Pilbara Craton and adjacent Ashburton Trough subdivided by major lithological types rather than formal stratigraphic nomenclature. The figure also indicates the approximate stratigraphic positions of major and minor deposits of the principal metallic minerals formed within the region.

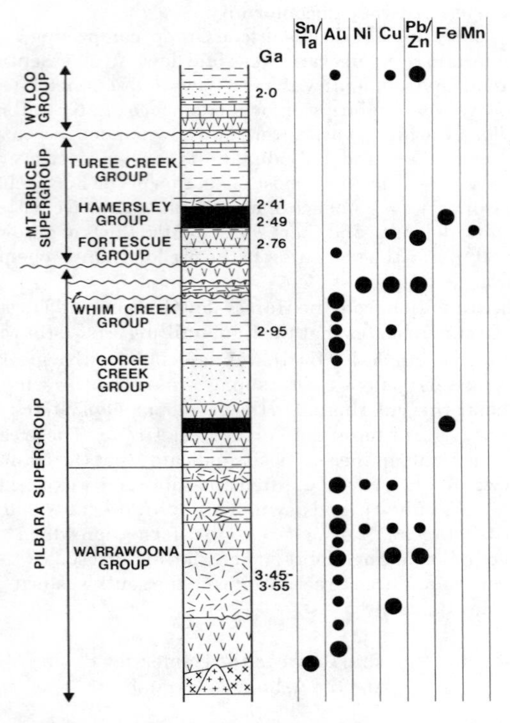

Fig. 8. Diagrammatic summary of the stratigraphic distribution of mineralization in the Pilbara Craton and Ashburton Trough.

Gold

In 1888 the Pilbara Block was the scene of one of the earlier Western Australian gold rushes. Activity quickly spread to the Ashburton Trough and various basement inliers within the Hamersley Basin. Altogether some 35 gold mining centres were established, with a total recorded production of about 19 000 kg of gold.

The greatest proportion of gold production has come from quartz veins intruding schistose mafic rocks within the Warrawoona Group at Marble Bar, Comet, Lalla Rookh, Bamboo Creek and numerous other centres (Figure 9). The age of this mineralization has been determined from lead isotope studies as about 3.2 Ga [Richards and others, 1981]. In these veins the gold is normally associated with minor pyrite, arsenopyrite, galena and chalcopyrite [Hickman, 1983].

Quartz veins in geosynclinal flysch have been important producers at Blue Spec, east of Nullagine, and Mallina, near Roebourne. Similar veins have also been the main source of gold in the Ashburton Trough. Deposits of this type are usually restricted to pelitic rocks, and are normally intruded parallel to either the cleavage or bedding of their hosts. Typically they contain concentrations of other metals such as antimony, copper or lead.

Gold concentrations within conglomerates of the Fortescue Group have attracted considerable interest from time to time because of supposed similarities to the Witwatersrand deposits. Such conglomerates have been mined at Nullagine and near Marble Bar but, apparently because of their immaturity, they do not contain the widespread mineralization found in their South African counterparts.

Many of the gold-bearing veins occur near granites and are probably of hydrothermal origin, while others appear to be metamorphic segregations. The comparative lack of metamorphism and absence of granite intrusion in the Fortescue Group may account for the lack of primary gold deposits in this sequence. However, the same cannot be said when comparing the lithologically similar flysch sequences in the Pilbara Block and Ashburton Trough, both of which are metamorphosed to similar grades, and neither of which is extensively intruded by granite. Overall, there appears to be a distinct tendency for gold mineralization to be concentrated within the earliest rocks of the region, i.e. prior to about 3.0 Ga.

Nickel, Vanadium and Chrome

The Pilbara Craton contains deposits of nickel,vanadium and chrome (Figure 10), but production has been limited to about 15 000 tonnes of chromite from Coobina in the Sylvania Dome. However, a nickel deposit at Sherlock Bay and a vanadium concentration at Balla Balla, although currently uneconomic, both have substantial resources.

Nickel, present as various sulfide minerals [Marston, 1984] and chromite, are associated with pyroxenites and serpentinite; while vanadium, contained within titaniferous magnetite, has formed as segregations within layered gabbro/anorthosite sills [Baxter, 1977]. Such mafic and ultramafic rocks are restricted to the Pilbara Supergroup and accordingly concentrations of these metals are not found in the Hamersley Basin or Ashburton Trough.

Base metals

Copper mineralization is widespread in the Pilbara Craton and Ashburton Trough (Figure 10). Styles of mineralization

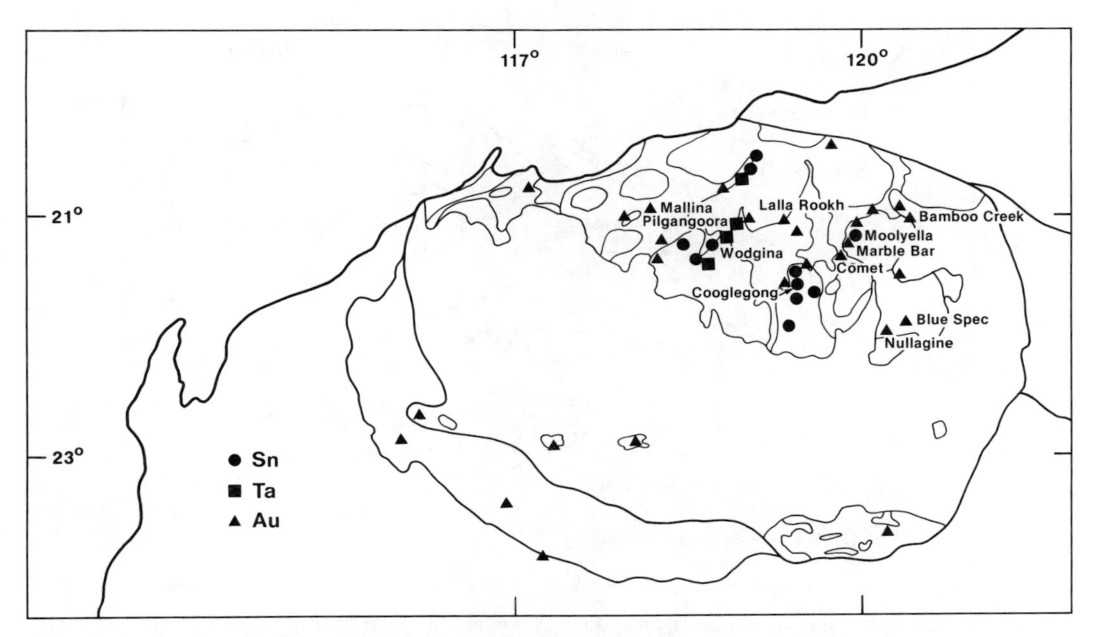

Fig. 9. Geographic distribution of gold, tin, and tantalum mineralization in the Pilbara Craton and Ashburton Trough.

range from volcanogenic and possible porphyry deposits, through sediment-hosted concentrations, to quartz veins and mineralized shears. Total production has amounted to about 17 250 t of contained metal [Marston, 1979].

Stratiform volcanogenic copper deposits are most commonly associated with the felsic Duffer Formation in the lower part of the Warrawoona Group. However, all such occurrences, which include those at Copper Hills and Whundo, have been small producers only. By far the largest concentrations, both in terms of past production and current resources, are contained in the Mons Cupri and Whim Creek deposits within the Whim Creek Group in the western part of the Pilbara Block. Here mineralization is found in a dome of rhyolite and within an overlying tuffaceous slate.

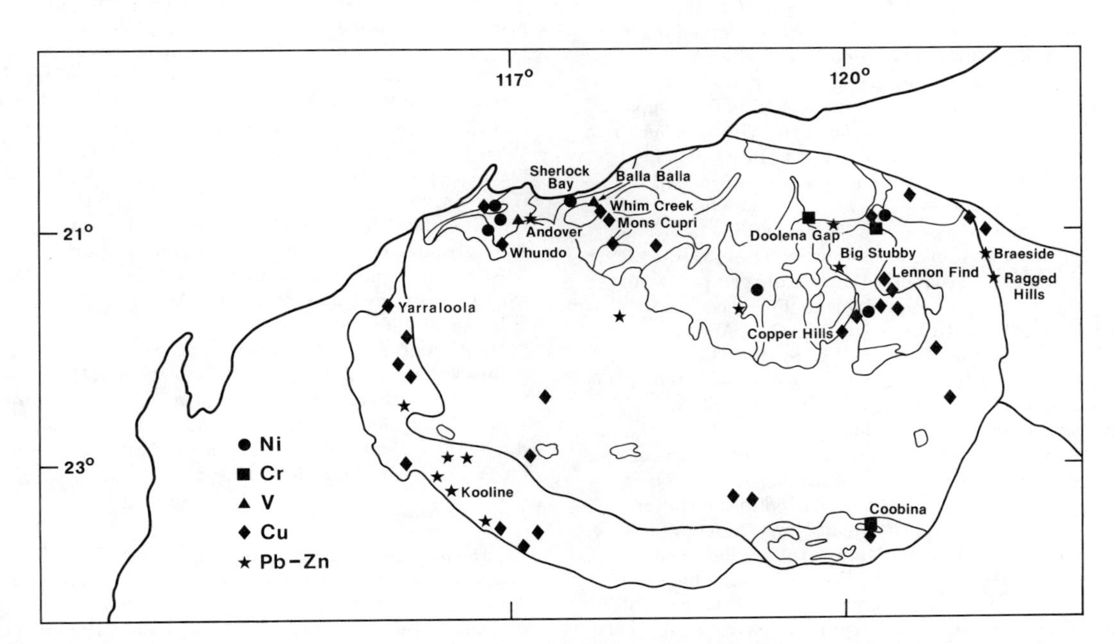

Fig. 10. Geographic distribution of nickel, vanadium, chromium, and base metal mineralization in the Pilbara Craton and Ashburton Trough.

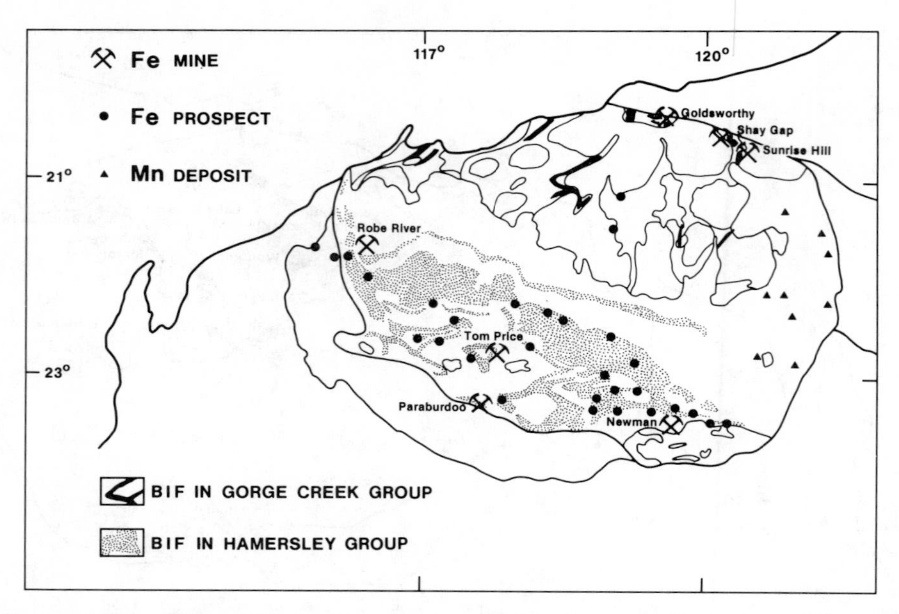

Fig. 11. Geographic distribution of iron and manganese mineralization in the Pilbara Craton and Ashburton Trough.

Copper-molybdenum stockworks have been investigated in a number of localities in the eastern Pilbara Block, but although substantial resources may be present, the grades are well below economic levels. These deposits appear to be a primitive form of porphyry copper formed at around 2.9 Ga ago.

Structurally controlled supergene concentrations of copper minerals have been mined in a number of places within a sulfide-bearing black shale unit (Roy Hill Member of the Jeerinah Formation) at the top of the Fortescue Group. Similarly, secondary copper minerals have been mined from fractures and caves within the Wittenoom Dolomite.

Copper-bearing quartz veins are ubiquitous within the region, with small commercial production coming from deposits within the Warrawoona, Fortescue and Wyloo Groups. Within the Fortescue Group, copper is typically associated with lead while the veins in the Wyloo Group may also contain lead, arsenic, bismuth and uranium.

About 6 300 tonnes of lead and 25 tonnes of zinc have been won from the Pilbara Craton and Ashburton Trough [Blockley, 1971]. Almost all of this production came from comparatively high-grade veins cutting the Warrawoona, Fortescue and Wyloo Groups at such localities as Doolena Gap, Andover, Braeside, Ragged Hills and Kooline (Figure 10). The veins follow major faults near the margin of the Pilbara Craton or lie parallel to the dominant cleavages in pelitic rocks of the Ashburton Trough. However, any long term potential of the region as a source of lead and zinc lies in its stratiform base metal deposits associated with felsic volcanic rocks. Such deposits occur at Big Stubby and Lennon Find in the Duffer Formation of the Warrawoona Group, at Mons Cupri in the Whim Creek Group, and at Yarraloola in the Wyloo Group.

Lead isotope studies reported by Richards and others [1981] and Richards [pers. comm.] on galenas from the Pilbara Craton and Ashburton Trough show a spectrum of ages from 3.47 Ga to 1.7 Ga with stratiform mineralization appearing at 3.47 Ga, 2.95 Ga and 2.0 Ga. It appears that lead-zinc mineralizing processes were active many times throughout the 1.8 Ga development of the region, with a pause only during the passive period which marked the deposition of the Hamersley Group.

Tin and Tantalum

The Pilbara Block has produced some 17 100 tonnes of tin concentrate and 1 500 tonnes of tantalite from pegmatites and derived alluvial deposits. Associated with the tin and tantalum are substantial concentrations of lithophile elements such as beryllium and lithium. Most of the primary tin-tantalum pegmatites are related to a suite of post-tectonic granites intruded at a late stage in the development of the Block [Blockley, 1980]. Early Rb-Sr dating placed the intrusion of these granites at about 2.6 Ga [de Laeter and Blockley, 1972], but more recent geochronology on their associated pegmatites (Pidgeon, 1978], and on the Fortescue Group basalts [Richards and Blockley, 1984], which locally overlie members of the suite, strongly suggests that their age is greater than 2.8 Ga.

The principal producing centres (Figure 9) are Moolyella (Sn), Cooglegong (Sn + minor Ta), Wodgina (Ta + Sn) and Pilgangoora (Ta).

Although within the context of the earth's history, tin and its associated metals are normally restricted to comparatively young rocks, there is no repetition of this style of mineralization within the region at times later than the +2.8 Ga event. This is due to the lack of suitable 'tin granites' after the intial period of cratonization of the Pilbara Block.

Iron and Manganese

Despite its diversity of mineralization, the Pilbara Craton is best known for its large iron ore deposits [MacLeod, 1966; Trendall, 1975]. These were first evaluated in the early 1960s and since production began in 1966, some 1 100 million tonnes of ore have been mined. The ore is associated with two widespread sequences of chert and iron-formation, the earlier appearing within the Gorge Creek Group and the later, and much better developed one, comprising the Hamersley Group. Most of the commercial ore has come from hematite-goethite enrichments formed during a weathering cycle which predated the deposition of the Wyloo Group, but significant deposits also formed during the Tertiary, in particular, the transported pisolitic goethites deposited as bog ores and mined at Robe River.

Of the total production to date, 90 per cent has come from ore associated with the Hamersley Group at Tom Price, Paraburdoo, Newman and Robe River, while the remainder was extracted from the Gorge Creek Group at Goldsworthy and Shay Gap (Figure 11). Of the estimated 33 billion tonnes of iron ore resources containing better than 55 per cent Fe in the Pilbara region, only 0.1 per cent is associated with the Gorge Creek Group. The extensive chert units within the Towers Formation of the Warrawoona Group are poor in iron and have produced no commercial iron ore. Similarly, thin units of iron-formation within the Wyloo Group have not yielded commercial concentrations of iron.

Manganese follows iron in being concentrated mainly in the Hamersley Group, with only a small proportion of the 2 million tonnes of ore produced coming from the Gorge Creek Group, and none from the Wyloo Group (de la Hunty, 1963).

The geological processess responsible for producing primary concentrations of iron and manganese therefore varied considerably with time during the evolution of the region, rising from negligible to a dramatic peak at around 2.5 Ga during the deposition of the Hamersley Group, and declining once again in Wyloo Group times.

Summary

In this brief review of the mineralization of the Pilbara Craton and Ashburton Trough it is apparent that gold, nickel, chrome and vanadium were preferentially concentrated in the older part of the geological column; iron and manganese deposits show a distinct peak in the Hamersley Group, corresponding to an age of about 2.5 Ga; while the base metals copper, lead and zinc are distributed throughout the whole time range, appearing wherever geological conditions were favourable for their deposition.

Tin and tantalum mineralization was confined to the period of craton-forming granite intrusion within the Pilbara Block and is particularly associated with the last stage of that event marked by the intrusion of a suite of differentiated post-tectonic granites.

References

Baxter, J.L., Molybdenum, tungsten, vanadium and chromium in Western Australia, West. Australia Geol. Survey Mineral Resources Bull. 11, 1977.

Blake, T.S. and N.J. McNaughton, A geochronological framework for the Pilbara Region in Muhling, J.R., Groves, D.I. and Blake T.S. (eds), Archaean and Proterozic Basins of the Pilbara, Western Australia: Evolution and Mineral Potential, Geol. Dept. and Univ. Extension, University of Western Australia Publ. 9, 1984.

Blockley, J.G., The lead, zinc and silver deposits of Western Australia, West. Australia Geol. Survey Mineral Resouces Bull. 9, 1971.

Blockley, J.G., The tin deposits of Western Australia with special reference to the associated granites, West. Australia Geol. Survey Mineral Resources Bull. 11, 1980.

Compston, W. and P.A. Arriens, The Precambrian geochronology of Australia, Canadian Jour. Earth Sci. 5, p. 561-583, 1968.

Compston, W., I.S. Williams, M.T. McCulloch, J.J. Foster, P.A. Arriens and A.F. Trendall, A revised age for the Hamersley Group, Geol. Soc. Aust. 5th Ann. Conv., Perth, Abstracts, 3, p. 40, 1981.

Daniels, J.L., The palaeogeographic development of Western Australia - Precambrian, in Geology of Western Australia, West. Australia Geol. Survey Mem. 2, p. 437-450, 1975.

de Laeter, J.R. and J.G. Blockley, Granite ages within the Pilbara Block, Western Australia, Geol. Soc. Australia Jour., 19, p. 363-370, 1972.

de la Hunty, L.E., The geology of the manganese deposits of Western Australia, West. Australia Geol. Survey Bull. 116, 1963.

Gee, R.D., Structure and tectonic style of the Western Australian Shield, Tectonophysics, 58, p. 327-369, 1979.

Hickman, A.H., Geology of the Pilbara Block and its environs, West. Australia Geol. Survey Bull. 127, 1983.

MacLeod, W.N., The geology and iron deposits of the Hamersley Range area, Western Australia, West. Australia Geol. Survey Bull. 117, 1966.

Marston, R.J., Copper mineralization in Western Australia, Geol. Survey West. Australia Mineral Resource Bull. 13, 1979.

Pidgeon, R.T., Geochronological investigation of granite batholites of the Archaean granite-greenstone terrain of the Pilbara Block, Western Australia; in I.E.M. Smith and J.G. Williams (eds.), Proc. 1978 Archaean Geochemistry Cong., p. 360-363, 1978.

Richards, J.R., I.R. Fletcher and J.G. Blockley, Pilbara galenas: precise lead isotopic assay of the oldest Australian leads; model ages and growth-curve implications, Miner. Deposita, 16, p. 7-30, 1981.

Richards, J.R. and J.G. Blockley, The base of the Fortescue Group, Western Australia: further galena lead isotope evidence on its age, Aust. Jour. Earth Sci., 31, p. 257-268, 1984.

Thorne, A.M. and D.B. Seymour, The sedimentology of a tide influenced fan-delta system in the early Proterozic Wyloo Group on the southern margin of the Pilbara Craton, W.A., West. Aust. Geol. Survey Prof. Pap. for 1984, p. 70-82, 1986.

Trendall, A.F., The Geology of Western Australia iron ore, in C.L.Knight (ed.), 1975, Economic Geology of Australia and Papua-New Guinea. 1. Metals. Austral. Inst. Min. Metall., Monogr., 5, p. 883-892, 1975.

Trendall, A.F., Geology of the Hamersley Basin, 25th Int. Geol. Congr., Sydney, Australia, Excursion Guide No 43A, 1976a.

Trendall, A.F., Striated and faceted boulders from the Turee Creek Formation - evidence for a possible Huronian glaciation on the Australian continent, West. Australia Geol. Survey Ann. Rept. 1975, p. 88-92, 1976b.

Trendall, A.F., A revision of the Mount Bruce Supergroup, West. Australia Geol. Survey Ann. Rept. 1978, p. 63-71, 1979.

Trendall, A.F., The Hamersley Basin, in A.F. Trendall and R.C. Morris (eds.), Iron Formation: Facts and Problems, Elsevier, Amsterdam, p. 69-129, 1983.

Trendall, A.F. and J.G. Blockley, The iron formations of the Precambrian Hamersley Group, Western Australia, with special reference to the associated crocidolite, West. Australia Geol. Survey Bull. 119, 1970.

PRECAMBRIAN SEDIMENTARY SEQUENCES AND THEIR MINERAL AND ENERGY RESOURCES

Ian B. Lambert

Elisian Resources Pty. Ltd., 2 Bonwick Place, Garran, Canberra, ACT 2605, Australia

Abstract. Evolutionary trends for mineral and energy resources in Precambrian sedimentary sequences are related to changes in tectonic environments and in compositions of the lithosphere, atmosphere, hydrosphere and biosphere.

Introduction

This paper outlines in turn the major trends in sedimentary rock abundances and depositional environments, secular geochemical trends, organic evolution, sediment-hosted metal deposits and fossil fuels.

Length constraints have necessitated presentation of most information in the form of summary diagrams, supported by short discussions of selected features with emphasis on Australian examples.

For further details the interested reader is referred to the reviews of geological evolution and mineral deposits by Veizer (1976), Windley (1977), and Lambert and Groves (1981).

Sedimentary Rock Distributions and Paleoenvironments

The generalized distributions of the major types of sedimentary rocks in the geological record are summarised in Figure 1.

Archean

Archean sedimentary strata occur in supracrustal sequences of high grade gneiss terrains and in greenstone belts, but they are subordinate to igneous rocks. Overall, greywackes, cherts and volcaniclastic sediments are much more abundant in the Archean than carbonates, mature clastics and evaporites (Lowe, 1980).

The oldest well-dated meta-sedimentary strata are in the 3.8 Ga Isua supracrustals of the North Atlantic craton (Allart, 1976). Supracrustal sequences here and in other Archean terrains contain quartzites, carbonates and banded iron formations (BIF), which have been interpreted as reflecting epicontinental platform or shelf conditions (e.g. Windley, 1977). However, high grade metamorphism makes it difficult to distinguish original quartz sandstone and limestone from exhalative cherts and carbonates. Associated clastic strata probably have major volcanic components.

Greenstone belts formed mainly between 3.6 and 2.6 Ga, although there are some 1.8 to 2.0 Ga examples. Different tectonic styles are reflected in the sedimentary components of greenstone belts (Groves, 1982). These will be illustrated by consideration of two extreme examples, viz. the east Pilbara and east Yilgarn cratons of Western Australia.

The early Archean east Pilbara craton (Hickman, 1981; Groves, 1982) contains arcuate greenstone belts which are not significantly disrupted by early structures or metamorphism. Within the predominantly volcanic lower part of the sequence there are intervals up to 500 m thick of ca. 3.5 Ga shallow water to subaerial sedimentary strata. These comprise chert and silicified arenites (Dunlop, 1978). At North Pole, in this region, the oldest microfossils and stromatolites yet identified occur in chert interbedded with barite (Schopf and Walter, 1983; Walter, 1983). Textures and structures in the chert imply it replaced carbonate, while barite crystals with gypsum interfacial angles suggest an original evaporite sequence (Dunlop, 1978; Lambert et al., 1978). Groves (1982) concluded these accumulated under conditions of low extension and slow subsidence. Above the basal sequence in this craton there are thick clastic deposits which accumulated mainly in platform (alluvial) to trough (submarine fan) environments (Eriksson, 1981), suggesting increasing extension. BIF in this upper sequence formed as pelagic sediments and as exhalative deposits associated with volcanism.

In contrast, the Archean east Yilgarn craton (Gee et al., 1981) appears to have had a relatively short period of development, mainly between 2.9 and 2.6 Ga. It is characterised by more linear greenstone belts with

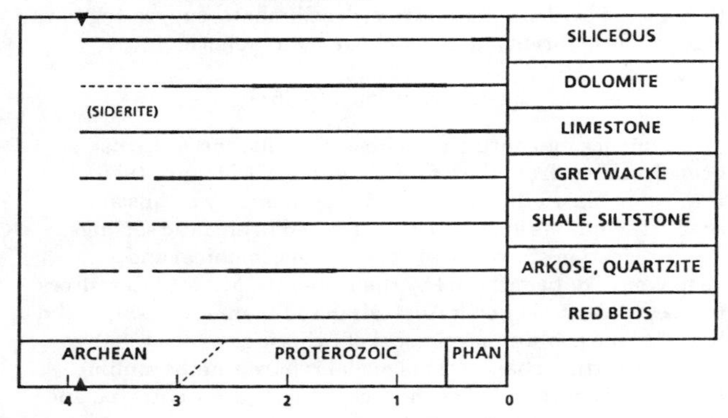

Fig. 1. Generalised abundance trends for major types of sedimentary rocks.

considerable stratigraphic disruption by early structures and metamorphism. Volcanics again dominate in the lower parts of the greenstone sequences; in a central basin these have intercalated sulfidic cherts and shales, while in peripheral belts oxide facies BIF is relatively common. Greywackes and associated volcano-clastic strata are most abundant at high stratigraphic levels. Groves (1982) concluded that the features of the east Yilgarn greenstone sequences indicate higher extension and more rapid subsidence than for the east Pilbara sequences.

Proterozoic

Proterozoic strata are preserved in large sedimentary basins and in relatively metamorphosed and structurally complex mobile belts, which formed on Archean basement.

As summarised in Figure 1, early Proterozoic sequences are characterised by the epicontinental assemblage of quartzite-orthoconglomerate-arkose-shale-carbonate.

Craton stabilisation did not occur at the same time everywhere. The oldest known cratonic sequence is in the ca. 3 Ga Pongola Supergroup in South Africa (Cloud, 1976). This is more basalt-rich than most Proterozoic sequences, but it differs from typical Archean greenstone sequences in having extensive shallow water quartzites and cryptalgal carbonates. It is succeeded by sequences of continental clastics and volcanics and the ca 2.5-2.75 Ga Witwatersrand Group which grades up from a probable marine deltaic assemblage to continental and marine coarse clastics (Pretorius, 1976). The oldest putative tillites occur in the Witwatersrand Group (Wiebols, 1955). The ca. 2.8 Ga Fortescue Group overlying the southern part of the Pilbara craton is another example of an early Proterozoic cratonic sequence which formed while Archean greenstone accumulation was occurring in zones of high extension elsewhere.

The earliest major developments of red beds and early diagenetic (primary?) dolomites were around 2.4 Ga; dolomites are more abundant than limestones throughout the Proterozoic. The oldest of the widespread early Proterozoic BIF are the 2.5 Ga (W. Compston and A. Trendall in Gee, 1980) examples in the Hamersley Basin of Western Australia (Morris and Horwitz, 1983).

Carbonate-pseudomorphed evaporite deposits have been documented from Proterozoic sequences of various ages. The oldest are of early Proterozoic age from the Great Slave Lake region, Canada (Badham and Stanworth, 1977); these contain the earliest evidence for halite-rich sediments, suggesting low NaCl concentrations in the Archean hydrosphere.

Basin Evolution 2.0-1.5 Ga

Evidence for continental margin subduction zones has been described from the Proterozoic (e.g. Hoffmann, 1980), but is uncommon. In this period, most sedimentary basins and orogenic domains appear to have formed in ensialic settings.

Integration of extensive geologic, geochemical and geophysical data gathered by the Bureau of Mineral Resources from ca. 2.0-1.5 Ga north Australian sedimentary basins is the basis for the evolutionary model of Etheridge et al. (1984), which is outlined below. This model is based on the similar timing and character of events across northern Australia, and even in other continents. In essence, there are two tectono-stratigraphic cycles (here termed lower and middle Proterozoic), separated by a major 1.8 Ga orogenic event.

Both the early and middle Proterozoic cycles comprise a lower quartz-rich clastic sequence which is commonly clearly fluviatile, and variable proportions of bimodal volcanics. These probably formed in initial phases of crustal extension and rifting of thin Archean crust. Both cycles also contain extensive middle sequences of finer grained clastics which include carbonaceous strata. There are also abundant dolomites and minor BIF in the middle sequences, which appear to have formed during transgressive phases associated with post-extension subsidence. Upper turbidite or molasse facies occur in both the lower and middle Proterozoic cycles, and they probably mark the beginning of orogenic phases.

Constraints from radiometric age determinations suggest that there was a major underplating of the north Australian crust around 2.2 Ga, probably concentrated above ascending small-scale convection cells in the mantle. Crustal extension is considered to have been focussed above the upwelling mantle cells, and it was followed by thermal subsidence. The widespread ca. 1.8 Ga orogeny, which may have been related to a rapid change in mantle convection patterns, resulted in production of I-type felsic magmas from the mafic underplated crustal layer. The further phase of stretching and rifting resulting in accumulation of middle Proterozoic sequences was initiated 20-50 Ma after this orogeny.

Late Proterozoic sedimentation, which is well represented in the central and southern parts of Australia, was initiated by another major period of rifting.

Geochemical Trends

As well as variations in the relative abundances of sedimentary rocks through geological time, there are secular geochemical trends. A major example is the increasing K/Na ratio in shales, siltstones and sandstones in the early Proterozoic (Engel et al., 1974). This is a reflection of the importance of K-rich igneous activity in the final stages of cratonisation, as is the sharp increase in $^{87}Sr/^{86}Sr$ ratios in carbonates during the early Proterozoic (Veizer and Compston, 1976). The rare earth element compositions of sedimentary rocks also change (Taylor and McLennan, 1981); Archean sediments are distinguished from younger strata by having generally lower REE abundances, lower La/Yb, and no Eu depletion, implying that the exposed Archean crust was considerably more mafic than in younger periods.

Cherts and carbonates display increasing $\delta^{18}O$ values with decreasing age in the Precambrian; this is interpreted by some as indicating decreasing ocean temperature (Knauth and Lowe, 1978).

At 2.5 ± 0.3 Ga, there is a marked divergence of sulfur isotope compositions of sulfide mineralisation and pyritic sedimentary strata from primitive $\delta^{34}S$ values of near $0‰$ (Lambert and Groves, 1981; Skyring and Donnelly, 1982). This implies increased sulfate concentrations in the hydrosphere, most likely related to a marked increase in the level of oxygen in the atmosphere. Therefore, sulfur isotopic data support geological arguments summarised by Cloud (1976) for evolution of an oxygenous atmosphere in the early Proterozoic.

Organic Evolution

A comprehensive treatment of Precambrian biology is available in Schopf (1983). Major features of organic evolution are sumarised in Figure 2.

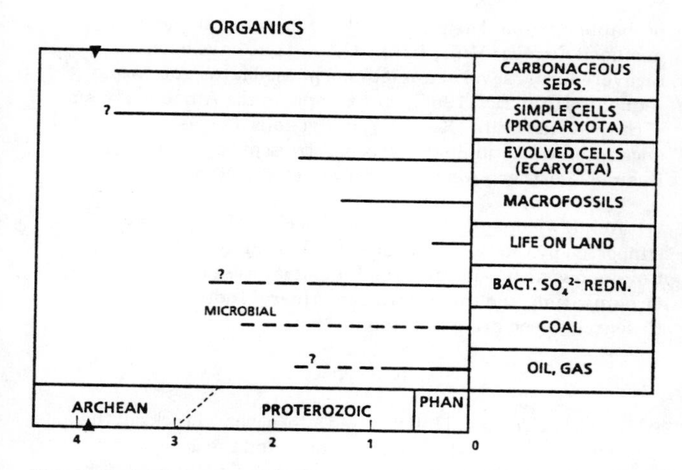

Fig. 2. Summary of some major aspects of biological evolution and of the distributions of fossil fuels through geological time.

Life on Earth must have evolved before 3.5 Ga, the age of the chert at North Pole which hosts the oldest generally accepted microfossils and stromatolites (above). Schopf and Walter (1983) studied the morphologies, distributions and physical organisation of these microfossils. They inferred that the microorganisms were photo-responsive, at least some were anaerobes, some used CO_2 as their immediate source of cellular carbon (autotrophs), and some cycled organic carbon within microbial mats (heterotrophs). It is noteworthy that these organisms occur in a sequence deposited in an unusual type of Archean environment - one in which sulfate evaporites formed. The sulfates are compatible with the activity of photosynthetic sulfur bacteria, but production of oxygen in this environment by biological or photochemical processes cannot be ruled out.

No microfossils have been found in greywacke-rich sedimentary sequences, which are much more common in the Archean record.

There is a marked increase in the abundance of stromatolites and microfossils in early Proterozoic sequences (Schopf, 1983). It is a reasonable conclusion from the geological and isotopic trends that there was proliferation of oxygen-producing and sulfate-reducing micro-organisms during this period.

Simple procaryotic organisms were joined by more evolved eucaryotes in the middle Proterozoic, possibly resulting in a further rise in oxygen levels. Macroorganisms appeared close to the Proterozoic-Cambrian transition.

Sediment-Hosted Metal Deposits

Figure 3 summarises the distributions of the major types of sediment hosted mineralisation through the geological record.

In the Archean, mineralisation formed as the result of igneous (including volcano-exhalative) or metamorphic processes is much more important than sediment-hosted mineralisation. There are the bedded barite deposits in Western Australia, South Africa and India, at least some of which formed by baritisation of original evaporite beds (Lambert et al., 1978). Also, a minority of Archean BIF occur in pelagic sedimentary sequences; the iron may have come from distal exhalative centres, or been concentrated by local water-rock interactions.

The important early Proterozoic BIF formed in predominantly sedimentary sequences and are generally larger than Archean examples. Detailed studies in the Hamersley Basin (Morris and Horwitz, 1983) indicate that shales interbedded with BIF are pyroclastic and chemical deposits and that there was very little terrigenous input into the marginal marine depositional basin. The sequence was deposited on a submarine platform or bank of largely volcanic, 2.8 Ga Fortescue Group. The main source of iron is likely to have been upwelling anoxic waters. These could have enriched Fe^{2+} mainly as a result of hydrothermal processes at rift zones and/or hot spots, although lower temperature seafloor weathering processes could also have been important. In the absence of silica-secreting organisms, these waters would also have been enriched in silica. The BIF would have precipitated as the upwelling waters entered oxygenated shallow water environments. Cloud (1976) has emphasised the abundance of BIF in the early Proterozoic and their subsequent dramatic decrease in abundance, arguing that this is consistent with oxidation of the hydrosphere/atmosphere system during this period.

Important Au and/or U deposits in pyritic quartz conglomerates, as exemplified by the huge Witwatersrand deposits (Pretorius, 1976), are restricted to early Proterozoic sequences. Materials eroded from Archean source regions were transported in high-energy fluvial systems, and uranium and gold were concentrated by physical and biological processes in near-shore environments. The stability of uraninite and pyrite during extensive physical reworking is another point favouring low oxygen levels in the contemporaneous atmosphere (Cloud, 1976).

Other types of U deposits formed close to early-middle Proterozoic unconformities; major examples of these occur in

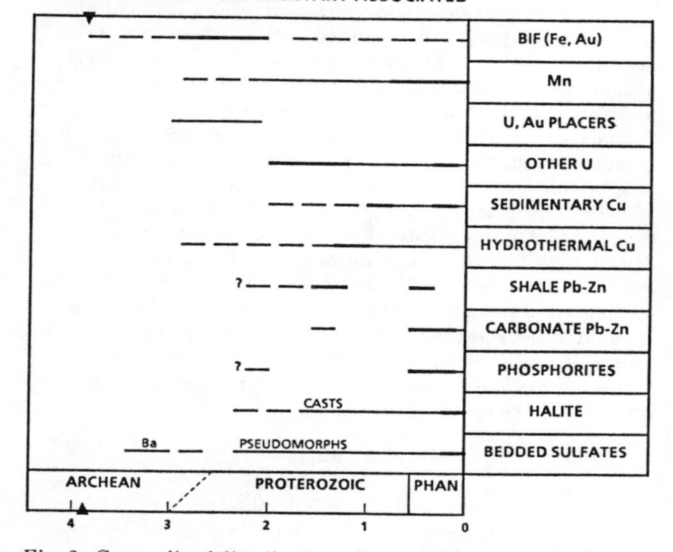

Fig. 3. Generalised distribution patterns for major types of metal deposits in the sedimentary rock record.

the Pine Creek Geosyncline of northern Australia, and the Athabasca Basin of western Canada (Ferguson and Goleby, 1980). The main sources of uranium were basement granitic rocks and early Proterozoic rocks derived from these. The deposits appear to have formed in multiple solution-precipitation events, after the build up of significant levels of oxygen in the atmosphere. Hexavalent uranium is very soluble in oxidised waters, which evidently migrated through faults, fractures and permeable zones. The uranium was precipitated by reductants, commonly organic matter.

The oldest major Pb-Zn deposits are the "shale-hosted" examples of middle Proterozoic age (Lambert, 1983), which are particularly well represented in Australia (Broken Hill, Mount Isa, McArthur River). Other major examples are Sullivan in Canada (Ethier et al., 1976), and Gamsberg-Aggeneys in Southwest Africa (Rozendaal, 1981). Upward enrichment of lead in the crust appears to have been a prerequisite for the formation of such deposits, and the ca. 1.8 Ga tectono-thermal event could have played an important role in this enrichment. The stratiform Pb-Zn deposits formed at and near the sediment-water interface in depressions in intracratonic troughs, well above lower Proterozoic basement. The ore-forming fluids were probably essentially basinal brines which leached metals from the trough sequences, possibly supplemented in some cases by metalliferous fluids from igneous centres. The sulfide came from hydrothermal and biogenic sources. No major stratiform Pb-Zn deposits are known in late Proterozoic sequences, but they reappear in the early Phanerozoic, for example the Selwyn Basin deposits of Canada (Morganti, 1981), and Rammelsberg (Hannak, 1982) and Meggen (Krebs, 1982) in Germany.

Hydrothermal copper mineralisation is also unusually well represented in Australian middle Proterozoic sequences (Lambert et al., 1985). Such mineralisation at Olympic Dam formed from fluids which were probably generated by magmatic activity in an active terrestrial rift, whereas the Mount Isa and Tennant Creek deposits apparently formed from fluids generated during late stages of regional metamorphism.

Major bedded copper mineralisation became abundant in the late Proterozoic, as exemplified by the African Copperbelt deposits (Fleischer et al., 1976) and White Pine (Brown, 1971), although the important Udokan deposits of the USSR are probably in early Proterozoic strata (Samonov and Pozharisky, 1977). The moderately oxidising conditions favouring copper transport in low temperature brines (Rose, 1976) would have become widespread only after the evolution of an oxygenous atmosphere. While there may have been syngenetic copper sulfide deposition in some cases, most deposits exhibit evidence for introduction of moderately oxidised cupriferous fluids into pyritic sediments during early diagenesis. Associated red beds are widely considered to have been the major sources of the copper, and other potential sources are basic volcanics and metal-enriched basement rocks. The sulfide was generated by bacterial sulfate reduction in the host sediments.

Fossil Fuels

Despite evidence for abundant microbial activity in early Proterozoic and younger sedimentary environments, few economic deposits of fossil fuels are known in Proterozoic basins (Fig. 3). There are "coal" (shungite) deposits of probable microbial origin in early Proterozoic strata of northern Europe (Muir, 1979). Oil and gas have been recovered from several unmetamorphosed late Proterozoic sequences (Murray, 1965), for example in the Amadeus Basin of central Australia. Recently, indigenous oil was encountered in middle Proterozoic strata of the McArthur Basin, of northern Australia (Powell et al., 1985).

Acknowledgements. The Baas Becking Laboratory is supported by the Bureau of Mineral Resources, the Commonwealth Scientific and Industrial Research Organisation, and the Australian Mineral Industries Research Association Limited.

References

Allart, J.H., The pre 3760 m.y. old supracrustal rocks of the Isua area, central west Greenland, and the associated occurrence of quartz-banded ironstone, in Windley, B.F. (ed), The Early History of the Earth, Wiley, London, 177-190, 1976.

Badham, J.P.N. and C.W., Stanworth, Evaporites from the Lower Proterozoic of the east arm, Great Slave Lake. Nature, 268, 516-517, 1977.

Brown, A.C., Zoning in the White Pine copper deposit, Ontonogan County, Michigan, Econ. Geol., 66, 543-573, 1971.

Cloud, P., Major features of crustal evolution. Geol. Soc. S. Africa, Spec. Publ., 78 (annex) 32 pp., 1976.

Dunlop, J.S.R., Shallow water sedimentation at North Pole, Pilbara Block, Western Australia, in Glover, J.E. and D.I. Groves (eds), Archaean Cherty Metasediments; Their Sedimentology, Micropalaeontology, Biogeochemistry and Significance to Mineralisation, Dept. Geol. and Extension Service, Univ. West Australia, Perth, 2 39-44, 1978.

Engel, A.E.J., Itson, S.P., Engel, C.G., Stickney, D.M. and E.J. Cray, Crustal evolution and global tectonics: a petrogenetic view. Geol. Soc. Am. Bull., 85, 843-858.

Eriksson, K.A.,Archaean platform-to trough sedimentation, East Pilbara Block Australia. In Glover, J.E. and D.I. Groves (eds), Archaean Geology Geol. Soc. Aust., Spec. Publ. 7, 235-244, 1981.

Etheridge, M.A., Wyborn, L.A., Rutland, R.W.R., Page, R.W., Blake, D.H. and B.J. Drummond, Workshop on Early to Middle Proterozoic of Northern Australia. Bur. Min. Res. Record 1984/31, 8 pp., 1984.

Ethier, V.G., Campbell, F.A., Both, R.A., and Krouse, H.R., Geological setting of the Sullivan ore body and estimates of temperatures and pressures of metamorphism. Econ. Geol., v. 71, 1570-1588, 1976.

Ferguson, J. and A.B. Goleby (eds), Uranium in the Pine Creek geosyncline. Int. Atomic energy Agency, Vienna, 760 pp., 1980.

Fleischer, V.D., Garlick, W.G. and R. Haldane, Geology of the Zambian copperbelt, in Wolf, K.H. (ed), Handbook of Stratabound and Stratiform Ore Deposits, V. 6, Elsevier, Amsterdam, 223-352, 1976.

Gee, R.D., Summary of the Precambrian stratigraphy of Western Australia, West. Aust. Geol. Surv., Ann Rept. 1979, 85-90, 1980.

Gee, R.D., Baxter, J.L., Wilde, S.A. and I.R. Williams, Crustal development in the Archaean Yilgarn Block, Western Australia, in Glover J.E. and Groves, D.I. (eds), Archaean Geology. Geol. Soc. Aust., Spec. Publ. 7, 43-56, 1981.

Groves, D.I., The Archean and earliest Proterozoic evolution and metallogeny of Australia, Revista Brasileira de Geosciencias, 12, 135-148, 1982.

Hannak, W.W., Genesis of the Rammelsberg ore deposit near Goslar/Upper Harz, Federal Republic of Germany, in Wolf, K.L. (ed.), Handbook of Strata-bound and Stratiform Ore Deposits, Elsevier, Amsterdam, v. 9.

Hickman, A.H., Crustal evolution of the Pilbara Block, Western Australia, in Glover, J.E. and D.I. Groves (eds.) Archaean Geology, Geol.Soc.Aust., Spec.Publ. 7, 57-70, 1981.

Hoffmann, P.F., Wopmay Orogen: a Wilson-cycle of early Proterozoic age in the northwest Canadian shield, in Strangway, D.W. (ed.). The Continental Crust and its Mineral Deposits, Geol. Assoc. Canada, Spec. Pap. 20, 523-549, 1980.

Knauth, L.P. and D.R. Lowe, Oxygen isotope geochemistry of cherts from the Onverwacht Group (3.4 billion years), Transvaal, South Africa with implications for secular variations in the isotope compositions of cherts, Earth Planet. Sci. Lett., 41, 209-222, 1978.

Krebs, W., The geology of the Meggen ore deposit. In Wolf, K.H. (ed) Handbook of Strata-bound and Stratiform Ore Deposits. Elsevier, Amsterdam, v. 9.

Lambert, I.B., The major stratiform lead-zinc deposits of the Proterozoic. Geol. Soc. Am., Mem. 161, 209-226, 1983.

Lambert, I.B., Donnelly, T.H., Dunlop, J.S.R. and D.I. Groves, Stable isotope compositions of early Archaean sulphate deposits of probable evaporitic and volcanogenic origins. Nature, 276, 808-811, 1978.

Lambert, I.B. and D.I. Groves, Early Earth Evolution and metallogeny, in Wolf, K.H. (ed) Handbook of Stratabound and Stratiform Ore Deposits, v. 6, Elsevier, Amsterdam, 339-447, 1976.

Lambert, I.B., Knutson, J. Donnelly, T.H. and H. Etminan, The diverse styles of sediment-hosted copper deposits in Australia, in Friedrich, G.H. et al., (eds), Geology and Metallogeny of Copper Deposits, Springer Verlag, Berlin, 540-558, 1986.

Lowe, D.R. Archean sedimentation. Ann. Rev. Earth Planet. Sci., 8, 145-167, 1980.

Morganti, J.M., Ore deposit models - 4. Sedimentary-type stratiform ore deposits: Some models and a new classification. Geoscience Canada, 8(2), 65-75, 1981.

Morris, .C. and R. Horwitz, The origin of the iron-formation-rich Hamersley Group of Western Australia - deposition on a platform, Precamb. Res., 21, 273-297, 1983.

Muir, M.D., Palaeontological evidence bearing on evolution of the atmosphere and hydrosphere, in Grant, P.R. (ed) Proceedings of a Meeting Evolution of the Atmosphere and Hydrosphere, Imperial College, London, 1979.

Murray, G., Indigenous Pre-Cambrian petroleum. AAPG Bull., 49(1), 3-21, 1965.

Powell, T.G., Crick, I, Jackson, M.J. and Sweet, I., Bureau of Mineral Resources Research Newsletter, No. 3, 1-2, 1985.

Pretorius, D.A., The nature of Witwatersrand gold-uranium deposits, in Wolf, K.H. (ed) Handbook of Stratabound and Stratiform Ore Deposits, Elsevier Amsterdam, 29-88, 1976.

Rose, A.W. The effect of cuprous chloride complexes in the origin of red-bed copper and related deposits. Econ. Geol., 71, 1036-1048, 1976.

Rozendaal, A., The Gamsberg zinc deposit, South Africa: A banded stratiform base metal sulfide deposit, Proc. Fifth IAGOD Symposium, E Schweizerbartsche vertagsbuchandlung, Stuttgart, 619-633.

Samonov, I.Z. and I.F. Pozharisky, Deposits of Copper, in Smirnov, V. I. (ed), Ore Deposits of the USSR, v. II, Pitman, London, 106-181, 1977.

Schopf, J.W. (ed), The Earth's Earliest Biosphere; Its Origin and Evolution, Princeton Press, Princeton, 543 pp., 1983.

Schopf, J.W. and M.R. Walter, Archean Microfossils: new evidence of ancient microbes, in Schopf, J.W. (ed) The Earth's Earliest Biosphere: Its origin and Evolution, Princeton Press, Princeton, 214-239, 1983.

Skyring, G.W. and T.H. Donnelly, Precambrian sulfur isotopes and a possible role for sulfite in the evolution of biological sulfate reduction, Precamb. Res., 17, 41-61, 1982.

Taylor, S.R. and S.M. McLennan, Evidence from rare-earth elements for the chemical composition of the Archaean crust, in Glover J.E. and D.I. Groves (eds), Archaean Geology, Geol. Soc. Aust., Spec. Publ., 7, 255-262, 1981.

Veizer, J., Evolution of ores of sedimentary affiliation through geologic history; relations to the general tendencies in evolution of the crust, hydrosphere, atmosphere and biosphere, in Wolf, K.H. (ed) Handbook of Stratabound and Stratiform Ore Deposits, v. 3, Elsevier, Amsterdam, 1-41, 1976.

Veizer, J. and W. Compston, $^{87}Sr/^{86}Sr$ in Precambrian carbonates as an index of crustal evolution, Geochim, Cosmochim, Acta, 40, 905-915, 1976.

Walter, M.R., Archaean stromatolites: evidence of earliest benthos, in Schopf, J.W. (ed) The Earth's Earliest Biosphere: Its Origin and Evolution, Princeton Press, Princeton, 187-213, 1983.

Weibols, J.H., A suggested glacial origin for the Witwatersrand conglomerates, Trans. Geol. Soc. S. Africa, 58, 367-382, 1955.

Windley, B.F., The Evolving Continents, Wiley, London, 385 pp., 1977.

MECHANISM OF FORMATION OF DEEP BASINS ON CONTINENTAL CRUST

E. V. Artyushkov

Institute of Physics of the Earth, Academy of Sciences, Moscow, USSR

M. A. Baer

Ministry of Geology of the USSR, Moscow, USSR

Abstract. An analysis has been undertaken of the sedimentary cover structure of a large number of deep basins on continental crust in the main Phanerozoic fold belts and in cratonic regions. This analysis reveals that most basins formed without significant stretching of the underlying crust.

Deep basins on continental crust can be related to two major types. Deep-water basins (1-3 km) arise from very rapid subsidence of short duration (1-10 Ma). A high rate of subsidence precludes thermal relaxation as a mechanism. The time and space relationships between subsidence and orogeny preclude a significant role of thrust loading. Data on crustal structure indicate destruction of the lower crust beneath deep basins formed by rapid subsidence. This is may be associated with a gabbro-eclogite transformation associated with asthenospheric upwelling to the base of the crust. Such upwelling is in agreement with slight uplift and volcanism that commonly preceded rapid subsidence.

The second basin type is represented by deep sedimentary basins (5-15 km) produced by slow sediment-loaded subsidence of long duration (500-1000 Ma). These basins are stable with respect to compression, which indicates that they form on a thick cratonic lithosphere. Cratonic basins should arise due to large density increases in the lower crust.

In the regions studied intense crustal shortening occurred only in deep basins formed by rapid subsidence. These are the classic "miogeosynclines". Thinning of the crust and lithosphere following rapid subsidence ensures later compression of the basins under convergent plate motions.

Introduction

Mechanisms of formation of deep basins on continental crust is one of the basic problems in geodynamics. Most scientists explain the formation of such deep basins by stretching or thrust loading. Stretching of continental crust was originally proposed for narrow rift valleys [Artemjev and Artyushkov, 1971]. Lithospheric stretching has been suggested for wide basins such as the North Sea, Aegean Sea and others [McKenzie, 1978a, b and others].

Intense stretching produces large deformations in the uppermost crust, ~10 km thick. Tilted blocks represent the most common deformation type. They can be observed in many continental rifts (Figure 1). Tilted blocks have been found in some present deep water basins, for instance, on the Armorican and Galicia continental margins of the Atlantic ocean [Montadert et al., 1979]. Many authors consider on this

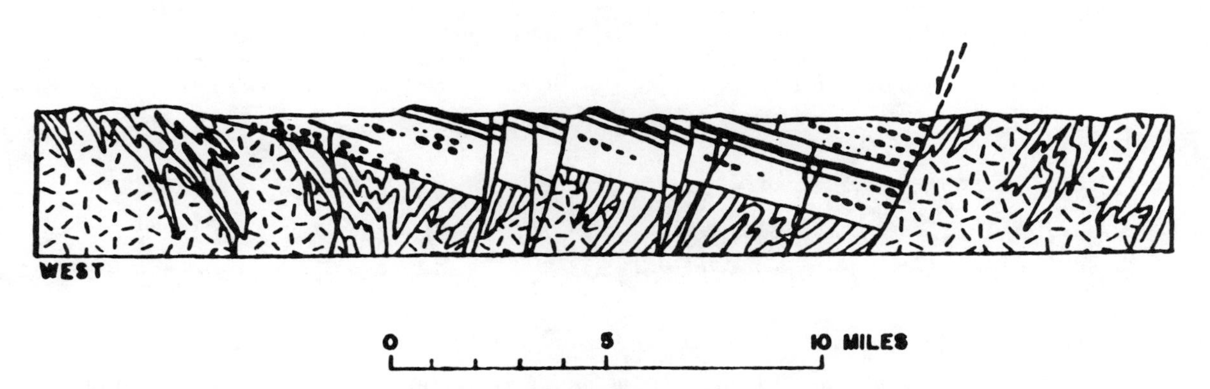

Fig. 1. Tilted blocks in the Late Triassic rifts in southern part of Northern Appalachians [King, 1959].

basis that most deep basins on continental crust also have been formed by stretching [Le Pichon and Sibuet, 1981; and others].

In order to evaluate the role of stretching on a global scale it is necessary to analyse the structure of a large number of deep basins underlain by continental crust. This is a difficult problem for the present deep-water basins, since most of them lack drilling information. Numerous deep basins on continental crust existed in the past, however. Most of them have been intensely shortened and have become incorporated into foldbelts. The sedimentary cover of these basins is exposed at the surface in thousands of geological sections or exposures that have been studied by generations of geologists.

In addition there are numerous undeformed or slightly deformed deep sedimentary basins on continental crust. Many of them have been extensively drilled and explored by seismic profiling, which permits an evaluation of the presence or absence of stretching.

We have analysed the structure of the sedimentary cover in numerous deep basins on continental crust, both strongly deformed and undeformed [Artyushkov and Baer, 1983, 1984a, b, 1986a, b, in prep.]. There are no known deformations indicating significant stretching over most of the territory of the basins. Here we briefly present some of the results of our analysis.

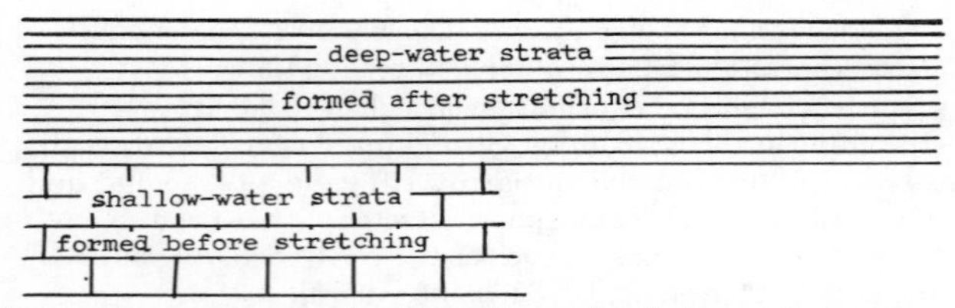

Fig. 2. Scheme illustrating the relationship between the areas covered with deep-water strata formed after intense stretching and shallow-water strata desposited before stretching.

Undeformed and Moderately Deformed Basins

The presence or absence of stretching can be easily revealed in undeformed or slightly deformed basins. After intense stretching the area covered with strata deposited following stretching should considerably exceed the area covered with strata deposited before stretching (Figure 2).

The initial water-depth after stretching by a factor of β is [LePichon and Sibuet, 1981]:

$$h_0 = 3.6\,(1 - 1/\beta)\,km \qquad (1)$$

For example, in order to produce a basin of an initial depth $h_0 \sim 1$ km by stretching, it is necessary to stretch the lithosphere by a factor of $\beta \sim 1.4$. Then deep-water strata will rest on the shallow-water strata deposited just before the subsidence on a part $(\beta-1)/\beta$ or ~70% of the area of the basin. On ~30 per cent of the area of the basin, deep-water strata should overlie much older strata or the crystalline basement.

Consider several examples of basins in the marginal parts of fold belts. The foredeep of the Ural Mountains was formed on the shallow-water shelf of the Russian Platform. After slight crustal uplift of short duration, rapid subsidence took place approximately between the Carboniferous and Permian [Pushcharovsky, 1959; Khvorova, 1961; Kamaletdinov, 1974]. A water-depth of $h_0 \geq 1$ km was reached very rapidly, in about 1 Ma. As a result shallow-water carbonates are abruptly overlain by deeper-water shaly limestones and marls. A very large number of wells have been drilled in the foredeep, which is 2000 km long and 50-300 km wide. Almost all the wells intersect deeper-water strata directly overlying shallow-water strata deposited just before the major subsidence (Figure 3). This means that deeper-water and shallow-water strata cover the same area, which precludes any significant stretching during subsidence.

The Martinsburg Basin of the Western Appalachians was formed on a shallow-water shelf by rapid subsidence in early Middle Ordovician time, following slight crustal uplift between the Early and Middle Ordovician [Cook and Bally, 1975; Shanmugan and Walker, 1980]. The initial water-depth was ≥ 1 km. The duration of subsidence was 1-2 Ma. The deposits of this basin were later slightly deformed by folding. The basin is well drilled and well covered by seismic profiling. A set of transverse profiles up to 130 km long has been compiled. These profiles show that deeper-water black shales rest everywhere on shallow-water carbonates (Figure 4). Both of these rocks units cover the same area, which precludes significant stretching.

The foredeep of the Caucasus formed in Early Oligocene time [Geology of the USSR, 1968; Danilchenko, 1970]. A water-depth of ≥ 1 km was reached in a few million years. A large number of wells indicate that strata deposited following the major subsidence rest everywhere on the strata deposited

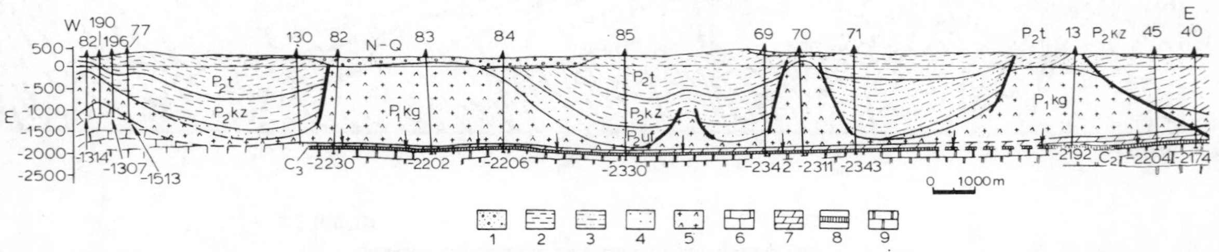

Fig. 3. A fragment of section across the outer unfolded part of the foredeep of the Urals [Kamaletdinov, 1974]. Transition from shallow-water rocks to deeper-water deposits is indicated by arrows. 1 - Cenozoic terrigenous deposits, 2-5 - upper Lower Permian-Upper Permian shallow water sediments (pelites, sandstones, clastics, evaporites), 6 - lower-mid Lower Permian shallow-water reef limestones, 7 - lower-mid Lower Permian deeper-water marls and marly limestones, 8 - deeper-water upper Upper Carboniferous cherty limestones, 9 - shallow-water Middle-lower Upper Carboniferous limestones.

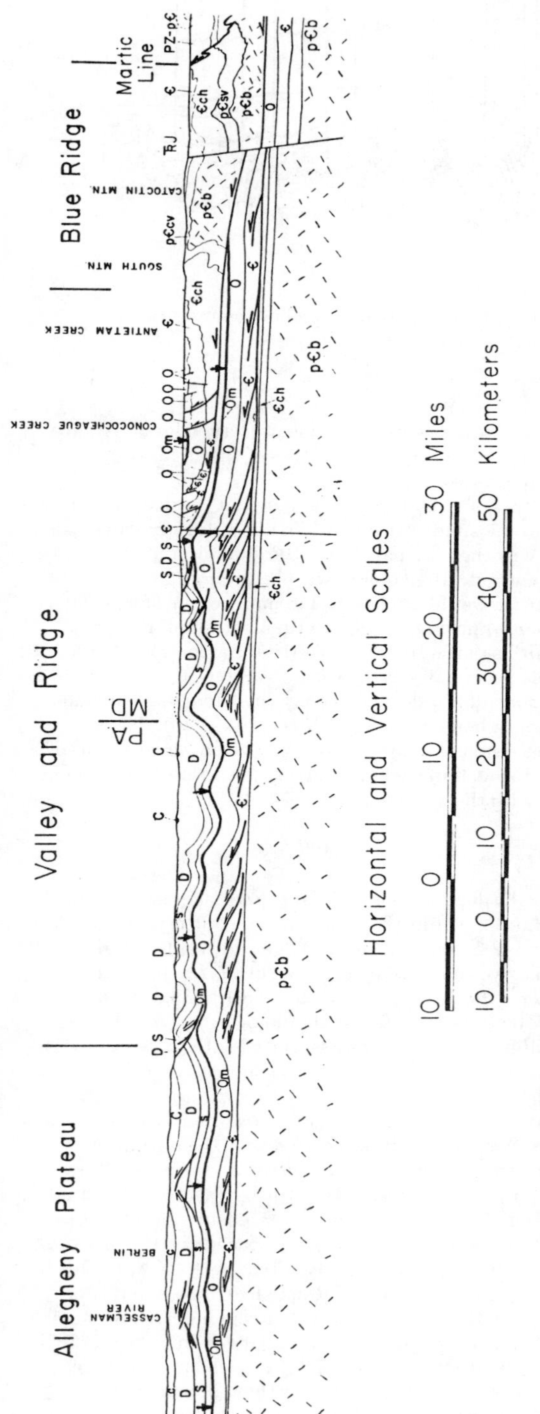

Fig. 4. Geological section across the Southern Appalachians in Pennsylvania and Maryland [Hatcher, 1981]. Transition from the Lower and lower Middle Ordovician shallow-water strata (O) to upper Middle and Upper Ordovician deeper-water strata (Om) is indicated by arrows.

just before this subsidence (Figure 5). Hence there was no significant stretching during the formation of this basin.

The West Siberian hydrocarbon basin can be taken as an example of a basin formed within a cratonic area. After slight crustal uplift of a shallow-water shelf in late Jurassic time, rapid subsidence followed[(Sonn, 1980; Surkov, 1982]. This formed a basin ~0.5 - 1 km deep. Beneath the region of rapid subsidence the crust is ~20% thinner than under the adjacent parts of the lowland [Karus et al., 1984]. Subsidence and thinning of the crust occurred without significant stretching. The Late Jurassic shallow-water strata that underlie the deeper-water Bazhenvsk Suite are continuous practically everywhere (Figure 6).

Thus an analysis of the sedimentary cover in a large number of present deep sedimentary basins shows that significant stretching occurred in no more than a small per cent of their area at most.

Intensely Deformed Basins

Let us now consider how to recognize stretching in intensely folded basins. A typical initial depth of such basins is ~2 km. It follows from (1) above, that in order to form a basin of this depth by stretching, the crust should be stretched by about two times ($\beta = \sim 2$). In this case deep-water rocks should overlie older rocks over at least one half of the basin.

Consider a case in which stretching produces tilted blocks in the uppermost crust (Figure 7a). Then deeper-water strata rest on considerably older strata at the boundary BC. On the top of the block AB deep-water strata overlie shallow-water strata formed just before the major subsidence. At this boundary, block tilting produces a large angular unconformity between the deep-water and shallow-water strata. The angle of tilting, θ, is related to the stretching factor β as [Le Pichon and Sibuet, 1981]:

$$\beta \approx \sin\phi / \sin(\rho - \theta) \tag{2}$$

Where ϕ is the initial dip angle of normal faults. Taking $\phi = \sim 50°$ we find that the angular unconformity between deep-water and shallow water strata should be $\theta = \sim 25\text{-}30°$ after stretching by $\beta = \sim 2$.

Block tilting by $\theta = \sim 25\text{-}30°$ should produce a saw-toothed relief of the boundary between shallow-water and deep-water strata, which will be preserved until folding occurs. This is observed very rarely, however. There are hundreds of long sections in foldbelts that include the transition from deep-water to shallow-water strata. Almost all of these sections demonstrate parallelism of shallow- and deep-water strata (Figures 3-5). Large angular unconformities resulting from stretching can be seen in no more than a few per cent of the sections.

A transitional layer from shallow-water to deep-water strata is usually from a few metres to a few tens of metres in thickness. It can be seen easily in small exposures ~100 m in size. Consider an exposure (Figure 7b) near the top of the tilted block in Figure 7a. The block is wider and it cannot be seen in the exposure. The angular unconformity resulting from stretching can be seen quite distinctly, however. This unconformity will be preserved even after rotation of the block in the process of folding.

Exposures with a large angular unconformity of the type shown in Figure 7b are very rare in fold belts. They consititute no more than a few per cent of the cases. We take

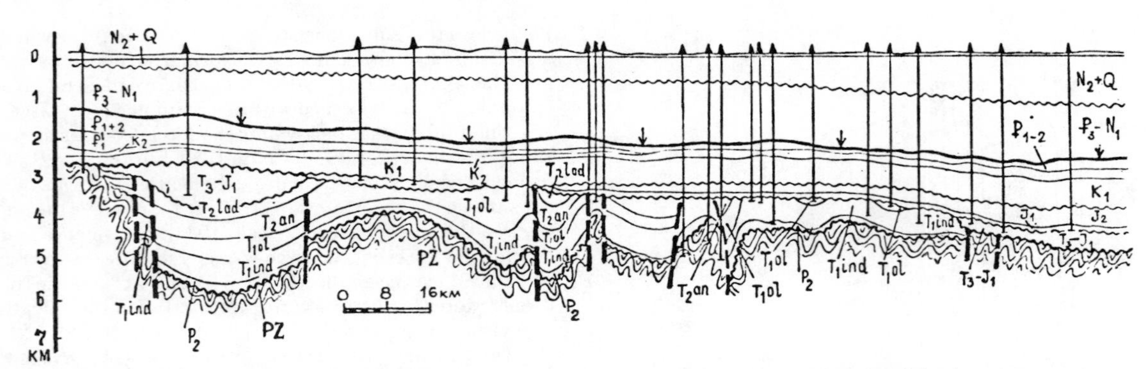

Fig. 5. Geological section across the foredeep of the Caucasus [Khain, 1977]. The transition between strata formed before and after subsidence in the Early Oligocene is indicated by arrows.

this to mean that crustal stretching with block tilting is a very rare phenomenon in such fold belts.

In some stretched regions broad blocks (≥ 10 km) are observed in the upper crust [Wernicke and Burchfiel, 1982]. Broad blocks slide without rotation along the detachment surface (Figure 8), and thus deep-water and shallow-water strata are parallel in such blocks.

Intense stretching should produce broad gaps between non-rotating blocks. It can be assumed that deep-water rocks may be deposited on the sialic basement in these gaps. Transitions of this type are unknown in fold belts, and thus this situation is a purely imaginary one.

Suppose that the gaps between broad blocks were filled with narrow, tilted blocks. In a deep basin, such blocks should cover more than one half of the basin. No angular unconformities will be observed on broad non-rotational blocks, whereas large angular unconformities should be observed on narrow, tilted blocks. Hence large angular unconformities should be observed in ≥ 50 per cent of the expsoures. As noted above, this is not the case for fold belts, as such features are unknown.

Stretching with the formation of broad blocks takes place now in the Basin and Range Province of the western United States [Wernicke and Burchfiel, 1982]. The Basin and Range Province, however, is not a deep basin. It is two kilometres above sea level. Intensely stretched rift valleys there are usually located on high plateaus. Very few of them are deep-water basins. It seems that narrow rifts and broad and deep basins on continental crust are structures of quite different types.

Consider some examples of intensely folded basins. Cratonic conditions were typical of the African Platform in the Paleozoic and Early Mesozoic. This area represented a shallow-water shelf in the Early Triassic [Aubouin et al., 1970; Argyriadis et al., 1980]. Rapid subsidence occurred in the Middle Triassic. This formed the large Dinarian-Taurian Basin [Artyushkov and Baer, 1984a] that spread over 3000 km from the Eastern Alps to Iran and had a width from several hundreds to 1000 km. In a few million years the basin floor subsided to a depths ≥ 1.5 km. During subsequent orogeny the sedimentary cover of the basin was emplaced on the adjacent regions in the form of large nappes. A typical section is shown in Figure 9. Parallelism of shallow-water strata can be seen easily. A system of closely spaced normal faults that should be typical of the stretched regions is absent here. The continuity of shallow-water strata is preserved for a distance up to 5-7 km. Thus there are no traces of significant stretching in this section.

Proponents of stretching can argue that the deep-water strata were deposited here on a wide, non-rotational block, and this is why they are parallel to the shallow-water strata. If this was correct, tilted blocks with large angular unconformities would be observed in many other places. There are, however, almost no sections with large angular unconformities in the Dinarian-Taurian Basin. The strata are conformable practically everywere.

Some normal faults developed in the Dinarian-Taurian Basin during subsidence [Argyriadis et al., 1980]. The distance between the adjacent faults was from several tens of kilometres to one hundred kilometres. An absolute extension associated with these normal faults is:

$$\Delta L \sim 2 h_0 \, cotan \, \phi$$

where h_0 is the depth of the basin and ϕ the average dip angle of normal faults. Taking h_0 to be ~ 1.5 km and ϕ to be $\sim 50°$ we obtain $\Delta L = \sim 2.5$ km. For the average width of the basin L~300-400 km this gives a relative extension of $\Delta L/L \leq 1\%$. On the other hand for the formation of a basin, ~1.5 km deep, by stretching the stretching factor should be $\beta = \sim 1.7$. Thus although slight extension was associated with the subsidence, it was quite insufficient to explain the subsidence by stretching.

The same situation is typical of many other basins. Subsidence was commonly associated with the formation of some normal faults which indicate crustal extension. It is usually supposed on this basis that the basins were produced by stretching. Normal faults, however, ensure an intense stretching only when they are closely spaced and the blocks bounded by the faults are strongly tilted (by ≥ 10-30°). This is observed, e.g., in the Basin and Range Province, the Bay of Biscay, eastern China, and some other regions. In most basins with tilted blocks the angle of block tilting is small ($\theta \leq 5\%$) and ensures relative extension of $\beta - 1 \leq 5$-10% [Artyushkov, 1987a]. This is considerably smaller than an extension $\beta - 1 \geq 30$-100%, which is necessary to form a deep basin by stretching. Thus thinning of the crust and the subsidence should be attributed mostly to other processes.

In the Tibetan Himalayas rapid subsidence occurred on a shallow-water shelf in Late Jurassic time [Gansser, 1964;

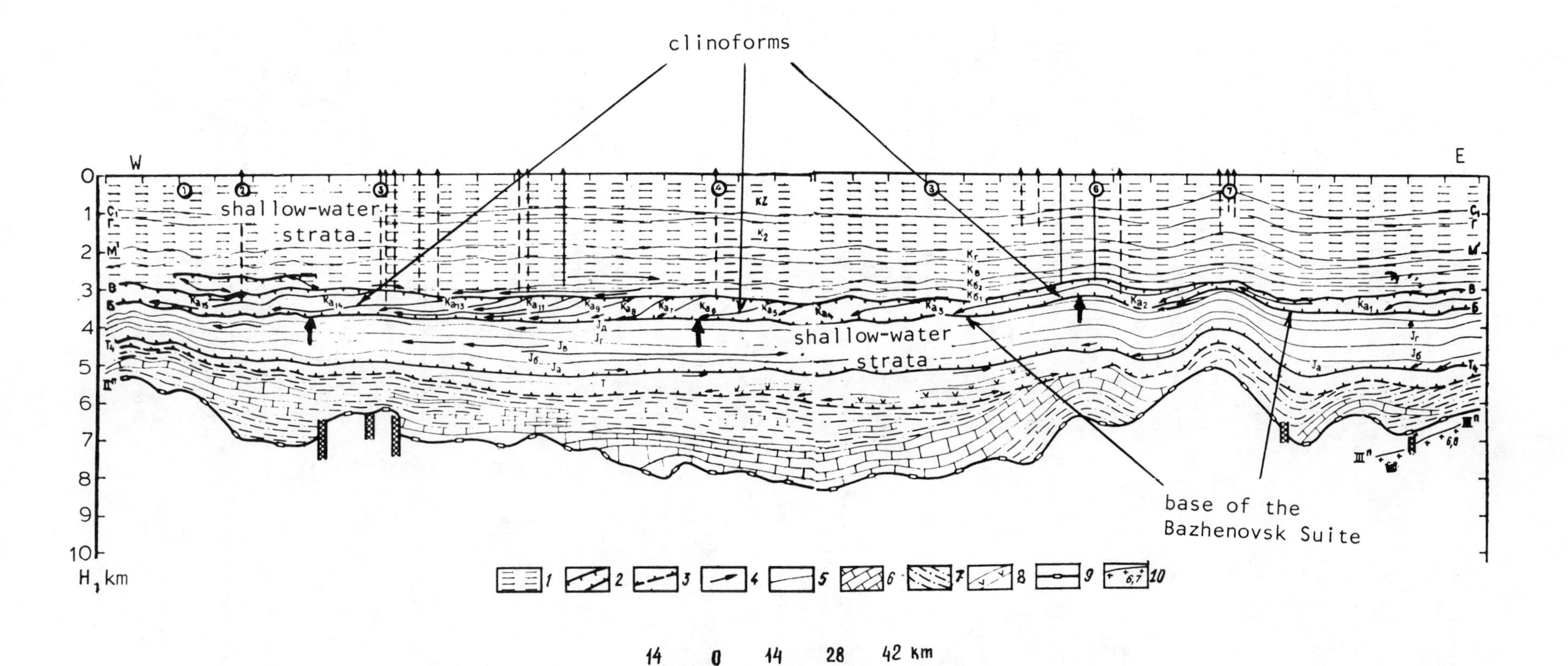

Fig. 6. Seismogeological profile across the northeastern part of the West Siberian Basin [after Kunin, 1983]. Transition from shallow-water to deeper-water strata is indicated by arrows. 1 - shallow-water strata of the upper Lower Cretaceous-Cenozoic, 2 - top and bottom of the upper Upper Jurassic - lower Lower Cretaceous deeper-water and deltaic deposits, 3 - base of the Triassic, 4 - seismically transparent layers, 5 - strong reflectors in the Triassic - Upper Jurassic shallow-water strata, 6-7 - Paleozoic carbonates (6) and terrigenous rocks (7), 8 - Triassic volcano-sedimentary rocks, 9 - top of the Upper Precambrian, 10 - top of the crystalline basement and boundary velocities.

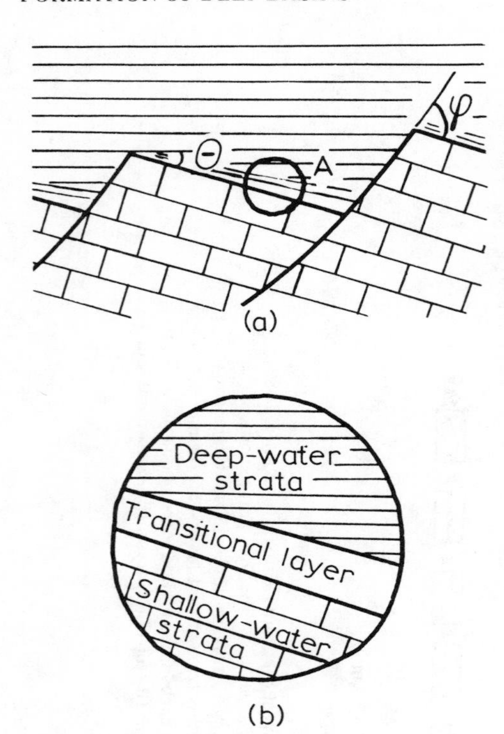

Fig. 7. The structures formed by intense stretching associated with blocktilting. a - Tilted blocks bounded by normal faults. b - An exposure near the upper boundary of a tilted block that includes the strata deposited both before and after stretching and subsidence.

Bassoulet et al., 1980; Srikantia, 1981]. This formed a deep-water basin, 2000 km long and 300 km wide. The duration of the subsidence was ~1 Ma. The transition from shallow-water to deep-water strata is exposed at the surface in many places. In the section of Figure 10 parallelism and continuity of shallow-water and deep-water strata can be seen in a block ~10 km wide. Similar sections are typical of the other parts of the region. This indicates that the basin was formed without significant stretching.

In the Verkhoyansk Fold belt the first rapid subsidence occurred in Early Carboniferous time [Geology of the USSR, 1970; Korostelev, 1982]. This formed a basin up to several kilometres deep. In numerous sections shallow-water and deep-water strata are conformable over most of the area (Figure 11). This means that the subsidence took place without significant stretching.

Rapid subsidence without significant stretching apparently occurred in all the miogeosynclinal zones of the Phanerozoic fold belts that covered tens of millions of square kilometres before compression.

Mechanism of Rapid Subsidence

The mechanisms that are commonly used for subsidence without significant stretching at the surface are thrust loading [Beaumont, 1981, and others], thermal relaxation [Sleep, 1971] and subcrustal erosion [Gilluly, 1963; Stegena et al., 1975]. The main parts of the basins considered, e.g. the Dinarian-Taurian, West Siberian and Verkhoyansk basins, were formed before there was nappe emplacement in the adjacent regions. A load of thick nappes must produce a basin with a depth that increases towards the edge of the nappes (Figure 12a). The depth of real basins, however, commonly decreased towards the edge of nappes (Figure 12b). This resulted in large block olistoliths derived from the nappes sliding down the slope into the basins. Basins of this profile cannot be produced by thrust loading. Among several tens of large basins considered, we found none that could have been formed by this mechanism.

Subsidence from thermal relaxation requires a previous heating of the crust and mantle. The subsidence considered here usually occurred in cratonic regions that were already cold when subsidence occurred. Furthermore, the characteristic time of thermal relaxation is much longer: ~50 Ma. Hence this mechanism cannot be used for subsidence of a duration of a few million years.

The present basins that were recently formed by rapid subsidence without significant stretching, e.g, the Pannonian Basin and the Aegean Sea, are underlain by attenuated continental crust ($h_0 = \sim 20\text{-}25$ km). Many past basins formed by rapid subsidence were intensely shortened at times of orogeny. For example, the magnitude of shortening of 2-3 is typical of the Caucasus and Alps. In most recently folded regions the crustal thickness is only ~40-55 km [Belyaevsky, 1974, 1981; Sologub et al., 1978]. Hence the crust in the basins was several times thinner, its thickness being $h_0 = \sim 10\text{-}20$ km before compression. Most basins were formed in cratonic areas where the crustal thickness is $h_0 \sim 35\text{-}45$ km. Thus rapid subsidence is associated with a strong crustal thinning.

Crustal thinning without significant stretching at the surface implies destruction of the lower crust. This can be produced by an intense stretching of this layer, or by its erosion by convective flows in the mantle. Concomitantly with the subsidence these processes should thicken the crust and produce an uplift in the regions surrounding the basin. No crustal uplift took place, however, around basins formed by rapid subsidence. For instance, the Dinarian-Taurian basin in the Triassic and the West Siberian Basin in the Jurassic were surrounded by a shallow-water shelf where no significant change in the depth occurred during subsidence.

Furthermore, in order to ensure subcrustal erosion or stretching of only the lower crust during ~1 Ma, the viscosity of this layer should be very low, i.e., its temperature should be near the melting point. Rapid subsidence was commonly

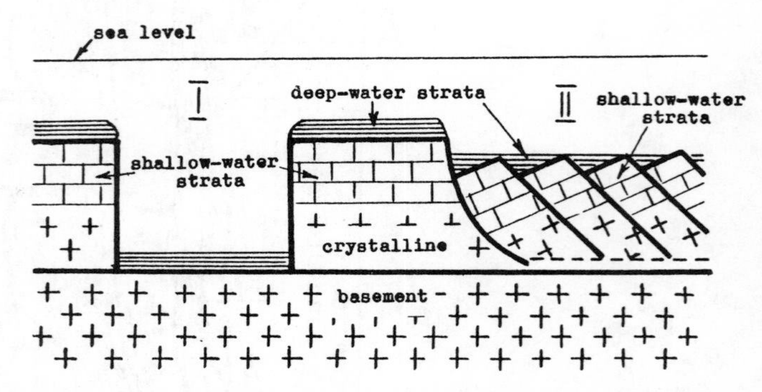

Fig. 8. The structures produced by stretching with the formation of broad non-rotational blocks.

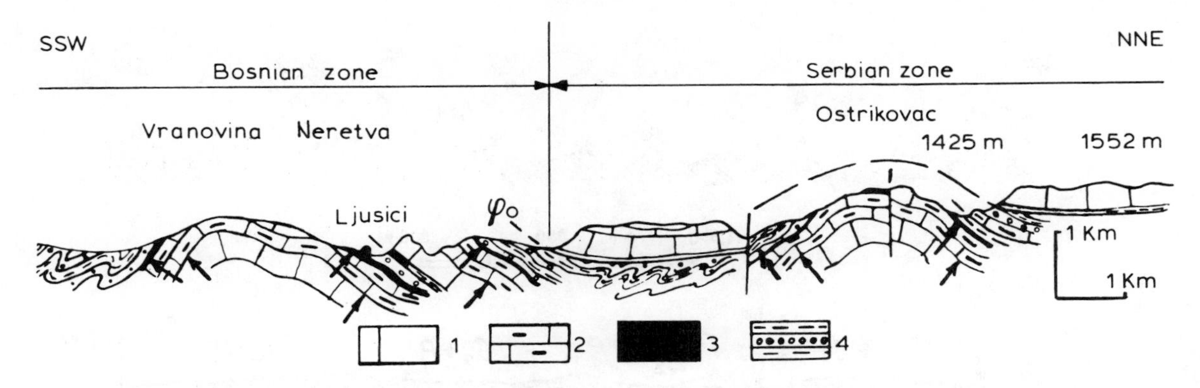

Fig. 9. Fragment of section across the inner zone of the Dinarides [Cadet, 1970]. 1 - Middle Triassic shallow-water limestones. 2-4 Ladinian-Cenomanian deep-water rocks (cherty limestones, radiolarites and flysch). Here and in the two subsequent figures the transition from shallow-water to deep-water rocks is indicated by arrows.

associated with the formation of some normal faults which indicate tensile stresses in the lithosphere. Under tensile stresses an intense heating of the lower crust should result in appreciable volcanic activity. Moreover, heating of the crust and mantle to high temperatures must produce significant uplift at the surface before the destruction of the lower crust. Large-scale crustal uplift and intense volcanism are common for high plateaus, e.g., for the Basin and Range Province. However, even in this region the intense stretching that is observed directly at the surface did not produce a deep-water basin.

Rapid subsidence that produced deep basins without significant stretching occurred under quite different conditions. This commonly took place in cool areas and was preceded by only slight crustal uplift of several hundreds of metres. The uplift and the subsequent subsidence were associated only with slight volcanism or no volcanism at all. These phenomena may indicate upwelling of the asthenosphere at moderate temperatures.

In our previous papers [Artyushkov and Baer, 1983, 1984a, b, 1986a, b] we suggested that rapid subsidence of continental crust without significant stretching occurs under conditions of upwelling to the base of the crust of a hydrous asthenosphere of moderate temperature T = ~800°C. An increase in temperature up to this value and water inflow strongly increase the rate of gabbro-eclogite phase transformation. In regions where the lower crust consists mostly of gabbro this may result in a rapid transformation of gabbro into eclogite. Dense eclogite ($\rho \sim 3.5$-3.6 g/cm^3) sinks into the underlying asthenosphere, which results in destruction of all or a considerable part of the basaltic layer. The attenuated crust subsides with the formation of a basin with a water depth (h_0):

$$h_0 = \frac{\rho_m - \rho_B}{\rho_m - \rho_0} h_B - \frac{\rho_m - \rho_{am}}{\rho_m - \rho_0} h_{am} \qquad (4)$$

Here ρ_m is the density of the mantle lithosphere, ρ_0 - the density of water, ρ_B, h_B - the density and thickness of the destroyed basaltic layer, ρ_{am}, h_{am} - the density and thickness of the asthenospheric anomalous mantle that upwelled to the crust. Taking $\rho_m = 3.35$ g/cm^3, $\rho_B = 2.9$ g/cm^3, $h_B = 20$ km, $\rho_{am} = 3.25$ g/cm^3, $h_{am} = 20$ km we find $h_0 = 3$ km. The depth of the sedimentary basin formed after cooling of the crust and mantle and filling of the initial depression by the sediments is

$$h_s = \frac{\rho_m - \rho_B}{\rho_m - \rho_s} h_B \qquad (5)$$

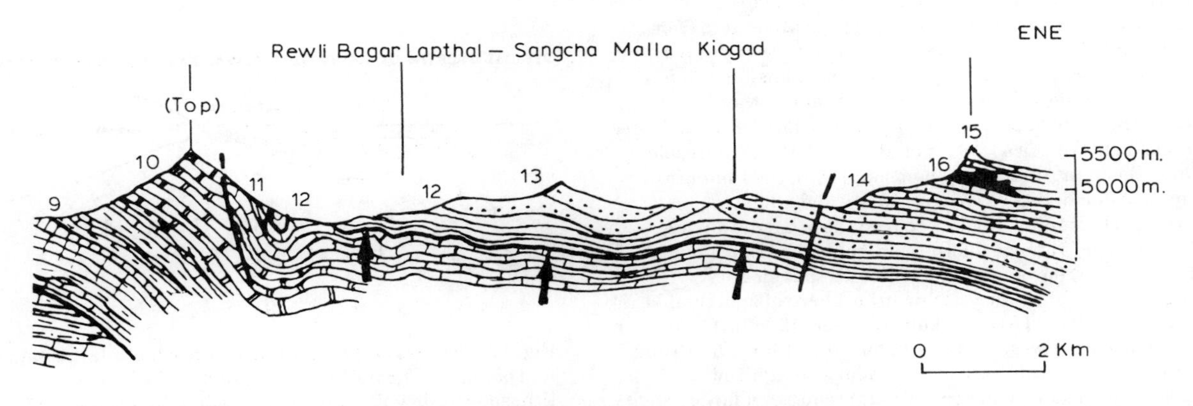

Fig. 10. Fragment of section of the "Tethyan" zone of the Western Kumaun Himalaya [Shah and Sinha, 1974].

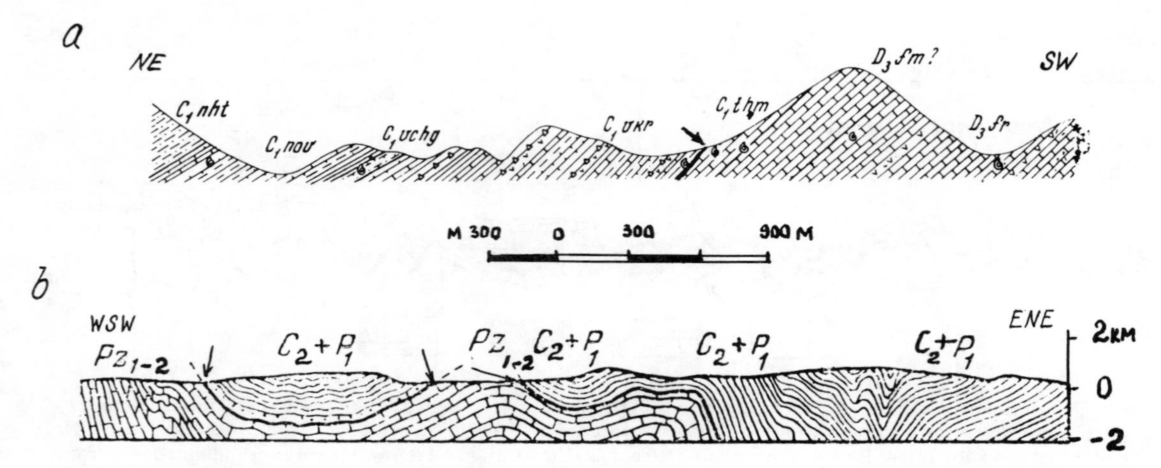

Fig. 11. Fragments of sections across the southern part of the Verkhoyansk Basin [Abramov, 1970; Mokshantsev et al., 1964].

where ρ_s is the sediment density. Taking $\rho_s = 2.50$ g/cm^3 we obtain $h_s = 11$ km.

The mechanism of eclogitization is very speculative and meets serious objections from the petrological point of view [Artyushkov and Baer, 1968a]. Hence we use this mechanism only as a working hypothesis. The other known mechanisms, however, meet more objections when applied to rapid subsidence without significant stretching in cool regions [Artyushkov and Baer, 1984a].

Cratonic Basins

Analysis of the structure of fold belts and of many present sedimentary basins reveals the wide occurrence of another class of deep basins on continental crust. There were and are many basins in which sediment-loaded subsidence took place at a low rate of ~10-100 m per million years for a very long period of time, ~500-1000 Ma. They are filled with shallow-water deposits up to 10-15 km thickness.

In basins formed by slow subsidence the strata at the top and base of the sedimentary sequence cover almost the same area, which indicates an absence of significant stretching. A typical geological section is shown in Figure 13. This is a part of the profile across the Turukhansk Basin in Siberia. Shallow-water strata, 5 km thick, were deposited there from Late Precambrian until Permian time. They are disrupted only by some normal faults that ensure an extension by a few per cent. The same structure is typical of many other basins, including the eastern part of the Proterozoic Belt - Purcell Basin in North America [Porter et al., 1982], the Proterozoic Adelaide Basin in Australia (Thomson, 1969), the Phanerozic Vilyuy Syncline in Sibera [Khain, 1979], and others.

Crustal shortening did not occur in these basins during slow subsidence. Intense folding commonly took place in adjacent regions, however. This indicates that slow subsidence takes place on cratonic lithosphere of appreciable thickness $d \geq 100$ km [Artyushkov and Baer, 1986b]. Deep basins formed by slow subsidence of long duration, ~500-1000 Ma, can be called "cratonic basins". A large magnitude of sediment-loaded subsidence (~5-15 km) requires a large density increase in the lithospheric layer. Thermal relaxation is unable to ensure such a density increase in cool cratonic areas, and has a comparatively small characteristic time (~50 Ma).

It has been suggested that slow subsidence of a large magnitude develops from a slow transformation of gabbro into eclogite or dense garnet granulite in the lower continental crust within a thick cratonic lithosphere [Yanshin et al., 1977; Artyushkov, 1983; Artyushkov and Baer, 1986b]. This is also only a hypothesis, however, as other known mechanisms are incapable of accounting for slow subsidence of large magnitude in cool cratonic areas.

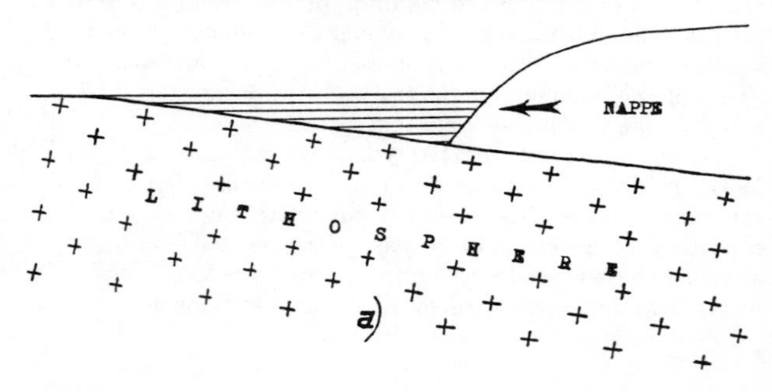

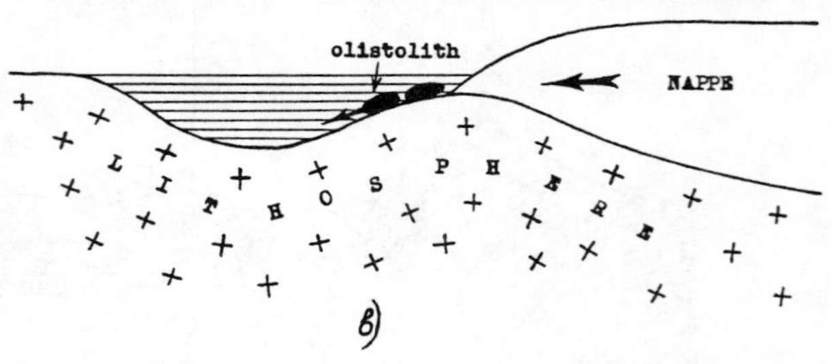

Fig. 12. Profiles of the basins near the edge of large nappes in fold belts. a - Theoretical profile of a basin formed by lithospheric downflexing under loading by a thick nappe. b - Typical profile of a deep basin near the edge of nappes.

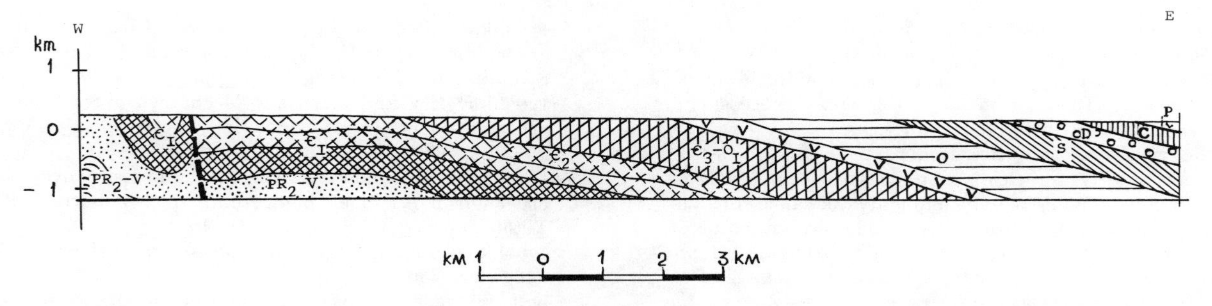

Fig. 13. The eastern part of the Turukhansk Basin on the western margin of the Siberian Platform [Votakh, 1976].

An absence of significant stretching in most deep basins on continental crust follows directly from the absence of the deformations typical of this process. This result is based on study of a very large volume of data and seems to be quite reliable. The occurrence of very rapid or very slow subsidence in most deep basins follows from study of numerous stratigraphic columns providing a high level of reliability. In contrast, the processes in the lower crust that produce the subsidence without significant stretching at the surface are much more difficult to test. Hence physical mechanisms accounting for the subsidence still represent a problem.

It should be noted that we do not reject stretching as a mechanism of formation for some basins (especially taking into account that it has originally been suggested by one of the authors [Artemjev and Artyushkov, 1971]). This phenomenon definitely occurred in some regions, and spectacular examples like that of the Bay of Biscay provide clear examples. Our analysis based on a large volume of data, however, shows that deep basins produced by stretching are not common phenomena. There are many more regions where subsidence deformed the sedimentary cover only slightly. We did not expect to discover this when we began our analysis.

Rapid Subsidence and Folding

Most scientists believe after Hall [1859] that any basin on continental crust can be intensely folded, if it is filled with a thick accumulation of sediments. Folded basins on continental crust are usually called "miogeosynclines".

An analysis of the development of fold belts shows that not all basins on continental crust can be intensely shortened. No cratonic basin formed by slow subsidence of large magnitude has been intensely folded in the fold belts studied. Intense shortening of continental crust occurred only in deep basins formed by rapid subsidence. Most of them were compressed soon after formation. Hence only basins of this type can be called "miogeosynclines".

Intense compression of miogeosynclines can be explained as follows [Artyushkov et al., 1982; Artyushkov and Baer, 1983; Artyushkov, 1987b]. After destruction of the basaltic layer the crustal thickness decreases to $h_c = \sim 10\text{-}20$ km (Figure 14). Larger additional stress Σ arises in the lithosphere due to lateral inhomogeneities in crustal thickness, inhomogeneties which produce large inhomogeneities in the potential energy of this layer [Artyushkov, 1973, 1983]. In order to remove this excess of energy, the crust tends to spread from elevated regions towards depressions. The additional stress in the lithosphere is

$$\Sigma = \frac{\rho_c - \rho_0(\rho_m - \rho_c) g h_c^2}{2(\rho_m - \rho_0)} + \Sigma_0 \qquad (6)$$

where ρ_c is the density of the crust and Σ_0 is a constant. In order to compress the crust under convergent plate motions it is necessary to apply to outer stress $\Sigma_{out} \geq \Sigma$. According to (6), Σ increases proportionally to the square of the crustal thickness h_c^2. This is why crustal thinning under a deep basin strongly decreases the force Σ_{out} necessary for compression.

After destruction of the lower crust the anomalous mantle, which is a part of the asthenosphere, upwells to the attenuated crust (Figure 14a). As a result the lithospheric thickness decreases to $d = \sim 10\text{-}20$ km. This thin lithosphere can be easily compressed and shortened under convergent plate motions

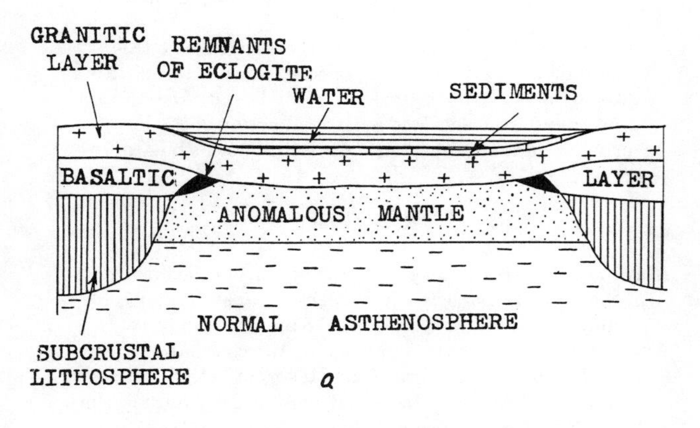

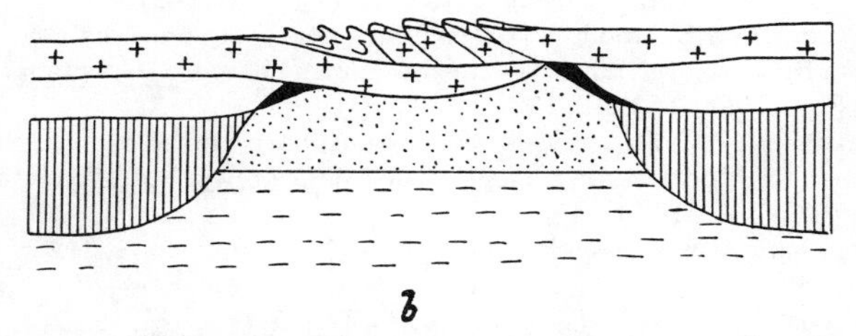

Fig. 14. Crustal shortening in a deep basin formed by rapid subsidence of continental crust. a- Thinning of the crust and lithosphere after destruction of the lower crust from upwelling of hydrous anomalous mantle. b - Compression of the attenuated lithosphere under convergent plate motions.

(Figure 14b). In contrast, the thick lithosphere of cratonic areas apparently cannot be shortened at all.

Thus rapid subsidence due to destruction of the lower crust is associated with a decrease in both the stress Σ necessary for compression and the lithospheric thickness d.

The crustal thickness in cratonic areas is $h_c = \sim 40$ km. There are, however, pratically no nappes of continental crust of this thickness in fold belts. Commonly the nappes include only granitic and sedimentary layers and represent plates of the thickness ≤15-25 km [Fountain and Salisbury, 1981, and others]. In some cases such nappes include a thin layer of mafic and ultramafic rocks at the base.

The plates probably represent continental crust in which the basaltic layer has been destroyed by eclogitization. Stacking of such plates by thrust folding in the process of crustal shortening can produce a thick sialic crust in some orogenic regions.

Conclusions

An analysis has been undertaken of the structure of a large number of deep basins on continental crust in fold belts and cratonic areas. The main results of this analysis are the following.

1. Most deep basins on continental crust were formed without significant stretching.
2. These basins are of two main types. The first type is represented by cratonic basins formed by slow subsidence during hundreds of million years or more. The second basin type is represented by miogeosynclines formed by rapid subsidence during a few million years.
3. Intense shortening of continental crust occurs only in miogeosynclines.
4. Slow subsidence of large magnitude occurs on a thick cratonic lithosphere. This is probably associated with a strong increase in the density of the lower crust. Rapid subsidence is associated with destruction of the lower crust under asthenospheric upwelling to the base of the crust.

References

Abramov, B.S., Biostratigraphy of the Carboniferous Deposits of the Sette-Daban (Southern Verkhoyansk Region), Nauka, Moscow, (in Russian), 178 pp, 1970.

Argyriadis, J., P.C. DeGraciansky, J. Marcoux, and L.-E. Ricou, The opening of the Mesozoic Tethys between Eurasia and Arabia-Africa, in Geologie des Chaines alpines issue de la Tethys, Mem. B.R.G.M., 115, 199-214, 1980.

Artemjev, M.E., and E.V. Artyushkov, Structure and isostasy of the Baikal Rift and the mechanism of rifting, J. Geophys. Res., 76, 1197-1212, 1971.

Artyushkov, E.V., Stresses in the lithosphere caused by crustal thickness inhomogeneities, J. Geophys. Res., 78, 7675-7708, 1973.

Artyushkov, E.V., Geodynamics, Elsevier, Amsterdam, 312 p, 1983.

Artyushkov, E.V., Rifts and grabens, Tectonophysics, 1986a.

Artyushkov, E.V., The forces driving plate motions and compression of the crust in fold belts. Mid-Term Report of the Interunion Commission on the Lithosphere, XXVII International Geological Congress, Symposium L06. Am. Geophys. Union, Washington, D.C., 1986b.

Artyushkov, E.V., and M.A. Baer, Mechanism of continental crust subsidence in fold belts,: the Urals, Appalachians and Scandinavian Caledonides, Tectonophysics, 100, 5-42, 1983.

Artyushkov, E.V. and M.A. Baer, Mechanism of continental crust subsidence in the Alpine Belt, Tectonophysics, 108, 193-228, 1984a.

Artyushkov, E.V., and M.A. Baer, Mechanism of continental crust subsidence in fold belts around the Northern Pacific, Tikohookeansk. Geol., 5, 3-16, 1984b.

Artyushkov, E.V., and M.A. Baer, Mechanisms of formation of deep basins on continental crust in the Verkhoyansk Fold belt: miogeosynclines and cratonic basins, Tectonophysics, 122, 217-245, 1986a.

Artyushkov, E.V., and M.A. Baer, Mechanism of formation of hydrocarbon basins: the West Siberia, Volga-Urals, Timan-Pechora basins and the Permian Basin of Texas, Tectonophysics, 122, 247-281, 1986b.

Artyushkov, E.V., and M.A. Baer, Mechanism of formation of hydrocarbon basins of the Middle East, Tectonophysics, in prep.

Artyushkov, E.V., M.A. Baer, S.V. Sobolev, and A.L. Yanshin, Mechanism of formation of fold belts, Sovetskaya, Geologia, 9, 22-36 (in Russian), 1982.

Aubouin, J., R. Blanchet, J.-P. Cadet, P. Celet, J. Charvet, J. Chorowicz, M. Cousin, and J.-P. Ramnoux, Essai sur la geologie des Dinarides, Bull. Soc. Geol. Fr., 7e Ser., XII(6), 1060-1093, 1970.

Bassoulet, J.P., J. Bulin, M. Colchen, J. Marcoux, G. Mascle, and Ch. Montenat, L'evolution des domains tethysiens au pourtour du Bouclier indien du Carbonifere au Cretace, in Geologie des Chaines alpines issue de la Tethys, mem. B.R.G.M., 115, 180-198, 1980.

Beaumont, C., Foreland Basins, Geophys. J. R. Astron. Soc., 65, 291-329, 1981.

Belyaevsky, N.A., The Earth crust of the territory of the USSR, Nedra, Moscow (in Russian), 280 p., 1974.

Belyaevsky, N.A., The structure of the earth's crust of the continents according to geological and geophysical data, Nedra, Moscow (in Russian), 432 p, 1981.

Cadet, J.P., Esquisse geologique de la Bosnie-Herzegovine Meridionale et du Montenegro occidental, Bull. Soc. Geol. Fr., 7e Ser., XII (6), 973-985, 1970.

Cook, T.D., and A.W. Bally (Editors), Stratigraphic Atlas of North and Central America, Princeton University Press, Princeton, 272, p. 1975.

Danilchenko, P.G., Main complexes of ichtiophauna of the Cenozoic Seas of the Tethys, in Fossil Vertebrate Fish of the USSR, Nedra, Moscow (in Russian), 175-183, 1980.

Fountain, D.M., and M.H. Salisbury, Exposed cross-sections through the continental crust: implications for crustal structure, petrology and evolution, Earth Planet. Sci. Lett., 56, 263-277, 1981.

Gansser, A., Geology of the Himalayas, Interscience, New York, 289 p, 1964.

Geology of the USSR, v. IX, North Caucasus, Nedra, Moscow (in Russian), 759 p., 1968.

Geology of the USSR, v. XVIII, Western part of the Takutsk ASSR, Nedra, Moscow (in Russian), I, 535 p., II, 256 p, 1970.

Gilluly, J., The tectonic evolution of the western United States, Q.J. Geol. Soc. London, 119, 133-174, 1963.

Hall, J., Description and Figures of the Organic Remains of the Lower Helderberg Group and the Oriskany Sandstone, New York, Geological Survey, Paleontology, 3, 532 p., 1859.

Hatcher, R.D., Thrusts and nappes in the North America

Appalacian Orogen, in Thrust and Nappe Tectonics, Geol. Soc. London, 491-499, 1981.
Kamaletdinov, M.A., The nappe structures of the Urals, Nauka, Moscow (in Russian), 232 p, 1974.
Karus, E.V., G.A. Gabrielyants, V.M. Kovylyn, and N.M Chernyshev, Depth pattern of West Siberia, Sov. Geol., 5, 75-84 (in Russian), 1984).
Khain, V.E., Regional Geotectonics. Non-Alpine Europe and Western Asia, Nedra, Moscow (in Russian), 259 p, 1977.
Khvorova, I.V., The flysch and lower-molasse formations of the Southern Urals, Geol. Inst. Acad. Sci. USSR Transactions, 37, (in Russian), 352 p, 1961.
King, P.B., The evolution of North America, Princeton, New Jersey, Princeton University Press, 190 p, 1959.
Korostelev, V.I., Geology and Tectonics of the Southern Verkhoyansk Region, Nauka, Novosibirsk (in Russian), 217 p, 1982.
Kunin, N.Ya., New possibilities of seismic stratigraphy investigations for regional works for oil and gas. Sov. Geol., II, 109-120 (in Russian), 1983.
Le Pichon, X, and J.C. Sibuet, Passive margins: a model of formation, J. Geophys. Res., 86, 3708-3720, 1981.
McKenzie, D., Some remarks on the development of sedimentary basins, Earth Planet. Sci. Lett., 40, 25-32, 1978a.
McKenzie, D., Active tectonics of the Alpine-Himalyayan belt: The Aegean Sea and surrounding regions, Geophys. J.R. Astron. Soc., 55, 217-254, 1978b.
Mokshantsev, K.B., D.K.Hornstein, A.S. Gusev, and others. Tectonic structure of the Yakutsk ASSR, Nauka, Moscow (in Russian), 291 p., 1964.
Montadert, L., D.A. Roberts, O.de Charpal, and P. Guennoc, Rifting and Subsidence in the Northern Continental Margin of the Bay of Biscay, Initial Reports of the Deep-Sea Drilling Project, 48, 1025-1060, 1979.
Porter, J.W., R.A. Price, and R.G. McCrossan, The Western Canada Sedimentary Basin, Phil. Trans. R. Soc. London, A305, 169-192, 1982.
Pushcharovsky, Yu.M., Marginal troughs, their tectonic structure and evolution, Geol. Inst. Acad. Sci. USSR Transactions, 28 (in Russian), 154 p, 1959.
Shah, S.K., and A.K.Sinha, Stratigraphy and tectonics of the "Tethyan" zone in a part of Western Kumaun Himalaya, Himalayan Geol., 4, 1-27, 1974.
Shanmugam, G., and K.R. Walker, Sedimentation, subsidence and evolution of a foredeep basin the Middle Ordovician, Southern Appalachians, Amer. J. Sci., 280, 479-496, 1980.
Sleep, N.H., Thermal effects of the formation of Atlantic continental margins by continental break up, Geophys. J.R. Astron. Soc., 24, 325-350, 1971.
Sologub, V.B., A. Guper, and D. Prosen (Editors), The structure of the Earth crust and Upper Mantle of the Central and Eastern Europe, Naukova Dumka, Kiev (in Russian), 272 p., 1978.
Srikantia, S.V., The lithostratigraphy, sedimentation and structures of Proterozoic-Phanerozoic formations of Spiti basin in the Higher Himalaya of Himachal Pradesh, India, in Contemporary Geoscientific Research in Himalaya, Dehra Dun, 31-48, 1981.
Stegena, L., B. Geszy, and F. Horvath, Late Cenozoic evolution of the Pannonian Basin, Tectonophysics, 26, 71-90, 1975.
Surkov, V.S. (Editor), Domanik Rocks of Siberia and their Role in Hydrocarbon Occurrences, Novosibirsk, 134 p. (in Russian).
Thomson, B.P., Precambrian basement cover - the Adelaide system, in L.M. Parkin (Editor), Handbook of South Australian Geology, Geol. Surv. S. Australia, 1969.
Votakh, O.A., Structural elements of the Earth, Nauka,Novosibirsk, (in Russian), 192 p, 1976.
Wernicke, B., and B.C. Burchfiel, Modes of extensional tectonics, Journal of Structural Geology, 4, 105-111, 1982.
Yanshin, A.L., E.V. Artyushkov, and A.E. Schlesinger, Main types of large structures on the lithospheric plates and possible mechanisms of their formation, Dokl. Akad. Nauk, SSSR, 234, 1175-1179 (in Russian), 1977.
Zonn, M.S., Paleogeographical conditions of accumulation of the Bazhenovsk Suite and its analogies in the northern regions of West Siberia, in N.A.Krylov (Editor), Hydrocarbons of the Bezhenovsk Suite of West Siberia, Inst. Geol. Prosp. Combust. Resource (IGIRFI), Moscow, 18-25 (in Russian), 1980.

NEOGENE-QUATERNARY PANNONIAN BASIN: A STRUCTURE OF LABIGENIC TYPE

V. G. Nikolaev

Geological Institute of the USSR Academy of Sciences, USSR

D. Vass

Shtur Geological Institute of Bratislava, Czechoslovakia

D. Pogacsas

Geophysical Exploration Company, Budapest, Hungary

Abstract. The Neogene-Quaternary sedimentary cover of the Pannonian Basin is subdivided into two structural complexes, the lower of Miocene age, and the upper one of Upper Miocene, Pliocene and Quaternary age. The lower complex shows typical narrow, extended structures, with synchronously developed linear zones of andesitic and rhyolitic volcanics. The upper complex shows gently sloping, irregularly shaped structures whose formation was accompanied by widespread basaltic volcanism. The thickest successions of the lower complex are associated with the basin's margins, whereas the thickest succesions of the upper complex are central to the basin. Miocene strata (up to the Sarmatian) were formed under conditions of extension, whereas the upper complex was dominated by vertical negative movements. Structural analysis of the deep-seated part of the earth's crust beneath the basin suggests that basin formation began in Middle Miocene time (main stage - Upper Miocene/Pliocene) due to the emplacement of a mantle diapir. The Pannonian basin, as well as basins of the internal Mediterranean belt seas, falls into a specific new class of structures herein identified as labigenic structures.

Introduction

The Pannonian Basin includes the whole of the Inner Carpathian area underlain by Neogene-Quaternary cover rocks. Many geologists today tend to regard the Pannonian Neogene-Quaternary basin as an inner structure resulting from subduction. This point of view was actively supported by Balla [1982], Horvath [Boccaletti et al., 1976; Horvath, Stegena, 1977] and some others. New data on the structure of the sedimentary cover, Cenozoic volcanics and deep properties of the region, however, now allow assignment of the Pannonian basin to a specific type of structure.

The map of the pre-Neogene Pannonian basement (Figure 1) makes use of current data, as well as considerable reference data [Nikolaev, 1981, 1983; Wein, 1973]. The pre-Neogene basement shows a widely developed Precambrian-Lower Paleozoic complex of schist-gneiss strata. The complex is commonly intruded by Upper Paleozoic bodies of granite-porphyrites and granodiorites. There are ubiquitous narrow linear zones of basic volcanics or, rarely, ultramafics. The zones range in age from Middle Paleozoic and Triassic to Jurassic-Lower Cretaceous. Platform strata (Triassic-Cretaceous, Upper Cretaceous and Paleogene) occur, as a rule, horizontally, although locally they may be part of nappes or deformed into folds. The area of the Pannonian basin became stable by mid-Late Cretaceous time, as verified by the extent of thin Upper Cretaceous sedimentary beds and other evidence [Belov et al., 1976; Knipper, 1975, etc.]. The youngest flysch trough continued until the Oligocene (Debretsen trough). At the same time, Paleogene sediments were being deposited. Thus, the basement that was to accommodate the Cenozoic Pannonian basin was heterogeneous in nature.

The structural map of the Neogene-Quaternary basement (Figure 2) shows, in the central part of the basin, major structures of the Great and Little Hungarian depressions, Zala depression and the Drava graben. Depth to the base of the cover varies from 3 to 7 km. Each of the structures can be divided into a number of smaller units. The structures have an irregular, non-linear shape. Though large-amplitude dislocations are rare here, there is a series of faults of a few metres amplitude. At the base of the Neogene successions, there are small overthrust structures. Precambrian sequences were thrust over Lower Miocene rocks along the surface of the overthrusts. The basin margins accommodate linearly extended structures: the Transcarpathian and Vienna depressions, and Sava graben, which includes several fold structures. The most warped parts are 3 to 5 km deep.

Analysis of a large volume of data suggests the presence of a regional angular unconformity at the base of the Pannonian stage (upper Upper Miocene) (Figure 3). Pannonian stage strata occur subhorizontally almost throughout the basinal

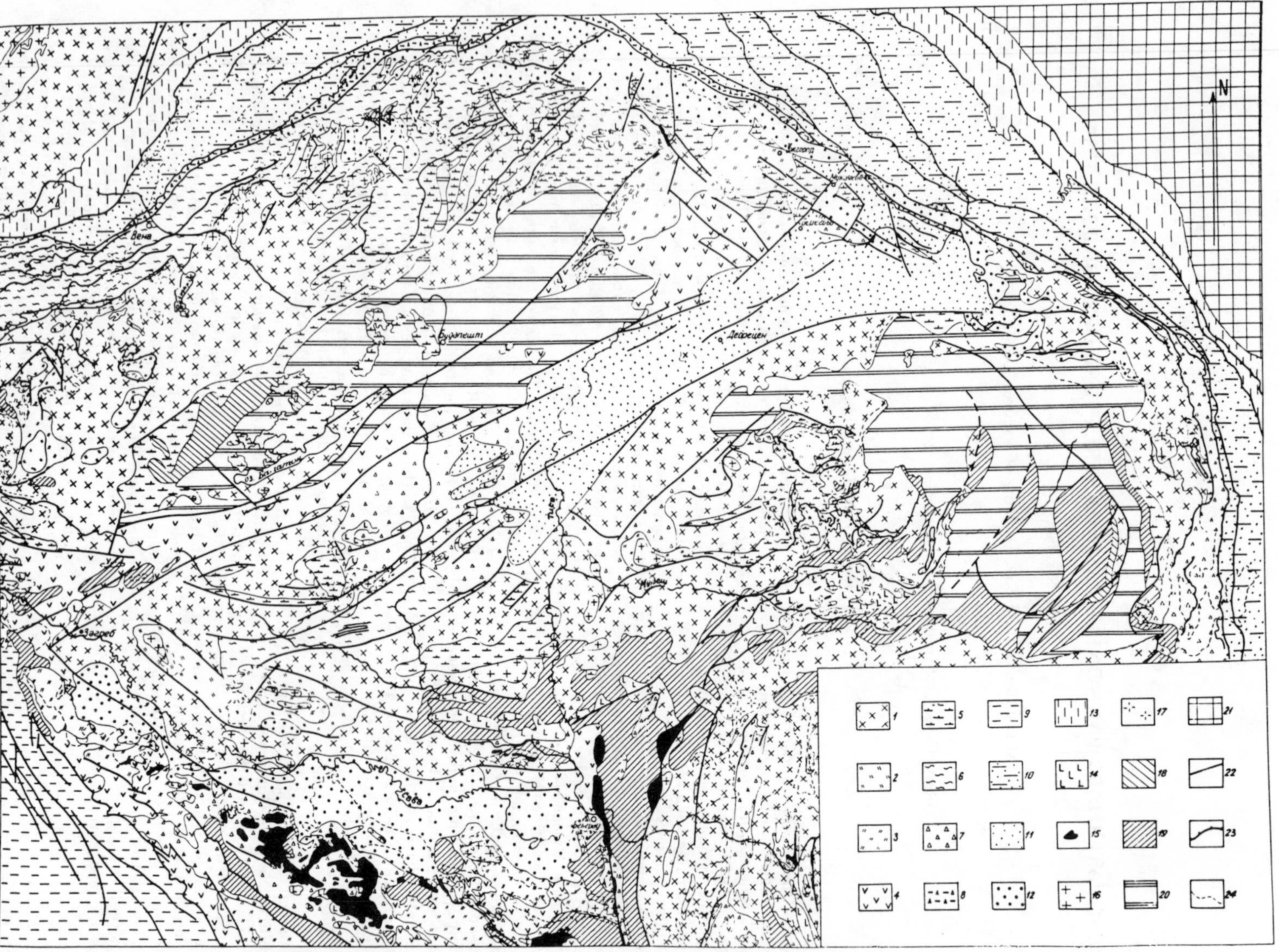

Fig. 1. Map of pre-Neogene structural-formational complexes of the Pannonian basin and adjacent regions. 1-17 - geophysical and orogenic fold complexes: 1- Precambrian-Lower Paleozoic, 2 - Lower-Middle Paleozoic eugeosyncline, 3 - Middle-Upper Paleozoic miogeosyncline and orogenic, 4 - Middle-Upper Triassic eugeosyncline, 5 - Triassic miogeosyncline, 6 - Triassic-Lower Cretaceous miogeosyncline, 7 - Jurassic-Lower Cretaceous eugeosyncline, 8 - Jurassic-Cretaceous eugeosyncline of the Klippe zone, 9 - Jurassic-Cretaceous miogeosyncline, 10-12 - flysch, 10 - Cretaceous-Paleogene, 11 - Upper Cretaceous-Paleogene, 12 - Paleogene, 13 - Oligocene (molasse of the Fore Carpathian trough), 14 - mafics, 15 - ultramafics, 16 - Middle-Upper Paleozoic granites, 17 - Cretaceous-Paleogene granites, 18-20 - complexes of sedimentary cover, 18 - Jurassic-Cretaceous, 20 - Paleogene. Other symbols: 21 - East European platform cover, 22 - major faults (shears, faults,overthrusts, etc.), 23 - major nappes, 24 - Neogene-Quaternary cover limits of the Pannonian basin.

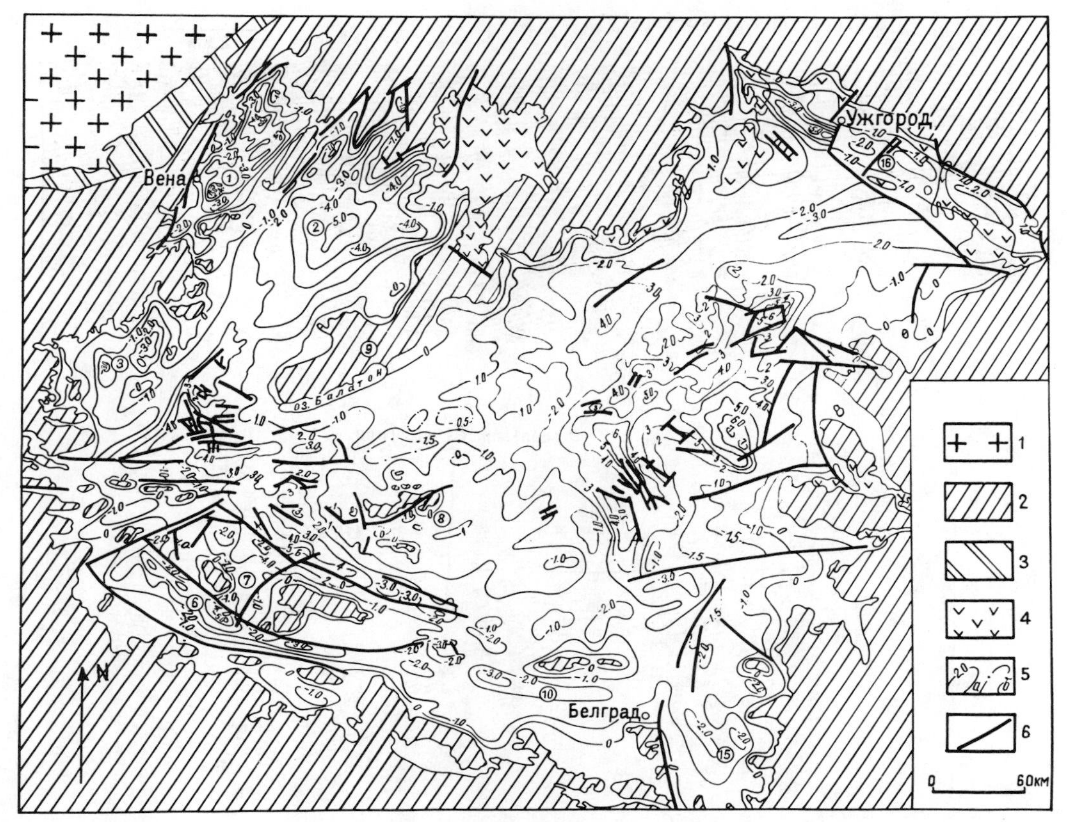

Fig. 2. Structural map of the base of the Neogene-Quarternary cover of the Panonian basin. 1 - Czechoslovak massif, 2 - outcrops of the pre-Neogene basement, 3 - Fore-Carpathian foredeep, 4 - Neogene volcanics outcrops, 5 - isolines of Neogene-Quaternary cover base (a - major, b - additional), 6 main faults. Figures on the map show: 1 - Hungarian trough, 2 - Little Hungarian (Fore-Danube) depression, 3 - Shtirian depression, 4 - Zala depression, 5 - Drava graben, 6 - graben of Sava, 7 - Slovenia-Croatia horst, 8 - Mechek uplift, 9 - Hungarian Central Mountain uplift, 10 - Slovenia-Sremian depression, 11 - Solonok trough, 12 - Keresh trough, 13 - Mako trough, 14 - Battonia uplift, 15 - Banat depression, 16 - Transcarpathian trough.

area, including the margins. The underlying, Sarmatian and older strata of the Neogene cover locally shows gently sloping folds tilted at the flanks to a few degrees. Occasionally, the slope angle may be 10 to 20°, and vary locally up to 60-70° at fault zones. Closely spaced seismic profiles and drilling control have made it possible to trace the unconformity throughout the area of the basin [Nikolaev, 1979; Lukacs et al., 1983]. The degree of angularity varies from a few to tens of degrees, while the unconformity is practically untraceable in the central parts of the most warped structures. This unconformity permits division of the Neogene-Quarternary cover of the Pannonian basin into two structural units or complexes. The lower complex comprises the Lower, Middle and a part of the Upper Miocene succession (throughout the Sarmatian stage). The upper complex is made up of Upper Miocene (Pannonian), Pliocene and Quaternary rocks. Both complexes show smaller unconformities, in particular at the base of the Quaternary. The lower complex shows a variety of rocks (marine siltstones, limestones, sands and sandstones, evaporites, etc.) with abrupt lateral facies transitions, whereas the upper one has a uniform section of lacustrine deposits, consisting mostly of siltstones, claystones and sands.

The Pannonian basin and adjacent regions show widespread Neogene-Quaternary igneous rocks [Konecny et al., 1969; Nikolaev, 1980; Panto, 1969, etc.]. They fall into three groups (Figure 4). The lower volcanic group is made up exclusively of acid varieties including ignimbrites and lavas of liparites and dacites. It is about 1500 m thick. The age of rocks is Eggenburgian-Middle Badian (Lower Miocene). They are developed in a narrow linear zone that has a NE-SW strike across the Pannonian basin.

The middle volcanic group is found over the northern and eastern margins of the basin, and is composed of acid and intermediate volcanics, including liparite-dacites and andesites. These two types of rocks often intercalate. The volcanics are associated with a group of hypabyssal rocks (granodiorite-porphyries, gabbro-diabases). The age of the middle complex is Upper Badian-Pannonian.

The upper volcanics are represented largely by basaltic lavas and their tuffs. The rocks are distributed in small areas

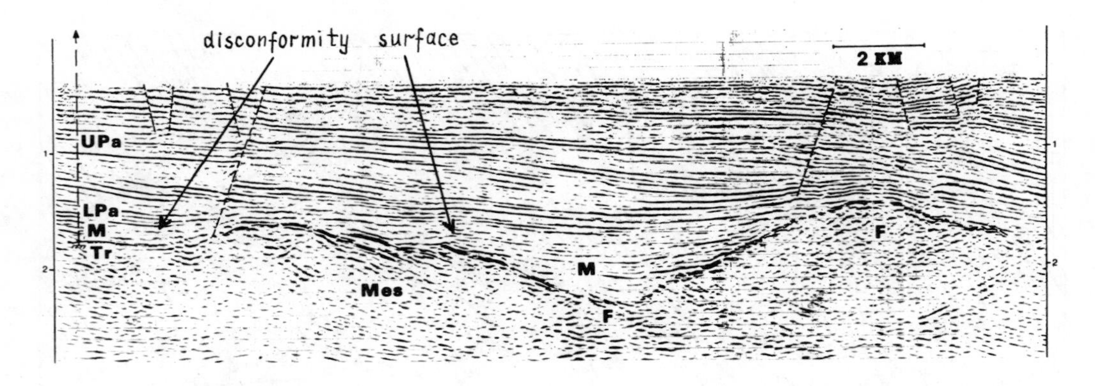

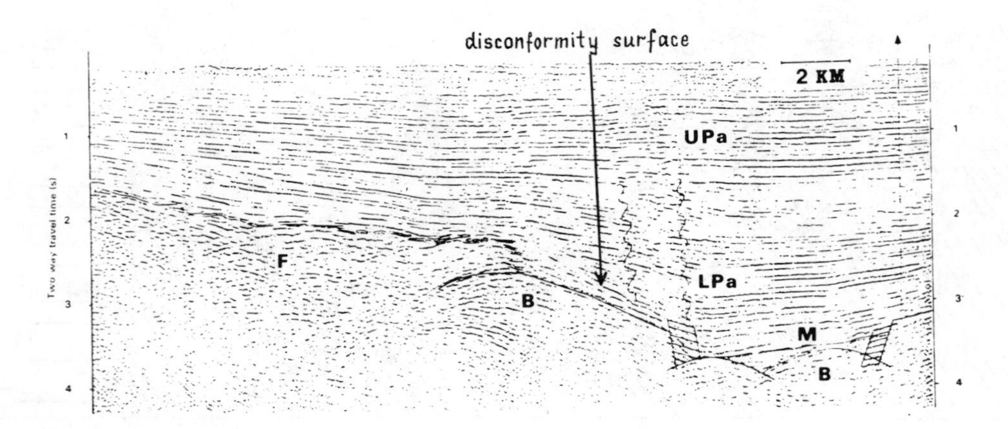

Fig. 3. Seismic profiles through some parts of the Pannonian basin showing the angular unconformities separating the Pannonian and the Mioicene. (a) central part of the Great Hungarian Plain, (b) East Hungary. Symbols: B- crystalline basement, Tr - Triassic rocks, Mes - Mesozoic sequence (in general), F - Flysch Formation, M - Miocene layers, LPa - Lower Pannonian, UPa - Upper Pannonian sediments. Available boreholes are also shown.

throughout the basin. The age of these basic volcanics is Pliocene-Pleistocene (1.5-4.0 m.y.).

Structurally, the lower and upper sedimentary complexes coincide in general although there are changes in the strike - e.g., the Middle Miocene sublatitudinal depression, and the Little Hungarian depression, the Chop anticline and other smaller features. The slope angles of the structures' limbs may vary rather abruptly, from 20-25 degrees in the lower complex to a few degrees in the upper one.

The thickest sequences of the lower structural complex are marginal to the basin (Figure 5). They range from 2000-3000 m in the Vienna, Transcarpathian and other depressions. In the western Pannonian basin the thickness of the lower complex varies from a few hundreds to one thousand metres. In the eastern part it changes from zero to a few hundreds of metres. There are large areas from which the lower complex is missing (almost the whole of the Great Hungarian depression and other sites).

In the lower structural complex, the lateral thickness of most structures changes gradually, so the structures must have developed synsedimentarily. There are some sites, however (e.g., Mako, Békés and other troughs), where Lower-Middle Miocene beds occur horizontally and onlap the surface of pre-Neogene rocks.

The thickness of the upper structural complex is distributed differently. It is at its maximum in the Mako trough, where it reaches almost 5000 m. The Great Hungarian depression strata are generally betwen 2000 and 3000 m thick, with some distinct structural elements. Similar values are observed in the Little Hungarian and Zala depressions. In the Drava graben, the thickness of the upper structural complex ranges from 2500 m (on the flanks) to 4000 m (in the central part). In the marginal zone of the Pannonian basin, the upper complex is much thinner. For example, in the Transcarpathian and Vienna troughs it is only about 1000 m thick.

Variations in thickness in the central part and on the margins are even more distinct at specific stratigraphic levels (Figure 6). Thus, there are 2 to 2.5 km of Lower Pannonian sequences in the Great Hungarian basin, contrasted with only 500 m in the Transcarpathian depression. The Quaternary strata are thick only in two areas: 800 m in the Great

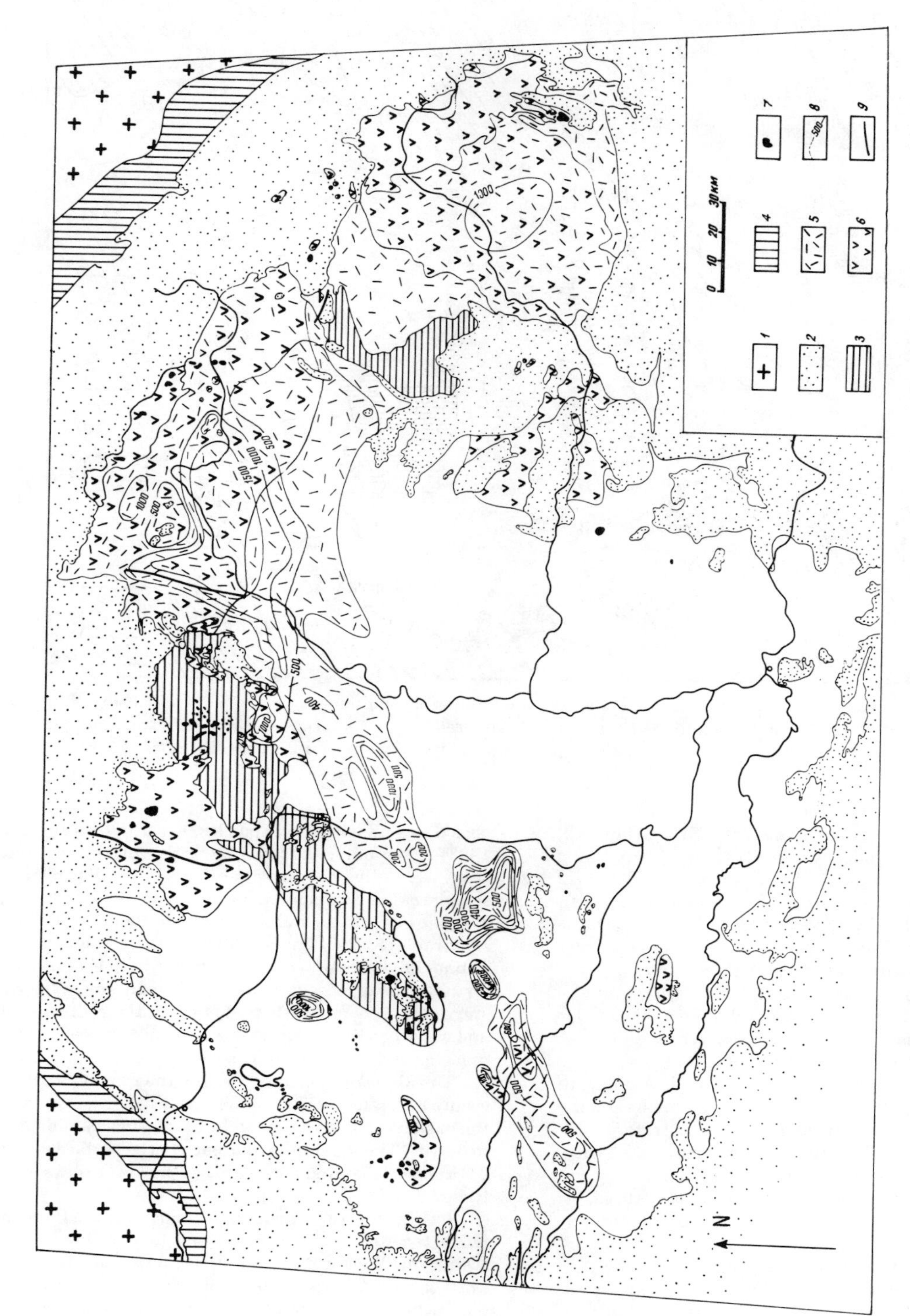

Fig. 4. Distribution of Cenozoic volcanics in the Pannonian basin. 1 - East European platform and Czechoslovak massif, 2 - Paleozoic-Paleogene fold structures, 3 - Paleogene sedimentary cover, 4 - Fore-Carpathian foredeep, 5-7 - Cenozoic volcanics, 5 - lower complex, 6 - middle complex, 7 - upper complex, 8 - isopachytes of valcanic depotists, m, 8 - major dislocations.

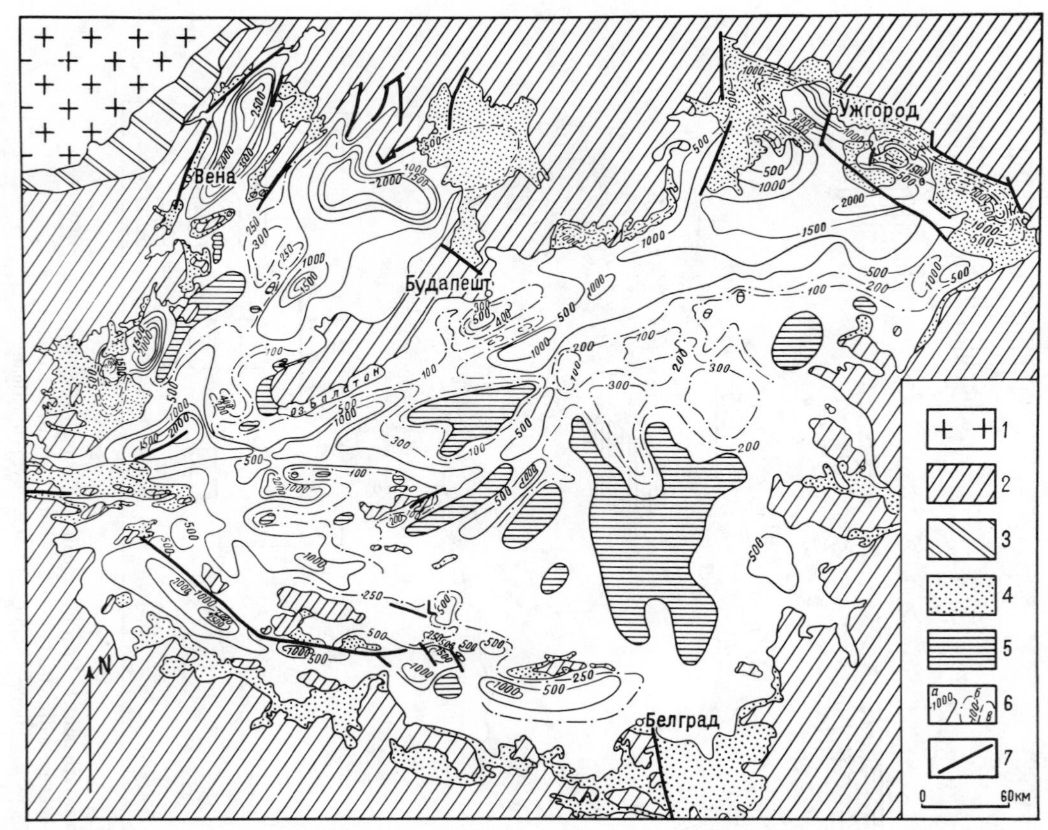

Fig. 5. Map of lower structural complex thickness in the Pannonian sedimentary cover. 1 - Czechoslovak massif, 2 - outcrops of pre-Neogene basement, 3 - Fore-Carpathian foredeeps, 4 - outcrops of lower structural complex, 5 - areas free of lower complex strata, 6 - isopachytes, m (a-major, b-additional, c-inferred), 7 - faults.

Hungarian depression and 300 m in the Little Hungarian depression. The general tendency is gradual thinning of the upper structural complex towards the peripheries.

The Neogene-Quaternary history of the Pannonian basin falls into two stages. The lower, Miocene (pre-Pannonian), stage is characterized by linear structures and diverse sediments. Associated with it are two linear volcanic complexes. Major warping took place at the margins, where the rate of sedimentation was almost 40-60 cm in a thousand years, whereas in the centre it was only a dozen or so centimetres per thousand years. The warping was differentiated and synsedimentarily compensated. Along with synsedimentary structures, the basin has synchronous small depressions whose beds onlap the underlying substrata. Troughs aligning the basin's margins must be associated with nappe generation in the surrounding fold structures. Frequent synsedimentary normal faults, linear structures, and volcanism point to conditions of extension at the Miocene stage.

The Pannonian-Quaternary stage is characterized by rapid sinking. Rates of sedimentation in the central basinal parts greatly increased (Figure 7). If the Miocene rate was about 2.5 cm per 1000 years, in the Pannonian-Pliocene it was about 50 cm per 1000 years [Horvath and Stegnea, 1977]. In marginal structures, the rate of sedimentation in the same period was under 5-10 cm per 1000 years [Vass and Cech, 1983]. Irregularly shaped troughs that emerged at the later stage were filled up synsedimentarily by uniform lake sequences. The associated basaltic volcanism is developed all over the basin area, and sometimes beyond it (Silesia, Elba zone, etc.). At the Pannonian-Quaternary stage, the Pannonian basin proper was formed whereas rear deeps virtually disappeared. Thus, the Pannonian-Quarternary stage was predominantly that of vertical movements. Faults and weak horizontal motions were concentrated on the basinal margins and neighbouring areas.

The Moho discontinuity within the Pannonian basin occurs at relatively shallow depths (Figure 8). On the whole, the basin is aligned by theMoho discontinuity at depths of 27.5 to 30 km. The maximum approach to the surface of the Moho (23 km) is registered in the area of the lower course of the Tisa River.

In fold structures around the margin of the Pannonian basin, the Moho discontinuity sinks to 35-40 km (Dinarides, West Carpathians), the maximum depth being 60 km (the Soviet Carpathians). Structurally, the Moho discontinuity shows irregular forms, corresponding in general to the base of the upper structural complexes of the Neogene-Quaternary cover. The "basaltic" layer within the Pannonian basin presumably is 10-15 km thick, increasing to 25-30 km

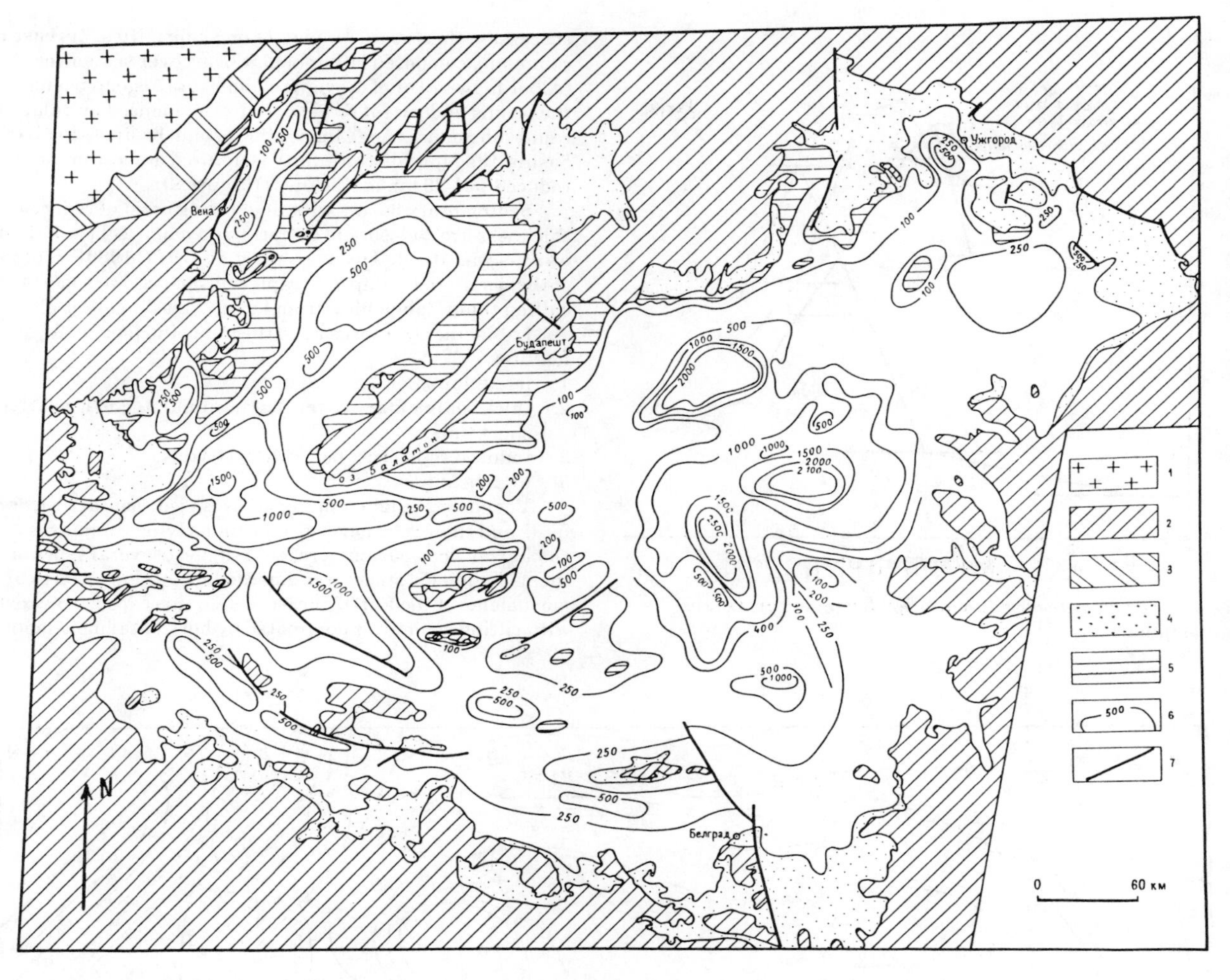

Fig. 6. Map of thickness of Lower Pannonian deposits in the Pannonian basin (lower part of the upper structural complex). 1 - Czechoslovak massif; outcropsof pre-Neogene basement, 3 - Fore-Carpathian foredeep, 4 - outcrops of pre-Pannonian deposits, 5 - areas free of Lower Pannonian strata, 6 - isopachytes of Lower Pannonian deposits, m, 7 - faults.

under the Carpathians and Dinarides. The "granite" layer in the Pannonian area is 10-15 km, showing no perceptible variations toward the neighbouring areas The upper mantle below the Pannonian basin is uplifted and has anomalous properties. Most likely, it is partly molten.

Thus the thickness of the solid earth's crust of the Pannonian basin, Neogene-Quaternary cover excluded, is 20-25 km. In surrounding fold structures, the value varies between 40 and 65 km, which is much higher than in the basin (Figure 9).

By the beginning of Miocene time, the Pannonian basin had become a stable area. As a result of comparative analysis of the present and paleostructures of a platform (which was the type of the Pannonian area at the Paleogene/Neogene boundary), the approximate thickness of the Pannonian basin's crust early in the Miocene was estimated [Nikolaev, 1983]. It was nearly 35-40 km, with almost equal quotas of "basaltic" and "granitic" layers. In the surrounding areas, the crust must have been slightly thicker as suggested by mountain-building there at the end of the Paleogene.

Throughout the Miocene (pre-Pannonian) stage, the Earth's crust must have remained stable, with local reductions in some places. Rapid thinning of the crust possibly took place in the Pannonian-Quaternary stage due to the reduced basalt layer thickness. Apparently, at the later stage, about 15 km of the solid crust was destroyed.

Thus, formation of the Neogene-Quaternary Pannonian basin is related to progressive emplacement of a mantle diapir and its effect on the solid crust. Evidently, the thinning of the crust and hence its sinking must be due to phase transitions between anomalous mantle and the lower parts of the crust (Figure 9) [Artyushkov, 1979].

Deep-sea basins of the Mediterranean belt (Black Sea, South Caspian, Aegean, Alboran, Tyrrhenian seas and others) have similar features. All of them have a sedimentary cover up to 16 km in thickness. Within this sedimentary cover

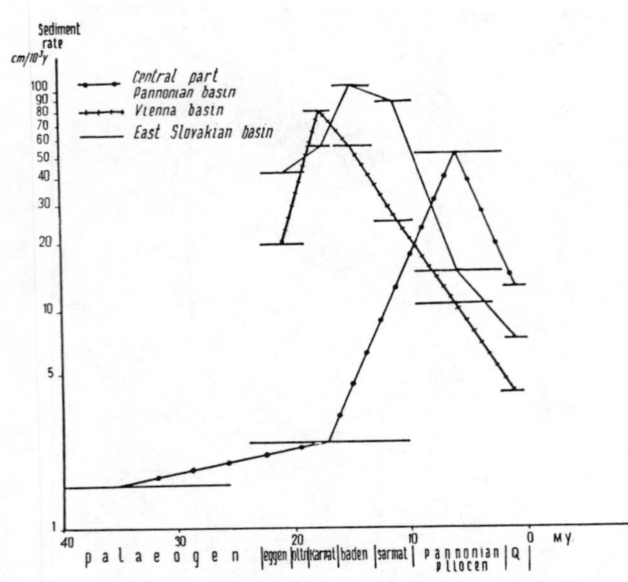

Fig. 7. Average rate of sedimentation for some parts of the Pannonian basin.

succession there are a number of unconformities. Its base is commonly composed either of platform cover sequences (Alboran basin) or of orogenic complexes (Aegean basins). All the basins underwent either collapse, evidenced by onlap, or rapid subsidence (Aegean, South Caspian basins). In all of the basins, the Moho discontinuity is uplifted and the crust is reduced to 4-20 km in thickness (Figure 9).

Sedimentary basins with underlying crust of reduced thickness are widespread on earth and make a class of their own, as was attested by Yanshin et al. [1977, 1980]. Since all of them suffered collapse or rapid subsidence, we suggest calling them "labigenic" (from Latin "labes" - collapse).

Labigenic structures have the following four main features:

1. irregular shape;
2. thick sedimentary cover that accumulated in a relatively short time;
3. reduced solid crust beneath the basins;
4. high heat flow.

Variations in labigenic structures (sedimentary cover units and their relations with the underlying complexes; volcanism; deep parameters, etc.) suggest several types of them. Labigenic structures emerge largely due to vertical movements of the Earth's crust, and in this aspect, contrast with rifting structures dominated by horizontal movement.

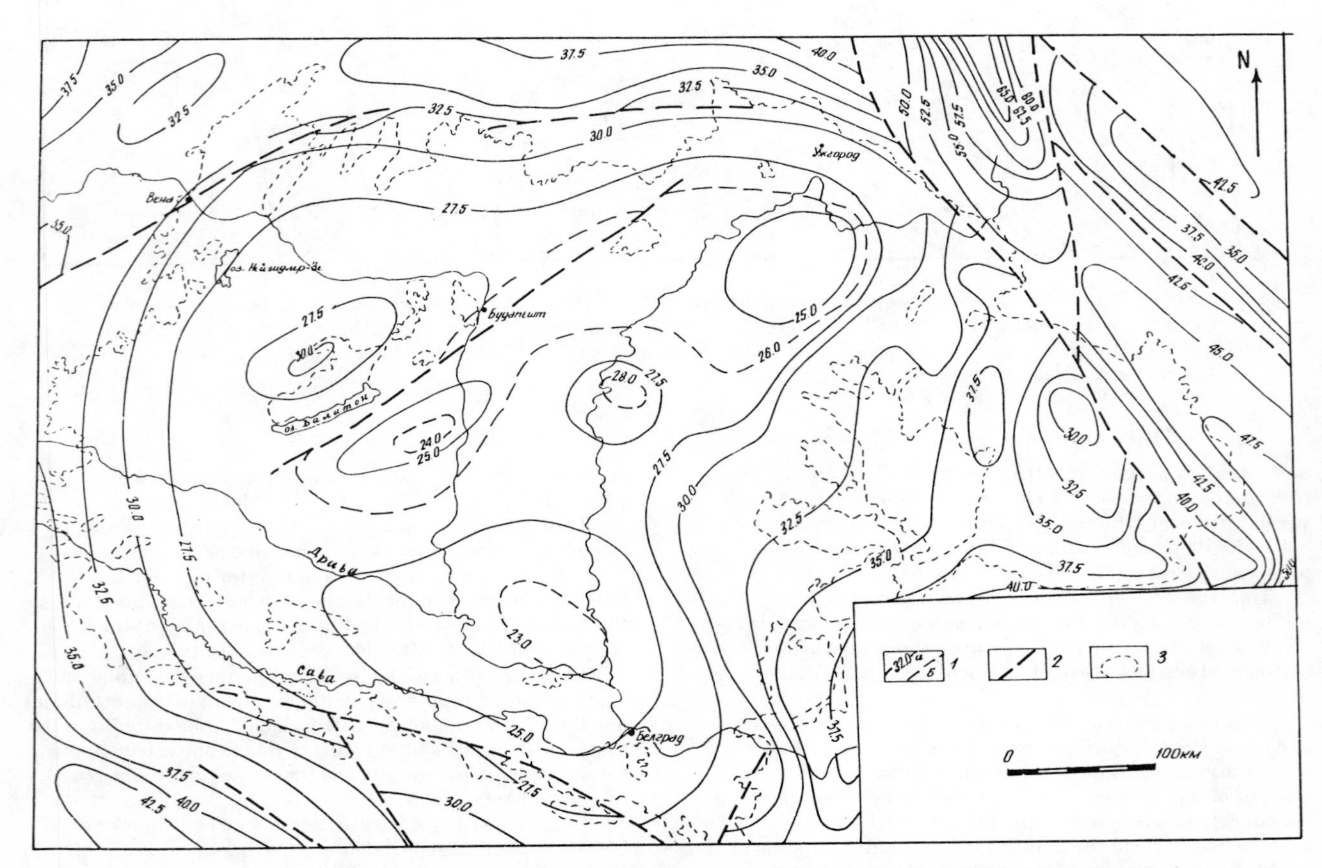

Fig. 8. Structural scheme of Moho discontinuity in the Pannonian Basin. 1 - isolines of Moho discontinuity, km (a-major, b-additional), 2 - faults, 3 - boundaries of Neogene-Anthropogene sedimentary cover.

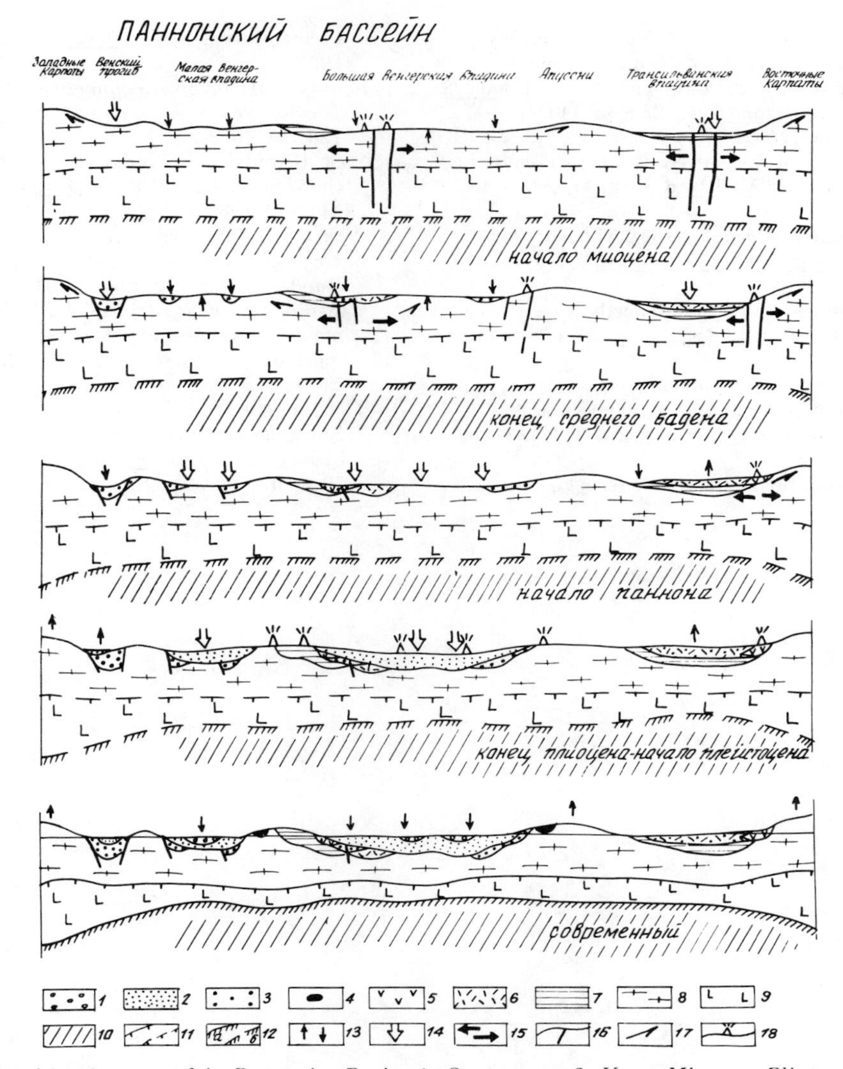

Fig. 9. Scheme of development of the Pannonian Basin. 1 - Quaternary, 2 - Upper Miocene - Pliocene, 3 - Lower-Middle Miocene, 4 - Pliocene-Quarternary basalts (upper volcanic complex), 5 - Miocene andesites, rhyolites (middle volcanic complex), 6 - Early Miocene ignimbrites (lower volcanic complex), 7 - Paleogene, 8 - granite geophysical layer, 9 - basaltic geophysical layer, 10 - assumed anomalous mantle , 11 - Conrad discontinuity, 12 - Moho discontinuity: a-established, b-assumed; 13 - weak vertical movements, 14 - intense submergence, 15 - assumed zones of extension, 16 - faults, 17 - zones of nappe emplacement and direction of their movement, 18 - volcanoes.

Acknowledgments. The authors thank the 9th Problem Commission of Multilateral Cooperation between the Academies of Sciences of socialist countries for providing the possibility of joint investigations. We are most grateful to Drs. L. Royden and E. Nelson for critical remarks that improved the paper.

References

Artyushkov, E.V., Geodynamics. "Nauka", Moscow, 327, 1979 (in Russian).

Balla, Z., Development of the Pannonian basin basement through the Cretaceous-Cenozoic collision: a new synthesis, Tectonophysics, 88, 1-2, 61-102, 1982.

Boccaletti, M., F. Horvath, M. Loddo, F. Mongelli, and L. Stegena, The Tyrrhenian and Pannonian basins: a comparison of two Mediterranean interarc basins, Tectonophysics, 35, 1-3, 45-69, 1976.

Horvath, F. and Stegena, L., The Pannonian basin: a Mediterranean interarc basin, In Symp. int. Hist. struct. basins mediterr., Split, 1976, Paris, 333-340, 1977.

Konecny, V., I.P. Bagdasarjan, and D. Vass, Evolution of Neogene volcanism in Central Slovakia and its confrontation with absolute ages, Acta geol. Acad. Sci. Hung., XIII, 1-4, 259-266, 1969.

Lukacs, Z., I. Pogàcsà, and I. Varga, Seismic facies analysis and stratigraphic interpretation of the unconformably dipping Pliocene features in the Pannonian basin. Trans. 28 Int. geophys., symp. Ballaton semes. 28 Sept. - 1 Oct. 1983, C.1, Budapest, 173-186, 1983.

Nikolaev, V.G., Structure of Neogene-Anthropogene cover in the Pannonian basin, Bull. MOIP otd. geol., 54, vyp. 6, 45-55, 1979 (in Russian).

Nikolaev, V.G., Neogene rear deeps in the peripheries of the Pannonian basin, Doklady AN SSSR, 245, 4, 915-917, 1979 (in Russian).

Nikolaev, V.G., Neogene volcanics in the northeastern Pannonian region, Izvest. AN SSSR, ser. geol., 4, 36-43, 1980 (in Russian).

Nikolaev, V.G., Pre-Neogene complexes and the deep structure of the Pannonian basin, In Problems of the Earth's crust tectonics, Nauka, Moscow, 263-371, 1981 (in Russian).

Nikolaev, V.G., Consolidation of the Earth's crust in the Pannonian basin, Veröff. des ZIPE, 77, Potsdam, 211-228, 1983.

Panto, G., Development of magmas and igneous rocks in the Tertiary volcanism of Hungary, Geol. Rundschau, Band 57, 128-143, 1968.

Vass, D. and F. Cech, Sedimentation rates in molasse basins ot the Western Carpathians, Geol. Zborn. Geol. Carpat, 34, 4, 411-422, 1983.

Wein Gy, Zur Kenntnis der tektonischen strukturen im Untergrund des Neogens von Ungarn, Jhar Geol. Bundesanstalt, 116, Vien, 85-101, 1973.

Yanshin, A.L., E.V. Artyushkov, A.E. Shlezinger, Main types of large structures of lithospheric plates and possible mechanism of their formation, Doklady AN SSSR, 234, 5, 1175-1178, 1977 (in Russian).

Yanshin, A.L., L.A. Esina, Ya.P, Malovitsky, and A.E. Shlezinger, Structure of sedimentary cover and formation of deep-sea basins, Geotektonika, 1, 72-85, 1980 (in Russian).

PROBLEMS OF PETROLEUM EXPLORATION UNDER PLATEAU BASALTS

Mahmoud Benelmouloud

Ministry of Energy and Petrochemical Industry, Algiers

Evgueni Zhuravlev

Gubkin Institute of Oil and Gas, Moscow, USSR

Abstract. Extrusive basalts are a specific feature of the Mesozoic-Cenozoic tectonic evolution of the lithosphere. In many sedimentary basins basalt flows cover oil- and gas-bearing strata. It is shown that basalt flows may serve as good seals for hydrocarbon pools. In some cases they produce thermal effects on the chemistry of hydrocarbons.

Introduction

Plateau basalts, also known as traps, are widely distributed on the continents. They are commonly found in sedimentary basins where they cover oil- and gas-bearing rocks.

Major sedimentary basins with associated plateau basalts are concentrated on the Eurasian continent, South America and Africa (see Figure 1). These rocks are mainly composed of basaltic covers and doleritic sills. Their thicknesses vary and sometimes exceed 3000 m. Sills and basalt flows are related to deep-seated feeder faults, which contain doleritic dykes. Plateau basalts are either directly superposed on each other or they are separated by sedimentary rocks or thin bands of tuff.

The formation of plateau basalts is attributed by some geologists [Dietz et al., 1970; Dewey, 1973] to the

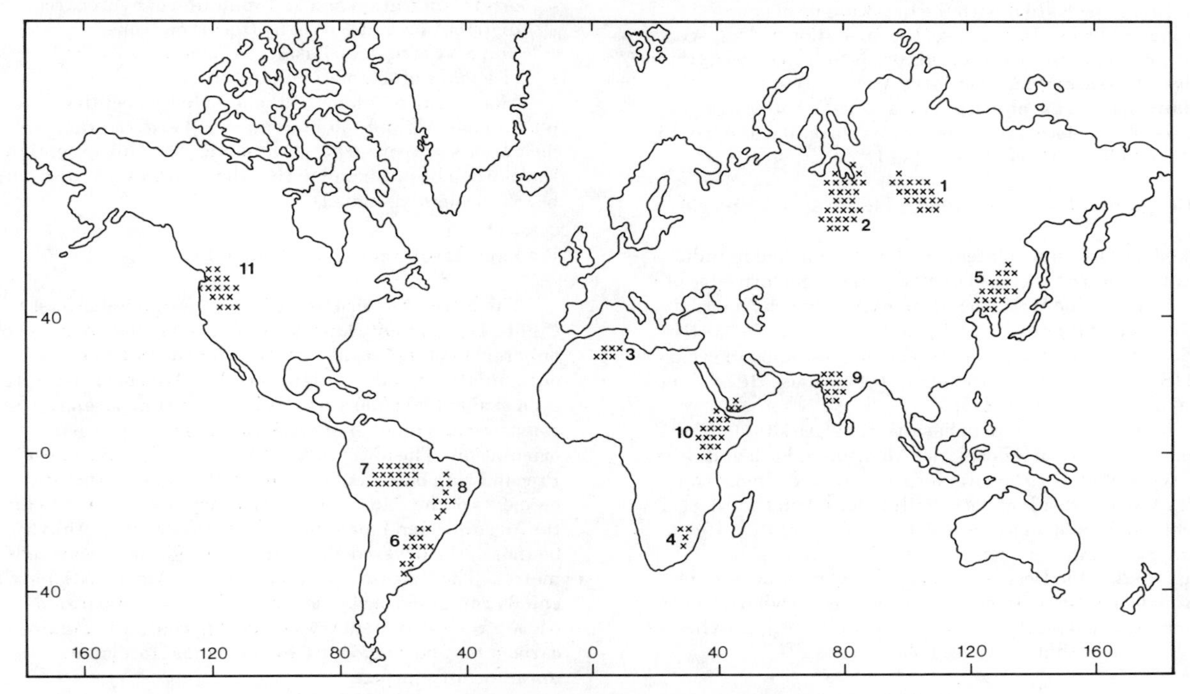

Fig. 1. Geographic location of plateau basalts in the world. (See Table 1).

Table 1. Major plateau basalt basins of the world

	Plateau basalts	Type of location	Area (km^2)	Age
1.	Tunguss	Exposed	3.5×10^5	Permo-Triassic
2.	Western Siberia	Buried	7.5×10^5	Permo-Triassic
3.	Algerian Sahara	Exposed	6.0×10^4	Triassic
4.	Karo	Exposed	1.0×10^5	Jurassic
5.	Mongolia	Exposed	3.5×10^5	Jurassic
6.	Parana	Exposed	5.0×10^5	Cretaceous (lower)
7.	Amazon	Buried	5.0×10^5	Cretaceous (lower)
8.	Maranhao	Buried	5.0×10^5	Cretaceous (lower)
9.	Deccan	Exposed	5.0×10^5	Paleocene-Eocene
10.	Ethiopia	Exposed	1.25×10^6	Oligocene-Miocene
11.	Columbia River	Exposed	5.0×10^5	Miocene-Pliocene

fragmentation of Pangea which began in the Mesozoic. Others [Gass, 1972] associate it with the breaking up of rising lithospheric blocks. In any case, the formation of deep-seated faults and rifts in the continental core is followed by rhythmic outflows of basic magma over large areas.

Many plateau basalts are of Mesozoic-Cenozoic age. On the basis of their locations they may be classified as exposed (visible) and buried types (see Table 1).

Geographic Location of Plateau Basalts in the World

The largest visible plateau basalts are located in India (Deccan), Eastern Siberia (Tunguss), on the eastern edge of the African continent (Karoo, Ethiopia), in South America (Parana) and in the region of the Columbia River in North America. These plateau basalts have been studied in great detail [Sobolev, 1936; Amarel et al., 1966; Ghose, 1976; Jones, 1976; Girod, 1978 and others].

As for the buried plateau basalts, they are situated in Western and Eastern Siberia, the Algerian Sahara and Brazil. These types of plateau basalts were discovered almost two decades ago as a result of deep drilling for oil and gas deposits [Mesner and Wooldridge, 1964; Khairi et Zhuravlev, 1982]. Drilling proved that buried plateau basalts in Parana, Tunguss and in the Deccan extend to great distances under Mesozoic-Cenozoic sedimentary rocks. On the whole, the area covered by buried plateau basalts appears to be approximately twice as large as that of exposed plateau basalts.

Tectonics

The tectonic position of plateau basalts is not the same in different parts of the world. While some of them are located in platforms others are located in the fold zones, and a third group is located in the foredeeps. The most widespread among them are the plateau basalts of the platforms. One may mention here the plateau basalts of the young epi-Hercynian Western Siberian platform and also the older platforms of Eastern Siberia, India, North Africa and South America. The total area covered by these plateau basalts is around 5 million square kilometres, i.e. approximately nine times the area of France.

Plateau basalts associated with orogenic zones occur more frequently but their dimensions are much smaller. There are the plateau basalts of Appalachia, the Morrocan Atlas and the Urals. The only exception is the huge basaltic cover of the North American Cordillera.

In West Bengal plateau basalts, most commonly known as traps, are located in the platform part of the pre-Himalayan foredeep [Tao, 1973].

Volcanism occurred extremely non-uniformly over geological time. There were six periods that experienced increased volcanic activity: Permo-Triassic, Jurassic, Lower Cretaceous, Paleocene-Eocene, Oligocene-Miocene and Miocene-Pliocene. Volcanism which occurred at the beginning of the fragmentation of Pangea formed the most extensive and thick plateau basalts [Zhuravlev, 1984].

Plateau basalts associated with volcanic activity over different time periods quite frequently discordantly overlie sedimentary rocks of different ages, which are of interest from the point of view of oil and gas prospecting. The unconformity surface often serves as a good seismic marker.

Prospecting in sub-plateau basalt sedimentary rocks for oil and gas is beset with a number of problems, the major ones among them being:

- origin, migration and accumulation of hydrocarbons,
- distribution and type of hydrocarbon traps,
- property of plateau basalts as "cap" rocks,
- efficiency of seismic exploration.

Most of the problems are related to the fact that, both the plateau basalts and sub-plateau basalt sedimentary formations have been poorly studied. The only exception is the plateau basalt-basin in the Algerian Sahara, which has been thoroughly studied.

Plateau Basalts of the Algerian Sahara

The North-Eastern part of the Algerian Sahara (see Figure 1), commonly known as the Triassic basin, is as yet the only region of the world with known hydrocarbon accumulations under plateau basalts. This basin occupies an area of about 200 000 km^2. Its Precambrian basement is composed of a number of blocks having a north-easterly orientation. The lowered and the elevated blocks of the Precambrian basement determine the major structural elements of the Paleozoic sedimentary cover. The thickness of the Mesozoic and Paleozoic sedimentary cover within the bounds of the lowered blocks may exceed some thousands of metres. The Paleozoic succession of the Algerian Sahara is chiefly composed of clastic rocks. These are separated from the Mesozoic sedimentary succession, containing mainly carbonates and evaporites, by the major Hercynian unconformity surface.

The base of the Mesozoic succession is composed of a Triassic clay-sand unit, the thickness of which reaches many hundreds of metres. This unit lies on Paleozoic sediments and in places where they have been completely eroded, directly on the Precambrian basement surface. Buried plateau basalts

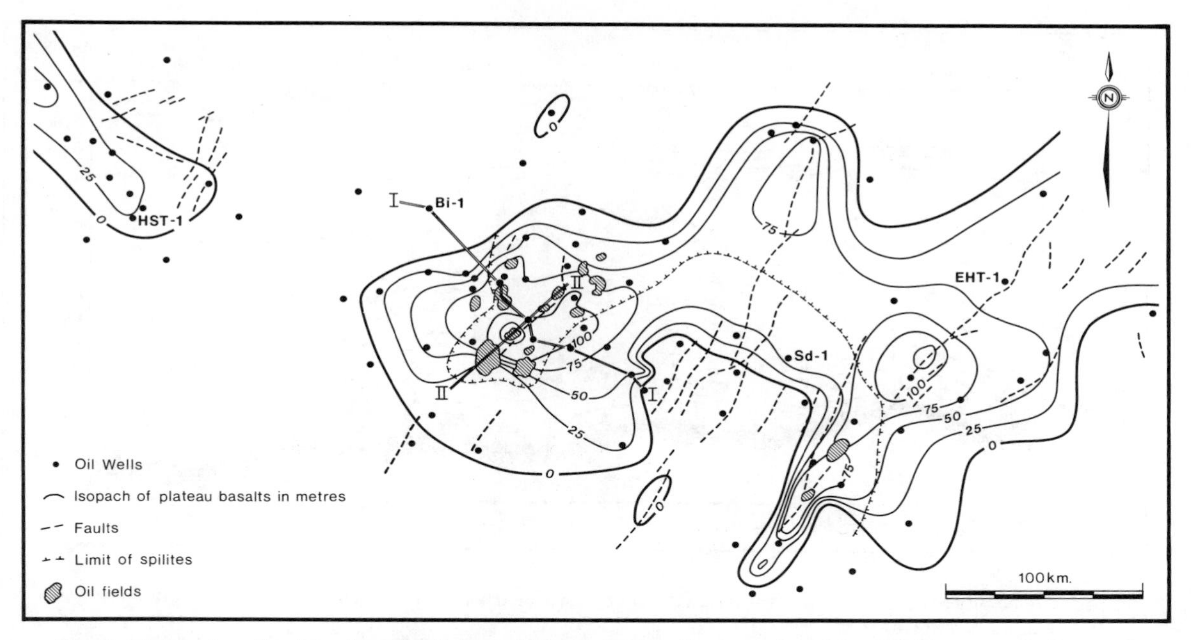

Fig. 2. Plateau basalts of the Algerian Sahara (Triassic basin). Cross sections I-I and II-II are seen in figures 3 and 4 respectively.

are, as a rule, located on this clay-sand unit. They have been discovered in more than 150 oil wells at depths of 1500 to 3500 m. These buried Triassic plateau basalts occupy an area of about 60 000 sq. km. (see Figure 2). Their average thickness is around 40 m and their maximum thickness is 140 m. The maximum thickness has been found close to major fault zones in fissures associated with deep-seated faults. The plateau basalts of the Triassic basin are composed of quartz and olivine normative basalts and dolerites of tholeiite facies. In the northeastern part of the Triassic basin, basalts and dolerites are partly spilitified, apparently due to the pouring out of lava into saline lagoons. The Triassic age (197 and 222 m.y.) of the dolerites was determined by the K-Ar method for two core samples (Khairi and Jouravlev, 1982). The Triassic plateau basalts contain up to ten basaltic covers. In the region of Rhourde El Baguel two main phases of basic volcanic activity are known, which are separated by a phase of relative calm recorded by alternating layers of clay, sandstone and dolomites. Triassic lava flowed on the peneplaned surface of the Hercynian unconformity (see Figure 3). In the shallow depressions, plateau basalts cover polygenic lower Triassic sediments of alluvial, deltaic and lagoonal origin.

Fourteen hydrocarbon pools have been discovered in the clastic Triassic sediments. In the underlying Paleozoic sediments, eleven oil and gas pools have been discovered below the Hercynian unconformity.

The majority of the hydrocarbon traps in the Triassic sediments are of structural type. Thin basalt covers did not interfere with the discovery of these hydrocarbon traps. In the geological cross-section it is clearly visible that oil

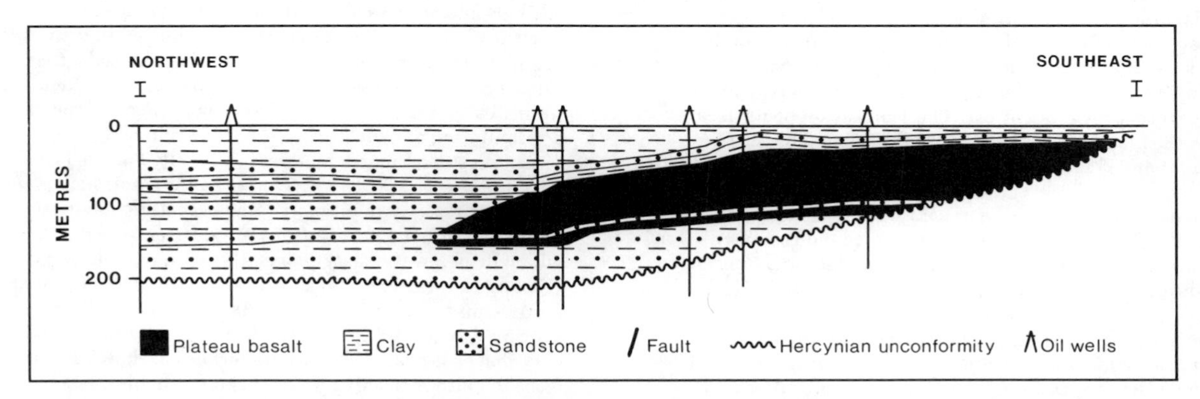

Fig. 3. Paleogeological cross section of the Triassic basin. (The reference line is the top of the clastic Triassic succession). Plateau basalts cover the surface of the Hercynian unconformity and polygenetic lower Triassic sediments.

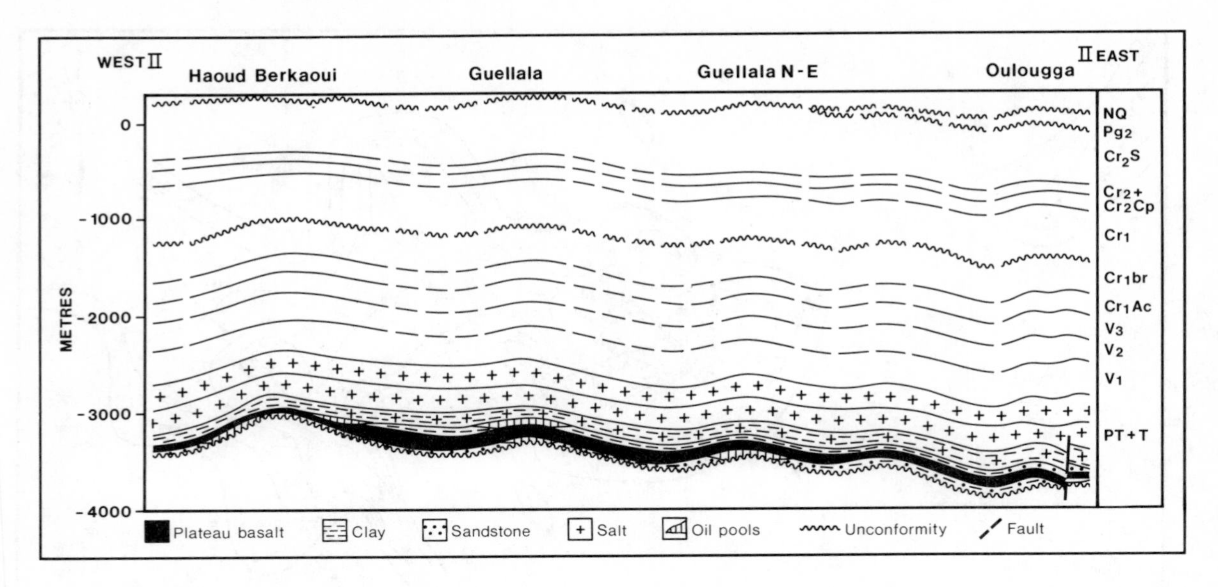

Fig. 4. Geological cross section of the Triassic basin. Hydrocarbon traps are controlled by the relief of the Hercynian unconformity.

accumulations are controlled by the relief of the Hercynian unconformity (see Figure 4). In the same structures oil pools may occur both below and above the plateau basalts themseleves and the overlying thin clay layers serve as an impervious cover. At higher stratigraphic levels this role is played by clay and evaporite sediments of Triassic and Jurassic-Cretaceous age. As a rule, igneous rocks untouched by secondary alteration processes serve as good impervious seals, whereas their weathered varieties are porous and permeable. In this case thin layers of weathered basalt do not serve as impervious seals. As for relatively thick basalt seals, where they are weathered only at the top, their central parts, composed of unaltered dolerites, remain impervious.

Tectonic and lithostratigraphic traps are less common. At the Ben Kahla oil-field, an oil pool has been discovered below the plateau basalts in a basal sandy layer, which experienced facies change across the dip. At the Oulougga oil field, the oil trap is intersected by faults (see Figure 4). Vertical displacement of layers, along these faults, caused sandy reservoir rocks to come into contact with impervious clay layers. As a result of this fact tectonic traps were formed which were later filled by oil. The Triassic oil pools, located under the plateau basalts, are small in size. The largest among them, Haoud Berkaoui, occupies an area of 9.5 x 19 km only.

Beneath the Hercynian unconformity, oil pools are also located in Paleozoic sediments. In the Haoud Berkaoui - Ben Kahla region, oil is found in Silurian sandstones, whereas in the southern and eastern parts of the giant oilfield Hassi-Messaoud, along the line Rhourde El Baguel - Nezla, oilpools are encountered in the sub-plateau basalt weathered quartzitic sandstones of Cambrian-Ordovician age, which are developed on the surface of the Hercynian unconformity.

The principal source rocks for the hydrocarbons confined in the sub-plateau basalt sedimentary complexes of the Triassic basin are thought to be black, Silurian clays (Tissot, 1980). Migration of hydrocarbons, from the source rock into traps, took place at the end of the Mesozoic.

Basalts and dolerites of the Algerian Triassic basin form a part of the huge Permo-Triassic plateau basalt province of the world, which was formed at the beginning of the breakup of Pangea.

Plateau Basalts in Other Parts of the World

In the Tunguss depression of Eastern Siberia (see Figure 1) extensive plateau basalt covers lie on carbonate, marl, clay and sandstone sediments of Triassic, Paleozoic and Eocambrian age. These sediments are now looked upon with great interest. Gas condensate deposits have been found under doleritic sills in Eocambrian and Cambrian sandstones at the Omorinskoye, Sobinskoye, Ayanskoye (see Figure 5) oilfields and others located in the southern part of Eastern Siberia. Odintsova and Drobot (1983) showed that the sills associated with plateau basalt formation exerted a thermal influence on the chemistry of hydrocarbons. This fact indicates that the intrusion of basic magma occurred after the accumulation of hydrocarbons in this region.

In Western Siberia plateau basalt fills Permo-Triassic rifts and covers adjacent blocks of the Paleozoic basement. These basalts create intensive linear positive magnetic anomalies. Carbonate and clay-sand Paleozoic sediments which lie below the plateau basalts are of marine origin. Carbonate rocks are characterised by a higher content of organic material. Within the rifts plateau basalt frequently alternates with red Triassic continental clay-sands and gray carbonaceous sediments. In numerous boreholes situated close to plateau basalts, hydrocarbon influx has been obtained from the top of the Paleozoic, which does not have good seals. It may be suggested that in regions of plateau basalt extrusion, basalt covers and dolerite sills serve as impervious

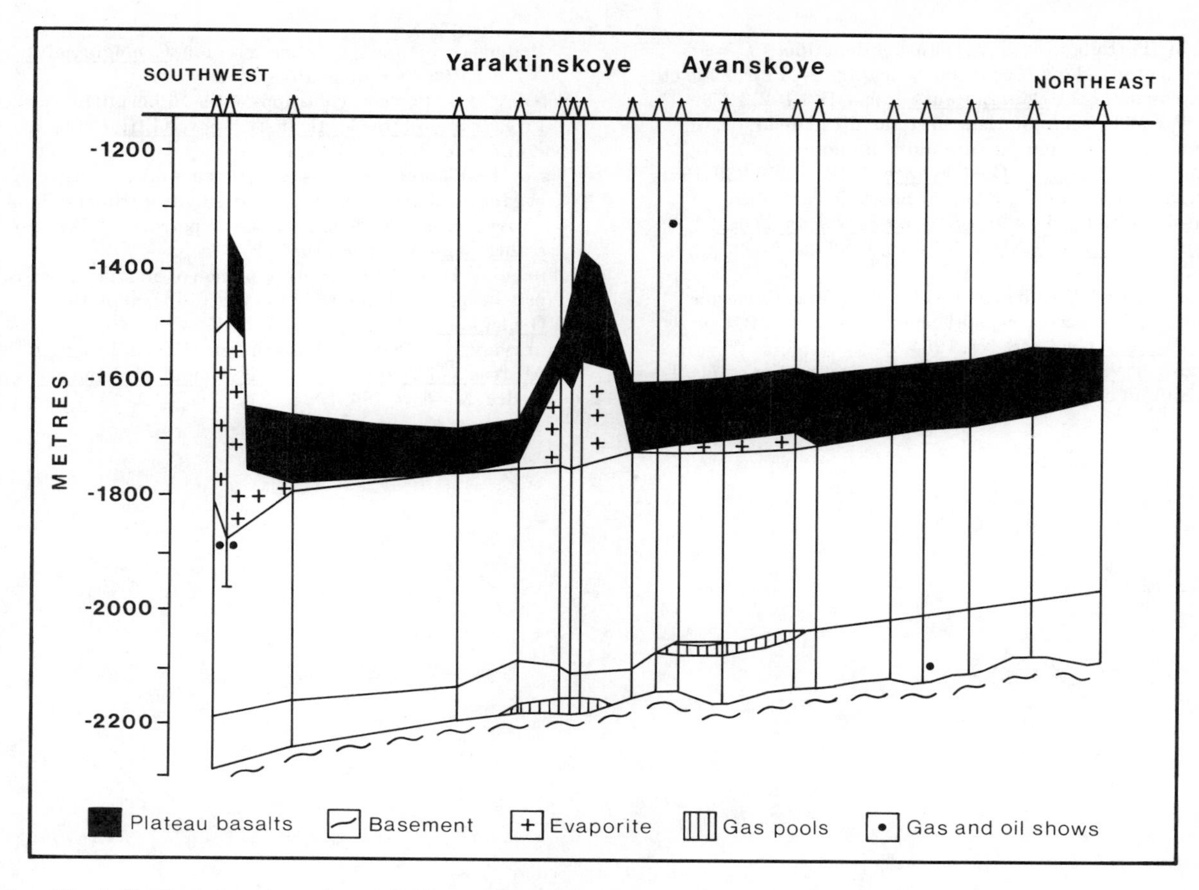

Fig. 5. Geological cross section of the Eocambrian and Cambrian strata of Eastern Siberia. After Odintsova and Drobot (1983).

covers for hydrocarbons, which are accumulated in the sediments lying below.

The thickness of the West Siberian plateau basalts is up to 3-4 km. Layers of basalts, dolerites and tuffs, comprising plateau basalts, have different acoustic resistance. On seismic profiles the thick stratified West Siberian plateau basalts are often mis-interpreted as petroleum sedimentary successions [Zhuravlev, 1984].

On the South American continent plateau basalts cover vast areas. Under plateau basalts of Cretaceous age lie Early and Middle Paleozoic marine clastic sediments. The thickness of these plateau basalts approaches 3 to 5 km.

In spite of the fact that at present plateau basalt-bearing basins of South America have very few small hydrocarbon deposits, they are of great interest for future exploration.

Conclusions

The presence of vast Mesozoic plateau basalt fields is a specific feature of the tectonic evolution of many sedimentary basins of the world. The data on the Triassic basin of the Algerian Sahara obtained by deep drilling indicate that sub-plateau basalt sedimentary complexes may contain significant hydrocarbon accumulations. Basaltic nappes and doleritic sills of the plateau basalt formation may serve as fluid-resistant layers for various hydrocarbon traps. Seismic exploration of the stratified plateau basalt regions is difficult.

Sedimentary basins containing plateau basalts occupy large, inadequately explored territories which form the last "white spots" on many continents, and may serve as important objects of hydrocarbon exploration.

References

Amarel, G., Cordani, U.G., Kavashita, K., Reijnolols, J.H., Postassium-Argon dates of Basaltic rocks from Southern Brazil, Geochim. Cosmochim. Acta, 30, 159-189, 1966.

Cox, K.G., The Kato volcanic cycle, J. Geol. Soc. London, 128, 311-336, 1972.

Dewey, J.F., Pitman, W.C., Ryan, W.B.F., Bonnin, J., Plate Tectonics and the Evolution of the Alpine System, Geol. Soc. of America Bull., v. 84, N 10, 3137-3180, 1973.

Dietz, R.G., Holden, J.C., Stroll, W.P., Geotectonic evolution and Subsidence of Bahama Platform, Geol. Soc. America Bull. 81, 1915-1928, 1970.

Gass, I.G., The role of lithothermal systems in magmatic and tectonic processes, Earth Sci. 9, 2, 261-273, 1972.

Ghose, N.C., Composition and origin of Deccan basalts, Lithos., 9, 65-73, 1976.

Girod, M., Pétrologie des laves dans les domaineds continentaux, dans "Les roches volcaniques. Pétrologie et cadre structural", Doin Editeurs, Paris, 166-192, 1978.

Jones, P.WW., Le magmatism au stade initial de la framentation des plaques arabique, nubienne et somalienne, Bull. Soc. Geol. France, 7, 18, 4, 829-830, 1976.

Khairi, S. and Jouravelv, E.G., Trapps de las province triassique d'Algérie, 4-ême Séminaire Nationale des sciences de la Terre, Alger, 5-7 juin, 1982, Resumes, 57, 1982.

Mesner, J.C. and Wooldridge, L.C.P., Maranhao Paleozoic basin and Cretaceosu coastal basins, North Brazil, Bull. Am. Ass. Pet. Geol., 48, N9, 1964.

Odintsova, T.V., and Drobot, D.I., Trapp magmatism and oil-gas content of the eocambrianclastic complex of the Prilenskiy oil and gas-bearing region, Geologia nefti i gasa, N7, 6-7, 1983. (In Russian).

Sobolev, B.S., Petrologyof trapps of the Silberian platforms, Proceedings of the Arctic Institute, v. XLIII, Geologia, Leningrad, 224, 1936. (In Russian).

Tissot, B., L'application des résultats des études de géochimie organique dans l' éexploration des hydrocarbures, dans "Noveaux aspects de la géologie du pétrole", 1, Edition francaise,SCM, 67-96, 1980.

Zhuravlev, E.G., Tectonic precondition of oil and gas search in pre-Jurassic sediment s of the West Siberia platform, Geologia nefti i gaza, N 3, 35-39, 1984. (In Russian).

Zhuravlev, E.G., Rifts and tapp magmatism of Pangea, Abstracts of 27th International Geological Congress, v. II, Nedra, Moscow, 484, 1984.